Andrea Strazzoni

Dutch Cartesianism and the Birth of Philosophy of Science

Andrea Strazzoni

Dutch Cartesianism and the Birth of Philosophy of Science

From Regius to 's Gravesande

DE GRUYTER

ISBN 978-3-11-056828-8
e-ISBN [PDF] 978-3-11-056969-8
e-ISBN [EPUB] 978-3-11-056862-4

Library of Congress Cataloging-in-Publication Data
Names: Strazzoni, Andrea, author.
Title: Dutch Cartesianism and the birth of philosophy of science : from
  Regius to 's Gravesande / Andrea Strazzoni.
Description: 1 [edition]. | Boston : De Gruyter, 2018. | Includes
  bibliographical references and index.
Identifiers: LCCN 2018030843 (print) | LCCN 2018033611 (ebook) | ISBN
  9783110569698 (electronic Portable Document Format (pdf)) | ISBN
  9783110567823 (print : alk. paper) | ISBN 9783110569698 (e-book pdf) |
  ISBN 9783110568264 (e-book epub)
Subjects: LCSH: Descartes, Rene, 1596-1650. | Logic. | Metaphysics. |
  Science--Philosophy--History.
Classification: LCC B1875 (ebook) | LCC B1875 .S88 2018 (print) | DDC
  199/.492--dc23
LC record available at https://lccn.loc.gov/2018030843

Bibliographic information published by the Deutsche Nationalbibliothek
The Deutsche Nationalbibliothek lists this publication in the Deutsche Nationalbibliografie;
detailed bibliographic data are available on the internet at http://dnb.dnb.de.

Typesetting: Integra Software Services Pvt. Ltd.
Printing and binding: CPI books GmbH, Leck
Cover image: Hendrick van der Burgh, The conferring of a degree at the University of Leiden
around 1650. WikimediaCommons, Public Domain.

www.degruyter.com

Acknowledgments

The preparation of this book began during my time as a doctoral student at the Erasmus University Rotterdam, between 2010 and 2015, and continued during my stays at the Higher School of Economics (Moscow), the University of Parma, and the Gotha Research Centre of the University of Erfurt. At the Erasmus University I conducted a substantial part of the research that led both to the completion of my doctoral dissertation, and then to the preparation of this book. I want therefore to thank Wiep van Bunge, Henri Krop, Han van Ruler, and Paul Schuurman who supported, helped and guided me in the essential part of my survey, as well as Christoph Lüthy and Theo Verbeek for their comments on my work. I also want to thank Stefan Heßbrüggen-Walter for his suggestions, and, with Darya Drozdova, for his hospitality at the Higher School of Economics in Moscow. In Parma (where I first approached the theme of Dutch Cartesianism during my undergraduate studies), I had the decisive support of Fabrizio Amerini, Stefano Caroti, Beatrice Centi, Roberto Pinzani, Mariafranca Spallanzani and Andrea Staiti. Eventually, the completion of this book was made possible by my post-doctoral scholarships[1] at the Gotha Research Centre, where I benefitted from the help and support of Martin Mulsow and Markus Meumann. Moreover, this book would not have been made possible without the fruitful conversations which I had, across Europe and beyond, with Igor Agostini, Siegrid Agostini, Vlad Alexandrescu, Érico Andrade, Peter Anstey, Delphine Antoine-Mahut, Roger Ariew, Fabrizio Baldassarri, Susanne Beiweis, Giulia Belgioioso, Jip van Besouw, Paul Richard Blum, Carlo Borghero, Erik-Jan Bos, Steven Vanden Broecke, Chiara Catalano, Andrea Cimino, Sorana Corneanu, Hannah Dawson, Adela Deanova, Antonella Del Prete, Giuliana Di Biase, Mihnea Dobre, Steffen Ducheyne, Marco Forlivesi, Daniel Garber, Alfredo Gatto, Benedino Gemelli, Guido Giglioni, Aza Goudriaan, Dana Jalobeanu, Martin Lenz, Brunello Lotti, Gideon Manning, Oana Matei, Dirk van Miert, Alberto Oliva, Riccardo Pozzo, Evan Ragland, Andrea Robiglio, Sophie Roux, Andrea Sangiacomo, Tad Schmaltz, Marco Sgarbi, Marco Storni, and Riccarda Suitner. Last but not least, I would like to thank Christoph Schirmer for his interest in the manuscript, and for his constant dedication to the editorial processes of peer-review and approval for publication.

---

[1] Herzog-Ernst-Stipendium der Fritz Thyssen Stiftung; Christoph-Martin-Wieland-Postdoc-Stipendium.

https://doi.org/10.1515/9783110569698-201

Contents

Introduction

When was philosophy of science born? And why? This book aims to answer these questions. Simply put, philosophy of science was born in seventeenth-century Dutch universities, where the introduction of Cartesian ideas called for philosophical reflection upon the validity, method, and concepts of natural philosophy. The disciplines which fulfilled this role were metaphysics and logic. The process was neither short nor straightforward, nor – admittedly – easily grasped through such a generalisation. As a matter of fact, philosophy of science has existed since antiquity, or for as long as philosophers have provided reflections on such topics. On the other hand, the institutional fragmentation of philosophical and scientific disciplines in the modern era has made it increasingly easy to distinguish between 'philosophy' and 'science', and for philosophers to provide reflections on the natural sciences. This book stands at the crossroads: namely, it disentangles the ways in which metaphysics and logic became functional for such reflection, and, forthwith, they began to be detached from what was labelled 'natural philosophy'. This is the reason why this book offers a study of Dutch Cartesianism. Dutch universities were the first where Descartes's ideas became official matter of teaching and came to reshape the function of metaphysics and logic. So that one can meaningfully ascribe the role of 'philosophy of science' to them, and to see this role at work in lectures, disputations, treatises. Again, this is a simplification. Philosophy of science did not appear abruptly, nor was it related only to a transformation of metaphysics, logic, and natural philosophy: its emergence coincided with a rethinking of the foundations of all philosophical disciplines. Philosophy of science appeared, first, as a reflection on philosophy itself, carried out in different disciplines. More precisely, it emerged as a renewed foundation of all philosophical disciplines, serving to guarantee the reliability of their methodologies and concepts, namely, to grant their status as indubitable *scientiae*.

The history of philosophy of science started with Cartesianism. This was because of the peculiar character of Descartes's own approach. Among the various revolutionary aspects of Descartes's theories – in physics, physiology, morals – a more fundamental one was his own use of metaphysics as a foundational discipline. Descartes was well aware of the groundbreaking impact of this approach. On 28 January 1641, for instance, he asked Mersenne – in decidedly naïve terms – not to reveal to anyone that his *Meditationes* were in fact the foundation of his physics:

> Je vous dirai, entre nous, que ces six méditations contiennent tous les fondements de ma physique. Mais il ne faut pas le dire, s'il vous plaît; car ceux qui favorisent Aristote feraient peut-être plus de difficulté de les approuver; et j'espère que ceux qui les liront

https://doi.org/10.1515/9783110569698-001

> s'accoutumeront insensiblement à mes principes, et en reconnaîtront la vérité avant que de s'apercevoir qu'ils détruisent ceux d'Aristote. (AT III, pp. 297–298)[1]

A few years later, when his philosophy was ready to become the philosophy of the schools in the form of his *Principia philosophiae*, he proposed a full-blown, new view of the order of the sciences:

> Ainsi toute la philosophie est comme un arbre dont les racines sont la métaphysique, le tronc est la physique et les branches qui sortent de ce tronc sont toutes les autres sciences, qui se réduisent à trois principales: à savoir la médecine, la mécanique et la morale [...]. (AT XI, p. 14)

Yet, Descartes alone could not have moved metaphysics – and logic – from their historical roles as the sciences of being and reasoning to that of a philosophy of science (or philosophy of natural philosophy, in its embryonic stage). This was the result of a century-long transformation shaped by acrimonious debates, quarrels, splits in the academic curriculum, which took place in Dutch universities, the first ones in which Cartesianism was officially taught. How did the story begin? Not surprisingly, it started with a quarrel – and a betrayal. The history of Dutch Cartesianism, indeed, was not only a history of 'Cartesians vs. Aristotelians'. It was also a Cartesian story, i.e. a long process of discussion on how to use Descartes, and finally on how to get rid of him. Decidedly, it was not a linear story; this notwithstanding, an evolution is recognisable. This book is intended as an account of the history of the foundation of scientific knowledge from the emergence of Cartesianism to its replacement by Newtonianism as the dominant scientific paradigm. This account sheds light on the evolution of foundational theories as these were handled by six major figures in Dutch and Dutch-related contexts: Henricus Regius, Johannes Clauberg, Johannes de Raey, Arnold Geulincx, Burchard de Volder, and Willem Jacob 's Gravesande. This evolution is analysed in the light of the polemics and debates over the uses of Cartesianism, raised by its introduction into the university: these were fuelled by the expounders of alternative world-views – viz. by Aristotelian, Hobbesian, Spinozist, experimental, and Newtonian philosophers – or were internal to Cartesianism itself. As such an evolution took place at the University of Leiden, where Cartesianism became the dominant philosophical and scientific paradigm in the seventeenth century before being replaced by Newtonianism. This book is mainly concerned with the debates and innovations expounded by the philosophers and scientists of this university. Yet, as foundationalism emerged from a broader context of dissemination of new ideas, this analysis will concern the Dutch and Dutch-related intellectual framework.

---

1 Here and hereafter, AT refers to Descartes's *Oeuvres* (Descartes 1897–1913).

Thus, after a historiographical appraisal of the problems of foundationalism in Cartesian philosophy (dealt with in the opening first chapter), the second chapter is devoted to the analysis of the first introduction of and quarrels over Cartesianism at the University of Utrecht, as determined by the teaching of a Cartesian natural philosophy and physiology by Henricus Regius. First, it is shown how his teaching gave rise to the *querelle d'Utrecht* (1641), in which two main issues were debated: the rejection of substantial forms, and the characterisation of man as *ens per accidens*. During the quarrel, questions were raised about the consistency of the new philosophy with theology and – insofar as it raised the problem of the individuation of bodies and species by substantial forms – of the unity of man. Secondly, it is shown how Regius's peculiar approach to natural philosophy (and physiology) led him to quarrel with Descartes himself in 1645–1648, with regard to two main questions: the nature of mind, and the method of natural philosophy, which Regius interprets in both cases from an empirical standpoint, as he maintains that the nature of mind may consist in matter (according to medical evidence), and that all knowledge has a sensory origin. Accordingly, he paved the way for 'radical' interpretations of Descartes which rejected his metaphysics, but also for an approach to a natural philosophy more open to the use of experience. In this chapter, I scrutinise the reasons behind Regius's approach to physics and metaphysics by paying attention, on the one hand, to his quarrel with Descartes, and on the other to the sources of his methodology, which he largely borrows from medical authors. In conclusion, I show that Cartesianism, in the hands of Regius, might serve as the basis for medicine but could not yet become a full-fledged alternative to Aristotelianism.

The third chapter gives an account of the debates over Cartesianism outlined below, which shifted from the University of Utrecht to Leiden, where the new philosophy was introduced by Adriaan Heereboord in the early 1640s, and was carried on by Johannes de Raey at the end of the decade. In Leiden, the quarrels over Cartesianism were prompted by the intervention of the theologian Jacob Revius, criticising Descartes's philosophy as a source of Pelagianism in 1647. This gave rise to a series of attacks, replies, and counter-replies which would dominate Dutch Cartesianism well into the 1650s: Revius's *Methodi cartesianae consideratio theologica* (1648), *Statera philosophiae cartesianae* (1650), and Cyriacus Lentulus's *Nova Renati Descartes sapientia* (1651) offered a full-blown critique of Descartes's philosophy, focusing on his metaphysics, method and on their uses in academia. Such critiques are analysed in this chapter as they brought about the first foundation of Cartesian philosophy after Descartes himself, namely, the development of a 'Cartesian Scholastic' by Johannes Clauberg, professor at Herborn and Duisburg. In fact, Clauberg's defence and foundation of a Cartesian academic philosophy was not the effort of one philosopher, but was coordinated with De Raey and with other members of the Cartesian network in the Netherlands and in Germany (including Abraham Heidanus, Tobias Andreae and Christopher Wittich), as a means to avoid the bans on Cartesianism and to provide a decisive answer to the theologians. This coordinated strategy of defence is revealed by two letters of De Raey to Clauberg: their contents shed light on the background of Clauberg's foundation, which constitutes the first, Cartesian reworking of the academic curriculum. In his works, indeed, Clauberg maintained a metaphysical foundation embodying rational-theological arguments, while providing at the same time a logical theory of the method for natural philosophy, as well as for law, theology and medicine. Yet, his concerns with the traditional structure of the curriculum led him to develop, besides a Cartesian first philosophy, also an *onto-sophia*, whose object is being as such. This is considered by Clauberg not only in the light of the 'first notions' of mind and body, but also of those abstracted from concrete realities, as 'unity', 'goodness', 'truth', and so on. This ultimately results in the reduplication of metaphysics, which will be reduced by Clauberg's followers – dealt with in the next chapters.

The fourth chapter analyses the establishment of Cartesianism at the University of Leiden in 1650s and 1660s. This was carried out by De Raey, who provided a defence and teaching of Descartes's physics in his *Clavis philosophiae naturalis* (1654), although not based on Descartes's metaphysics: physical principles, indeed, are presented by De Raey as self-evident truths, and consistent with Aristotle's theory of *scientia* or universal and necessary knowledge. This was not the only peculiar characteristic of Leiden Cartesianism, as De Raey also provided a differentiation between philosophical and practical knowledge (including medicine and revealed theology), as these are based on different sources of knowledge, namely, intellect and sensory experience. In the hands of Christopher Wittich, the separation thesis became the standard in conceiving the place of Cartesianism in the university, which was confined to natural philosophy in 1650s and strictly secluded from revealed theology. At the same time, the need to develop a moral philosophy consistent with the Reformed creed became the centre of the debate between Revius and Heidanus, a Reformed theologian who saw in Cartesianism a philosophy more consistent with Calvinism than Scholasticism. Accordingly, he was eager to support the appointment in Leiden of Arnold Geulincx, who was developing a philosophical ethics independent of revealed theology but consistent with the Reformed creed. For this purpose, besides the relation of body, soul and world, Geulincx considered those relations of man, world, and God from which moral duties follow. Accordingly, he provided his ethics with a foundation in rational theology. In turn, this foundation entails a reflection on the type of knowledge that constitutes physics and determines its very method. Given the inscrutability of God's reasons in creating the world, Geulincx could claim that physics has to proceed by hypotheses based on experience rather than by a deduction of natural laws from metaphysical principles. In this way, the epistemic consequences of his foundational theory refuelled a reflection on the method of natural philosophy itself, making his foundation a sample of the transformation of Cartesian foundationalism – dominated by Descartes's metaphysical physics – into a reflection on physics itself. In other words, Geulincx provided, together with De Raey, a de-metaphysicisation of physics.

The fifth chapter is a study of the emergence of 'radical Cartesianism' as an actor's category in 1660s and 1670s, which prompted a further development of foundationalism as a reflection on the limits and proper method of philosophy. The key figure in this double process was De Raey, who in the late 1660s started to develop a new logic or metaphysics, intended to counter, on the one hand, the uses of Descartes outside natural philosophy and metaphysics itself, and on the other, the erosion of Descartes's metaphysical tenets. His writings thus turn out to be relevant as he offers a mapping of the interpretations of Cartesianism covering its uses in theology (as by Meijer and Spinoza) and the spreading of Hobbes's materialist philosophy in the Low Countries. Moreover, his logical-metaphysical theories embody a novelty in determining the aim of these disciplines, both with respect to the Cartesian tradition (i.e. to Clauberg and Geulincx) and to the Scholastic one, and start to assume a reflective role as they assess the limits of scientific knowledge. In his hands, logic and metaphysics become one discipline, as they have the same function, i.e. the definition of the principles of scientific knowledge (viz. the notions of physical and mental realities), and of its scope and applications. On the one hand, therefore, he can overcome the reduplication of metaphysics and ontology of Clauberg, and the introductory or didactic function that logic played both in Clauberg's and Geulincx's philosophy. On the other, his logical and metaphysical systematisation can be read as the result of a long-standing debate over the objects of these disciplines (either mere concepts, or entities existing outside mind), in which De Raey places himself by dealing with the theories of Burgersdijk and Ramus. This chapter thus works as a junction in the account of the establishment of Cartesianism as the philosophy of the university, and its challenges by emerging, alternative worldviews, interpretations and uses. Together, these converging theoretical and historical factors shaped a new, reflective function for logic and metaphysics.

The sixth chapter focuses on the evolution of Cartesianism in the last quarter of the seventeenth century in Leiden and Amsterdam, against the background of the emergence of alternative views in natural philosophy capable of replacing it as a dominant paradigm, namely, the experimental philosophy of Robert Boyle and the mathematical-experimental approach of Huygens and Newton. The last evolution of Cartesianism is reconstructed in this chapter by considering the

'Cartesian empiricism' of Burchard de Volder, and the reflections on the language of philosophy and practical disciplines by De Raey in Amsterdam, where he moved in 1669. In 1675 De Volder established the Leiden *Theatrum physicum*. There he performed experiments in order to teach the principles of a mechanical philosophy largely inspired by Descartes but open to the use of experimental and mathematical evidences in the formulation of natural laws. A similar approach was in fact assumed in the same years by Wolferd Senguerd, who used the *Theatrum* to perform experiments in pneumatics to teach some of the principles of his eclectic worldview, encompassing some Cartesian principles (such as that of the circularity of movement), but also rejecting the vortex theory. Yet, only De Volder developed a foundational theory for the basic principles of mechanicism, namely, the assumption that every phenomenon can be explained by the notions of matter and movement alone. This was the result of a movement internal to Cartesianism, as De Volder not only reacted (positively) to the emergence of an experimental-mathematical natural philosophy, but was also involved in the defence of Cartesianism against the *Censura philosophiae cartesianae* of Pierre-Daniel Huet (1689), in which Descartes's metaphysics is rejected as inconsistent given its very foundation on doubt and *cogito*. The intermingling of these different issues resulted, in De Volder's hands, in a further de-metaphysicisation of physics – as metaphysics cannot provide a justification for the laws of motion – and in the narrowing of the scope of foundationalism, which can only sanction the psychological character of clarity and distinction as a criterion for internal truth, defined in terms of indubitability only. Accordingly, for De Volder metaphysics cannot demonstrate, on a Cartesian basis, that phenomena are actually ruled by the principles of mechanism: insofar as, for him, metaphysics has a prominent reflective role, and loses its status as justification of the absolute truth of scientific statements. This process can be labelled as the transition from foundationalism to philosophy of science and does not characterise only his 'Cartesian empiricism'. In Amsterdam, De Raey was, over the same years, developing his *Cogitata de interpretatione* (1692), embodying one of the first, self-standing philosophical considerations of language. Still maintaining his separation thesis, and attacking Hobbes and the radical Cartesians, in this text he aimed to clarify how words meaning sensory data and abstract notions (such as those of mathematics) can be used in philosophy. Rather than setting a method for how to use mental faculties, De Raey aimed, at the end of his career, to provide an updating of the linguistic meanings of scientific vocabulary, namely, a reflection rather than a justification of science.

The seventh chapter focuses on the aftermath of the decline of Cartesianism as a leading force in the Dutch academic context. After De Volder and De Raey, indeed, only Ruardus Andala in Franeker carried on the teaching of Cartesian physics (which he taught by commenting upon Descartes's *Principia*) and metaphysics, mainly for the sake of contrasting Spinozism and other forms of radical Cartesianism. Thus, Descartes's philosophy came a dead end on the eve of the eighteenth century. Yet, Leiden Cartesianism and the Leiden experimental tradition (which could include, after De Volder and Senguerd, the activities of Herman Boerhaave), favoured the introduction of Newtonianism as the standard in teaching natural philosophy. This was carried out by 's Gravesande, who used logic and metaphysics (including rational theology), that is, the chief foundational disciplines in the Cartesian tradition, to introduce students to and justify the assumptions (methodological and ontological) of a Newtonian approach in natural philosophy. This had two outcomes. On the one hand, his arguments in metaphysics (i.e. ontology) have the function of clarifying which objects natural philosophy can investigate. These are natural laws, whose ultimate source in substances or modes cannot be ascertained by intellect alone, as they depend on the power of God. On the other, given the fact that experience is our only means of grasping such laws (which are then mathematically handled) he provides a rational-theological justification of experiential evidence, as capable of providing us with a degree of certainty equal to mathematical evidence. As for the Cartesians, this is still the paradigm of scientific knowledge. Accordingly, with the introduction of Newtonianism foundational disciplines such as logic and metaphysics (including rational theology) served as a justification for and a reflection on a given scientific methodology. Thus, philosophy started to be actually detached from the natural sciences, and to assume a subservient role with respect to them, as once it was the *ancilla theologiae*.

1 The quest for a foundation in early modern philosophy: A historical-historiographical overview

1.1 HPS, &HPS, HOPOS (and history of philosophy)

Since the 1960s the integration of the history of science and the philosophy of science has been substantiated by the presence of university departments offering a curriculum of studies catering to both disciplines. At Princeton University, Charles Gillespie established the first curriculum of studies in the history and philosophy of science – henceforth HPS – in 1960, with the purpose of attracting students to the study of the history of science. In Princeton, history of science was taught by John E. Murdoch, while Hilary Putnam gave courses in philosophy of science. At Indiana University it was a historian of philosophy, Norwood Russell Hanson, who established the first department of HPS, hosting the teaching of historian of science Alfred Rupert Hall. Eventually, the actual integration of history and philosophy of science was brought about by the appearance of Thomas Kuhn's *Structure of Scientific Revolutions* (1962), which, drawing attention to the notion of paradigm in scientific theories, favoured the marriage between historical and philosophical analysis of science. Since 1960s, HPS has flourished, and nowadays at least five universities in the United Kingdom and sixteen in the United States offer degrees in HPS (Mauskopf/Schmaltz 2012, pp. 1–10). More recently, kindred approaches have emerged, such as the 'integrated HPS' (&HPS), endorsed in some international conferences, which aims at merging the two disciplines rather than offering philosophical insights into the history of science and vice versa:

> It must be good history of science and philosophy, in that its claims are based on a solid grounding in appropriate sources and are located in the relevant context. And it must be good philosophy of science, in that it is cognizant of the literature in modern philosophy of science and its claims are, without compromise, articulated simply and clearly and supported by cogent argumentation. (Stadler 2014, p. 761)

Another approach is the 'history of philosophy of science' (HOPOS), established by the birth of the *The International Society for the History of Philosophy of Science* (1996) and of its *Journal* (2011), aiming to "construe this subject (*HOPOS*) broadly, to include topics in the history of related disciplines and in all historical periods, studied through diverse methodologies," and to "explain the links among philosophy, science, and mathematics, along with the social, economic, and political

https://doi.org/10.1515/9783110569698-002

context, which is indispensable for a genuine understanding of the history of philosophy."[1] Both &HPS and HOPOS are forms of HPS which include the study of the reflections scientists had on their own method and theories of understanding nature, and bring the history of philosophy into HPS itself. The condition of these approaches is that philosophy and science, intended as physics or natural philosophy, had been until the eighteenth century a branch of philosophy as such: philosophers were scientists and, in most cases, vice versa. As philosophers, moreover, they often provided reflections (either descriptive or normative) on their 'scientific method' i.e. their ways of proceeding in the study of nature. Accordingly, HPS is more and more linked to history of philosophy and to the historical appraisal of philosophy of science broadly intended as reflection on natural philosophy.

Given their nascent character, the method and the subject matter of &HPS and HOPOS – and by extension of HPS – are still unclear. The flourishing of these disciplines, or, more precisely, of the attempts at defining subject matter and method for HPS through an exploration of its potential scope and interconnections with other disciplines, has prompted meta-philosophical discussions on the status of HPS itself. Thomas Uebel – a philosopher of science – has explained in a collection of essays on *The Present Situation in the Philosophy of Science* the rapid growth of this field as a consequence of a

> change in methodological attitude that late 20th century philosophy of science prided itself on, a change sometimes characterised as a naturalistic turn or even a turn to scientific practice: either way it involves the self-conscious rejection of *a priori* reflection about grand philosophical themes related to science and instead demands detailed knowledge of current scientific theories and experimental practices. (Uebel 2010, p. 13)

Accordingly, HPS is a reflection on scientific practices in their historical development, i.e., it is philosophy of science applied to history of science, and can be more properly defined as "philosophy of science by other means." For Thomas Morman, commenting upon Uebel, this definition has the positive effect of forestalling "a profusion of undesired meta(meta)disciplines which threaten the conceptual unity of an interdisciplinary research dealing with the history and philosophy of scientific culture" (Morman 2010, p. 30).

On the other hand, a historian of science, Peter Dear, has signalled how HPS is today more and more oriented to its integration with history of philosophy: "where once philosophers were taken to lead historians in setting the agenda and questions for HPS, contemporary philosophers of science who look to history do so by following the lead of historians. [...] Perhaps the chief movers in this

---

1 www.hopos.org, cover page, visited on 9 June 2017.

endeavour are Daniel Garber and Roger Ariew, both specialists in Cartesianism"
(Dear 2012, pp. 68–69). In fact, Uebel's definition seems to exclude the possibility of a study of the self-reflection of past philosophers (and scientists) on their
methods and the conceptual premises of their theories, which Ariew and Garber
have successfully accomplished in the last two decades.[2] It is not by chance,
indeed, that the initiators of the history of philosophy of science are specialists
in Cartesianism. In recent years, the historiography of Cartesian philosophy has
been boosted by renewed attention to the problem of the foundation of philosophical knowledge. Instead of a crystallised dualistic approach to metaphysical problems – on the one hand – and to natural-philosophical method – on the other –
the philosophy of René Descartes has been increasingly analysed in light of the
metaphysical problems entailed by his methodology and vice versa. It is now a
consolidated historiographical result that the breakthrough of the philosophy of
Descartes brought about a reflection on the method, assumptions, and functions
of philosophy, vindicated in the development of foundational theories as premises of philosophy as such.[3] His tenet that the source of philosophical knowledge
is to be found in the clear and distinct ideas of reason rather than in sensory experience called for a reformulation of the principles of philosophy, and caused an
unprecedented interest in the foundation of philosophical knowledge both as the
justification of the very possibility of acquiring a certain, indubitable, and secure
knowledge of philosophy (*scientia*), and as the definition of the main concepts,
method, and first principles of the different philosophical disciplines (*scientiae*).[4]
Accordingly, the peculiar character of Descartes's philosophy as a reflection on
natural philosophy seems to justify HPS's focus on Cartesianism today. Moreover,
the developments of scientific theories which mark the early modern age itself,
that is, the gradual overcoming of the Aristotelian worldview determined by the
installation of Copernicanism and the collapse of the Scholastic system of knowledge based on Aristotle's corpus, support the possibility of an HPS that regards
science as we may conceive it today: that is, the mathematical-experimental
study of natural phenomena. Cartesianism, in sum, seems to entail a philosophy of science (either normative or descriptive) which concerns (early-)modern
science, and to mark at the same time an important moment in the process of
the detachment of science from philosophy, which can be acknowledged in the
gradual differentiation between 'professional' philosophers concerned with

---

2 See Ariew 2011 (revised edition of Ariew 1999), Garber 1992, Garber 2001.

3 On the problem of foundation in early modern philosophy, see Hatfield 1990, Garber 1992,
Garber 2001, Garber 2006, Fichant 1998. See also Burtt 1932, Buchdahl 1969.

4 On the meaning of 'scientia' in the history of early modern philosophy, see Achinstein 1968,
Sorell/Rogers/Kray 2010.

metaphysical, logical and moral problems, and 'scientists' concerned with the study of nature, which came to be fully visible in the course of the eighteenth century. In this sense, HPS may emerge as a historical and philosophical study of science independent of the history of philosophy and, at the same time, capable of taking into account the philosophical reflections of scientists at the time they were still 'natural philosophers'. Yet, if Descartes's philosophy may be labelled as a philosophy of science, i.e. as aimed at providing the study of nature with a foundation, the impact of his 'revolution' is to be assessed through its reception. That is, what is to be considered is the function that different parts of philosophy assumed before and after Cartesianism. This book is intended to offer such a consideration: it is a study in the history of philosophy showing why and how Descartes's foundationalism, in the hands of his followers and successors, became an essential part of philosophy throughout the seventeenth and eighteenth centuries. In particular, it will show that the Cartesian quest for a foundation of philosophy shaped a new function of metaphysics and logic as forms of reflection upon the principles of knowledge of natural philosophy, determining a change in the function of philosophy itself – which gradually became a reflection on science – and, in doing so, enhanced the development of philosophy and science as distinct bodies of knowledge.

1.2 Descartes's foundationalism: A historiographical appraisal

Descartes aimed to provide his whole philosophy with a foundation through metaphysical and rational-theological arguments, these being the core of his *Discours de la méthode* (1637), *Meditationes de prima philosophia* (1641), and the first part of his *Principia philosophiae* (1644). His metaphysical foundationalism has been thoroughly surveyed by Garber, who has pointed out the new role of metaphysics with respect to the Aristotelian tradition which dealt with the notion of being as such (special metaphysics) and immaterial entities such as God and angels (special metaphysics or rational theology):

> In its strict Aristotelian meaning, metaphysics was usually taken to be the science of being qua being, the science of being as such. In addition, metaphysics was often taken to include an account of God, separated (i.e., immaterial) substances, and substance in general. [...] Although the view that physics depends in some substantive way on metaphysics was not completely unheard of among medieval Aristotelian schoolmen, physics was generally held to be a discipline largely independent of metaphysics, and as a more concrete discipline dealing with sensible things, it should be studied before the student took up metaphysics. Therefore, in this strict sense, for an Aristotelian, one could not properly talk about the metaphysical foundations of physics. (Garber 2006, p. 21)

Garber has shown the development of the notions of matter and motion in Descartes's metaphysics, interpreting 'foundation' as (i) the cluster of basic notions and principles dealt with in Descartes's metaphysical writings, and hence employed in his natural philosophical theories, and as (ii) the deduction of the principles of motion and laws of impact from the idea of God (Garber 1992, chapters 5, 6 and 9). On the other hand, Daniel Flage and Clarence Bonnen give a slightly different interpretation of what is a foundation in Cartesian philosophy, that is, (iii) the preparation of the mind for the acknowledgment of the first principles of philosophy, and (iv) the justification of the reliability of our faculties in acquiring the truth. Given the possibility of doubt, a foundation ensures the appraisal of the first principles in Descartes's natural philosophy.[5] In fact, the problem of scepticism plays a crucial role in early modern philosophy. As shown by Richard Popkin's classic study, modern scepticism is to be traced back to the 'intellectual crisis' of the Reformation, and to the rediscovery of the arguments of the ancient sceptics.[6] Descartes embarked on an "intellectual crusade" against the scepticism of his time, attempting to overturn the sceptical arguments as the basis of his metaphysics, which starts with radical doubt (Popkin 2003, pp. 144–145). The quest for a foundation, characteristic of Cartesianism, was motivated by the need to reach an evident knowledge through a new method that could prevail over the Aristotelian introduction to philosophy, which was vulnerable to the arguments of the sceptics. As signalled by Flage and Bonnen:

> Consistent with the shift away from an empiricist epistemology, first principles are known by reason, not by abstraction from experience [...]. This epistemological shift marks a significant departure from the Aristotelian tradition, which held that even the truths of mathematics and metaphysics are abstracted from experience. (Flage/Bonnen 1999, p. 112)

Accordingly, a main reason for Descartes's foundationalism was, along with the overcoming of scepticism, the justification of a way of practising philosophy more professionally than in what he labelled as the commonsensical, juvenile Scholastic approach. Descartes's foundation shows why his philosophy, based

---

5 "Descartes gives voice to another theme that recurs throughout the *Meditations*, namely, that the discovery of false beliefs is the motive for systematic doubt. If knowledge is based on the wrong foundations, error can easily follow. [...] The philosopher at the beginning of the *Meditations* is understood as one in a state of philosophical naiveté, one having all the natural biases toward the reliability of sense perception. By weaning oneself from sense experience, one becomes aware of those axioms that are 'in us from birth' [...], and one is in a position to recognize their truth by the natural light," Flage/Bonnen 1999, p. 112.

6 See Popkin 2003, third revised and expanded edition of Popkin 1960. On Descartes and scepticism, see Verbeek 1993b, Grene 1999, Lennon 2008.

on an alternative source of knowledge and upholding the differentiation between the deceiving 'evidences' of sensory experience and the mechanical causes of phenomena, could replace a four-centuries old-Aristotelian tradition in philosophy. The theoretical need for a foundation of knowledge, therefore, went along with the necessity of demonstrating the reliability of Descartes's groundbreaking methodology with respect to the established Aristotelian paradigm.

The relation between foundationalism and the 'rationalist' aim of acquiring evident knowledge has been noted by Tom Sorell in the area of ethics as well, as this is a matter deriving from Descartes's metaphysics and physics, the *scientiae* of soul and body.[7] In fact, he interprets foundation as the derivation of philosophical theories from self-evident principles, as does Garber.[8] However, the problem of foundationalism seems to go beyond the entailments of Descartes's rationalism. Indeed, in recent years the rationalism/empiricism dichotomy has been partially replaced by the categories of 'speculative' and 'experimental' philosophy (Anstey 2005). This has allowed a broader approach to the history of Cartesian philosophy, appreciated in its different experimental-empirical aspects, taking into account the establishment of Newtonianism (Dobre/Nyden 2013a). Yet, this replacement has not disproved the inner connection of Descartes's foundationalism and his search for philosophical principles in pure reason, that is, by a method alternative to the ways of discovery and demonstration established in European universities at the beginning of seventeenth century, such as the methodologies of Aristotle, Galen, and the more recent ones of Ramus and Zabarella.[9] Moreover,

---

7 "Descartes is a rationalist in ethics, but he always regarded ethics as a highly derivative science, partly because he thought that what was worthwhile to do in life depended on *scientia* about the perfections of mind and body (metaphysics) and *scientia* about the workings of the human body (physics)," Sorell 2005, p. 83. I will deal with the relations between Cartesianism and Protestant ethics in chapter 4.

8 "Rationalism [...] can be associated with foundationalism, the idea that there are a small number of self-evident truths in the light of which all or most other truths are evident, or from which other truths can be derived by self-evident reasoning. Cartesian rationalism extends to ethics and the conduct of life, where it asserts that detachment from the appetites is sometimes necessary for distinguishing genuine from merely apparent goods, and for identifying an order of priority among the genuine goods," Sorell 2005, p. xii. See also chapter 3, *The belief in foundations*, pp. 57–84.

9 On the Ramist tradition in logic and its derivations, Ong 1958, Hooykas 1968, Ashworth 1974, Schmidt/Biggemann 1983, Bruyère 1984, Leinsle 1985, Ashworth 1988, Robinet 1996, Feingold 2001, Verbeek 2001, Skalnik 2002, Meerhoff/Magnien 2004, Hotson 2007, Reid/Wilson 2011, Vasoli 2007, Angelini 2008, Goulding 2010, Pozzo 2012, Sgarbi 2012, Strazzoni forthcoming a. On the Paduan tradition in logic, see Poppi 1969, Poppi 1970, Poppi 1972, Bottin 1972, Papuli 1983, Schmidt/Biggemann 1983, Berti 1983, Risse 1983, Mikkeli 1992, Reiss 2000, Sgarbi 2012, Sgarbi 2014. For an introduction to these themes, see Gilbert 1960, Risse 1964, Risse 1970, Dear 1998, Nuchelmans 1998.

questions have been raised by the analysis of the Scholastic background of Descartes's foundation: John Cottingham has shown how "supporting the trunk of his physics by unearthing its metaphysical roots, gradually overwhelmed Descartes by its complexity; and that in attempting to complete the task, he was drawn, little by little, to fall back on the very Scholastic apparatus that he so derided in his scientific work" (Cottingham 2008a, p. 58). Descartes's use of the Scholastic tradition in metaphysics has been thoroughly unveiled by Roger Ariew, particularly where Descartes's relation to the Scotist notions of objective being, idea, and the self-substantiality of pure matter is concerned (Ariew 2011). In sum, we are now in possession of an image of Descartes which takes into account his awareness of the conceptualisation of past philosophy, the groundbreaking novelty of his approach, and of the interrelation of physics and metaphysics, as well as his blind spots in his own theories.[10]

The use of metaphysics in the foundation of physics by Descartes is, in itself, a foremost example of the use of a philosophical discipline as a form of reflection upon the functioning of human faculties in attaining clear and indubitable knowledge (*scientia*) and on the basic notions of natural philosophy. However, it still not evidence for the change in function of philosophy as a collective enterprise and in relation not only to Cartesian (or Cartesian-inspired) physics. If the problem of the foundation of Descartes's natural philosophy has been exhausted, the reception of Descartes's foundationalism has not received the same attention. Accordingly, we are still not in possession of a clear picture of the historical development of Descartes's foundationalism which may corroborate – or provide with a foundation – contemporary approaches to philosophy and science from the point of view of an HPS integrating the history of philosophy. On a more historical level, we do not have a view of the changes in function of philosophy as a discipline practised and taught in European academies in the early modern age. As a main hypothesis, I will assume that present-day methodological obscurities are directly related to our deficient recognition of Cartesianism as a historical factor reshaping the relation of philosophy and science.

Whilst some studies on the problem of a foundation of Cartesian philosophy in different contexts have appeared – such as the doctoral dissertations of Mark Aalderink, concerning Descartes's theory of knowledge and its reception by the Flemish philosopher Arnold Geulincx – himself a professor at Leiden – (Aalderink 2009) and Mihnea Dobre (Dobre 2010, Dobre 2013a, Dobre 2017), who has focused on the foun-

---

10 Not surprisingly, the study of the self-awareness of Descartes as a philosopher went along with a growing interest for the 'biographic' development of his philosophy: see Gaukroger 1995, Clarke 2006, Condren/Gaukroger/Hunter 2006.

dation of natural philosophy in French Cartesianism – the topic has been generally neglected. The scholarly tradition on the dissemination and reception of Descartes's philosophy and science – on which I address the reader to Dobre/Nyden 2013b for an up-to-date, full-blown account – is nowadays substantiated by a ponderous body of literature which had its initiators in the classic studies of Ast, Damiron, Bouillier, Cousin, Ueberweg, Monchamp, Bohatec, Brulez, Brunet, Mouy, Thijssen-Schoute and Dibon.[11] Cartesian studies have mostly focused on those geographical contexts in which Cartesianism inspired the curriculum of universities – as in the Dutch and German areas – and of learned circles and scientific academies – as in France.[12] As to the Dutch context, Cartesian studies have been largely dominated by an approach focusing on the institutional clashes between philosophers and theologians, and on the interrelations of Cartesianism and Spinozism.[13] In Germany, Cartesianism has been studied especially with regard to its uses in medicine,[14] to its influence on seventeenth and eighteenth-century *Schulphilosophie*,[15] and to its contribution to German radical Enlightenment.[16] In France, Cartesian studies have focused mostly on the uses of Cartesianism in natural philosophy,[17] theories of matter and mind, and the polemics these brought about.[18] Furthermore, contexts in which Cartesianism did not constitute a dominant paradigm (as the emergence of new paradigms was embodied by more experimentally-oriented bodies of knowledge) – such as in the English[19] and Italian[20] ones – have enjoyed remarkable attention. A view of the transnational development of Cartesian philosophy, eventually, is at a mature stage of development (for a comparative treatment, see Schmaltz 2016a, and the collec-

---

11 Ast 1807, Damiron 1846, Bouillier 1854, Cousin 1866, Ueberweg 1866, Monchamp 1886, Bohatec 1912, Brulez 1926, Brunet 1926, Mouy 1934, Thijssen-Schoute 1954, Dibon 1954.

12 On the history of French scientific institutions, see Brown 1934, Brockliss 1981, Brockliss 1987, Hahn 1971, Hirschfield 1981. On Dutch and German ones, see *infra*.

13 Brunet 1926, Dibon 1954, Dibon 1990, Thijssen-Schoute 1954, Ruestow 1973, De Hoog 1974, McGahagan 1976, Verbeek 1992a, Van Ruler 1995, Van Bunge 2001. See also Dijksterhuis 1950, Vanpaemel 1985, Frijhoff/Spies 2004, Schmaltz 2005a, Schmaltz 2016a.

14 Rothschuh 1953, Rothschuh 1968, Trevisani 1992, Trevisani 2012, Smith 2013, Omodeo 2017, Theis/Ferrari/Ruffing/Vollet/Guenancia 2009. See also Angyal 1941.

15 Althaus 1914, Wundt 1939, Freedman 1984, Wollgast 1988, Mulsow 2009. See also Blum 1998.

16 Mulsow 2002, Mulsow 2015, Suitner 2016.

17 Mouy 1934, McClaughlin 1977, McClaughlin 1996, McClaughlin 2000, Vanpaemel 1984, Clarke 1989, Brockliss 1995, Des Chene 2002, Goldstein 2008, Shank 2008, Dobre 2010, Dobre 2013a, Dobre 2017, Borghero 2011, Roux 1998, Roux 2006, Roux 2013a, Roux 2013b, Ariew 2014.

18 Watson 1966, Watson 1998, McClaughlin 1979, Lennon 1993, Schmaltz 2002, Schmaltz 2005b, Shelford 2007, Lennon 2008, Nadler 2011, Ariew 2013, Ariew 2013.

19 Webster 1969, Gabbey 1982, Hutton 1990, Fouke 1997, Atherton 2005, Jesseph 2005, Dessì/Lotti 2011, Jalobeanu 2011, Hatfield 2013.

20 Belgioioso 1985, Belgioioso 1992, Belgioioso 1999, Belgioioso 2005, Armogathe 2005.

tions of essays of Lennon, Sorell, Schmaltz, Borghero, Del Prete, Garber, Roux, Antoine-Mahut, Gaukroger).[21] And yet, notwithstanding the fact that we now have a detailed view of the various declinations of Cartesian philosophy throughout the seventeenth and eighteenth century, various factors have prevented the writing of a history of foundationalism as a topic underlying the history of Cartesian philosophy and of early modern philosophy as such. First, the failure of Cartesian physics in the second half of the seventeenth century, that is, its speculative claims, the absence of a mathematical formulation of natural laws, and the ambiguous use of experience have brought about an image of Cartesianism as a dead branch in the history of early modern philosophy and science. Accordingly, scant attention has been paid to the interrelation of physics, metaphysics, and other branches of philosophy apart from the case of Descartes. As a consequence, Descartes's followers have mostly been labelled as uninteresting figures in early modern philosophy and science, whereas their master brought about an actual novelty in the history of philosophy. Secondly, the dichotomy of Cartesianism and Newtonianism as different scientific alternatives has hindered the appreciation of philosophical foundationalism as a topic entangled with the birth of modern science as such. Insofar as foundationalism has been considered essential to Descartes's 'rationalism', its relevance in early modern philosophy has not been systematically dealt with. Thirdly, the ongoing dichotomy of empiricism and rationalism has prevented foundationalism from prevailing over such perspectives on the history of early modern philosophy.[22] As a result, even though it has been assessed why Cartesianism called for a foundation of philosophy – i.e., as an answer to the 'sceptical crisis' and as a defence of a novel way of reasoning in philosophy – it is yet to be understood where the construction of the philosophical edifice started, and what the reasons were for the different solutions given to the problem of a foundation after Descartes. An answer to these questions will show why (and how) a process of internal transformation of Cartesian foundationalism gave rise to a philosophy of science in the early modern age. Such internal transformation is to be studied by a survey of the problems emerging during the reception of Cartesian ideas, that is, from the clash between some issues underlying Descartes's philosophy and its adaptation to particular historical demands. Moreover, it must be made certain that the solutions to such problems created a philosophical framework capable of accepting scientific renovation beyond Cartesian philosophy itself. This book is intended to contribute to the solution of these questions. In this, it focuses on

---

21 Lennon/Nicholas/Davis 1982, Sorell 1993, Lennon 2003, Schmaltz 2005a, Borghero/Del Prete 2011, Garber/Roux 2013, Kolesnik-Antoine 2013a, Antoine-Mahut/Gaukroger 2016, Schmaltz 2016a.
22 For a more detailed account of contemporary perspectives and method of historiography early modern philosophy, see Laerke/Smith/Schliesser 2013, Lenz/Waldow 2013.

the geographical, cultural, and institutional context where Cartesianism impacted and confronted the established Scholastic paradigm: i.e. Dutch and Dutch-related – i.e. German – universities of the mid-seventeenth century, as those of Utrecht, Leiden, Amsterdam, Groningen, Herborn, Duisburg, where philosophers who acknowledged themselves as Cartesian or relied on Cartesian notions were active throughout the seventeenth century, and whose work contributed to the creation of a unified philosophical background in the universities. In fact, the use made of Descartes in the Netherlands and Germany was determined by the needs of the university. The idea of a 'Dutch Cartesianism' is nowadays accepted as a historiographical category in the scholarly context, insofar as Cartesian philosophy had its main dissemination in the universities and soon became a 'philosophy of the Schools', and was shaped through a constant confrontation with Scholastic theories, leading to clashes, attempts to create syntheses, or to fit with new content into old ways of teaching and organising philosophy. Scholastic philosophy, taught in Dutch academies mainly through the textbooks of Franco Burgersdijk (1590–1635), did not include a foundation of philosophy as a justification of its reliability, but only an introduction to logical instruments of philosophy. Metaphysics, on the other hand, was among the last disciplines to be taught, and included a science of being – *metaphysica generalis* – and the basics of a rational theology – *metaphysica specialis* (Bos/Krop 1993, Krop 2003a). With the emergence of Cartesianism, this order of disciplines would change: Cartesianism brought about a replacement of the disciplines hitherto based on the Scholastic paradigm as parts of the official curriculum of the universities, calling upon a reflection on philosophy as such. Thus, foundationalism became crucial for the justification of the adoption of Cartesianism and Newtonianism and for their development into full-fledged philosophical alternatives. Even if with the rise of Cartesianism a foundation was provided in different contexts – as Dobre has demonstrated – the rise of foundationalism can be appreciated through a study of the Dutch and Dutch-related academic context, where the Cartesian revolution led to a substantial change in the official academic culture. Yet, although until now historians of philosophy and intellectual historians have devoted several works to the history of philosophy in Dutch and German universities[23] in the seventeenth century, they have not provided a comprehensive view of how the role of philosophy in the universities changed as a consequence of the introduction of the new philosophy. Moreover, even if the partial dismissal of the

---

23 On the relevant institutional context in Dutch and German universities, see Dibon 1954, Dibon 1990, Kuiper 1958, Menk 1981 Van Berkel 1985, Smit/Jensma 1985, Trevisani 1992, Trevisani 2012, Verbeek 1992a, Cerrato 1999, De Haan 1993, Van Rijen 1993, Hotson 1994, Hotson 2007, Freedman 1999, Feingold/Freedman/Rother 2001, Otterspeer 2001, Wiesenfeldt 2002, Van Miert 2009, Reid/Wilson 2011, Huussen 2013.

aforementioned historiographical dichotomies has recently led to the appearance of some work in the history of philosophy of science in the Dutch and Dutch-related context, one can still see difficulties in discerning a common element underlying the development of philosophy in this context in the early modern age through Cartesianism and Newtonianism.[24] According to this book, this common element is the appropriation of metaphysics (including rational theology) and logic as foundational disciplines which provide reflections (both prescriptive and descriptive) on natural philosophy up to the emergence of Newtonian mathematical-experimental science. This book will bring the analysis of 'Dutch Cartesianism' to the next level, and will fill a gap in current historiography. Yet, this will be incidental to the main purpose of the book, which will not be a history of Dutch Cartesianism *qua* Dutch: but *qua* characterised, more than in other contexts of dissemination, by a rethinking of the use of philosophy as this constituted the foremost part of academic education. Accordingly, this is not a rewriting of the history of Dutch Cartesianism, but a new chapter in the narration of such history. Moreover, it is part of a history of Dutch philosophy in the period between Cartesianism and Newtonianism. Indeed, foundationalism not only characterised Cartesian philosophy, but Newtonian science, too, was provided with a philosophical defence and introduction in Dutch universities.

1.3 From foundation to philosophy of science: Leading problems

As to the problems which caused Cartesian foundationalism to bring about a change in function of philosophy as a reflection on science, these are revealed by the history of Dutch Cartesianism and will be made clearer in the course of this book. Since Descartes aimed to replace Scholastic philosophy, the foundation of Cartesian philosophy must be considered as a comprehensive corpus of academic disciplines: logic, moral philosophy, metaphysics as science of being, and with respect to the higher arts (medicine, theology, and law). So the first question must be whether the foundation of philosophy as a purely rational enterprise was consistent with a plan of reform of the whole course of philosophy, which included disciplines consistently based on experience. Secondly, as a main aim of Descartes was to develop a moral philosophy as the foremost among the sciences (Rutherford 2013), it must be clarified why a Cartesian ethics needed a foundation. In the case of Spinoza, for instance, one finds a rational ethics developed without a foundation, as Spinoza started his *Ethica* with a series of axioms the

---

24 See Schliesser 2011, Janiak/Schliesser 2012, Jorink/Maas 2012, Dobre/Nyden 2013a, Ducheyne 2014a, Ducheyne 2014b.

evidence for which is not justified.[25] Thirdly, needing to be verified is the impact on the foundational theories of the emergence of a mathematical-experimental science which debunked the principles of Cartesian physics, such as the discovery of the laws of impact by Huygens, Wren, and Wallis in the 1660s. The question must be asked whether a foundation of physics could enable an integration of such science in a Cartesian framework, in terms of the weight of experience in the formulation of the principles of philosophy and the use of experiments as a means of discovery and teaching.

In the course of the survey, different kinds of foundations of philosophical knowledge will be confronted as answers to such problems. As a general thesis, I will maintain that the foundation of philosophy – first and foremost, of natural philosophy – was carried out through logical and metaphysical arguments (including rational-theological ones), which fitted the need to answer the introductory requirements of new paradigms in the university, demonstrated the truth of their principles, and assessed the reliability of the new methodology in leading the mind to grasp such truth. A rational-theological foundation relies on a conception of God in order to ensure that our faculties do not deceive us; a logical foundation consists of a survey of the ways we deal with the contents of our mind, and a metaphysical foundation is an examination of the basic concepts of philosophy and science. Such solutions can be traced back to Descartes's writings, where the foundations of his philosophy are embodied by i) the notions of metaphysics matching a dualistic worldview, i.e., the basics of a Cartesian ontology; ii) the drawing of physical laws from metaphysical principles, such as the notion of body, soul and their modes, and the perfections of God; iii) the preparation of the mind for philosophy as the cleansing of the mind from the Aristotelian errors; iv) the justification of the right functioning of mental faculties: as to a) evidence as the criterion of truth, provided by means of doubt and *cogito*, and to b) the truth of past demonstrations, by the appeal to the goodness of God.[26] Descartes located

---

25 "Often this metaphysics served (as for Descartes) not only to give a foundation for the new sciences but especially to defuse the threat that this new science seemed to pose to traditional religious culture. Spinoza's metaphysics constitutes a totally different approach. Not only is there no need for a separate justification of the new, scientific way of thinking – certainly not one arrived at through the proof of a personal, benevolent God – but, for Spinoza, it also is possible to reflect about being and man [...] and even about God," De Dijn 1996, p. 10.

26 These points have been thoroughly addressed in secondary literature: on Cartesian ontology, see Garber 1992, Marion 1993, Watson 1998, Schmaltz 2008, Anstey/Jalobeanu 2011. On Descartes's deductive physics, see Clarke 1982, Garber 1992 (chapters 7, 8), Garber 2001, Schuster 1993, Gaukroger/Schuster/Sutton 2000, Gaukroger 2002, Lüthy 2006, Zittel 2009, Sorell 2010, Schuster 2012. On the introduction to philosophy, see Williams 1998, Flage/Bonnen 1999, Broughton 2002, Sorell 2005, Clarke 2006, Curley 2008. On evidence as truth criterion, see Alanen 2003,

the actual place of the consideration of these issues in metaphysics, which he conceived as the roots of the whole tree of philosophy in the French edition of his *Principia* (1647), constituted, as to the other parts, of the trunk of physics and the branches of medicine, mechanics, and moral philosophy. Thus, metaphysics is the foundation of Descartes's physics, and includes a consistent use of rational-theological arguments. On the other hand, he excluded any use of logical considerations in his philosophy, as these are, according to him, not provided with any function of discovery (Gaukroger 1989). Dutch philosophers would address all these strategies and develop – or reject – Descartes's arguments according to their own interests and standpoints. Accordingly, this book will show why and how Descartes's arguments were used, rejected, or survived in 1) the foundation of philosophy as a comprehensive corpus of academic disciplines and as a rational enterprise, 2) the foundation of a philosophical ethics, 3) the Cartesian foundation of an empirical science.

As a matter of fact, these three foundational strategies cannot be clearly distinguished at all. An analysis of the foundation of philosophy and science raises a large number of questions, both historical and methodological. The problem of the definition of the basic concepts to be used in such research – as well as in any study of historical topics, since these are defined by our pre-conceptions – reveals the interplay of these two kinds of issues. Accordingly, historical analysis may help in solving a methodological problem, which, in turn, can shed light on the very history of philosophy, resulting, however, in a circle between history and theory. Such circularity entailed by a historical approach aimed at solving methodological problems and vice versa can be successfully addressed by paying attention to the problems expressly faced by early modern philosophers: in this case, to the problem of a foundation of philosophical knowledge, the different solutions to which can be regarded as interpretative categories in the history of philosophy. The main problem emerging from my study, in fact, is that of the definition of the meaning of 'foundation' itself, which I will regard as a heuristic concept which can be clarified through a historical analysis. By 'foundation' I will primarily refer to those arguments aimed at providing philosophy and science with a demonstration of their reliability in acquiring the truth, that is, as the demonstration of the reliability of human faculties in philosophical and scientific reasoning. Secondarily, I will refer to foundation as the theory providing

---

Clarke 2005, Ayers 1998, Datson 1998, Gaukroger 2008, Patterson 2008, Boyle 2009, Curley 1993. On Descartes's use of rational theology as a guarantee for rational truths, see Della Rocca 2005, Cottingham 2007, Cunning 2010, Cunning 2014, Lennon 2014. On Descartes's theory of knowledge (in its uses in rational theology and physics), see Perler 1996, Perler 1998, Goudriaan 1999, Wohlers 2002, Schmidt 2009, Barth 2017.

philosophy and science with its ontological apparatus, as to the notions of mind, body, and God. Thirdly, foundation is the introduction of students and scholars to new ways of thinking, and – fourthly – the deduction of the first principles of various branches philosophy and science from logical, metaphysical, and rational-theological notions. Since this book is aimed at developing a history of philosophy of science as a study of the self-reflection of philosophers on philosophical and scientific knowledge the main focus will be on the first kind of foundation. It is however to be noted that this notion is twofold: in fact, this kind of foundation can have both a prescriptive and a descriptive role with regard to the methodologies to be used in philosophy and in the investigation of nature as such. Namely, a foundation can be developed as a justification of a given method or of actual scientific practices, or as the setting for this method. However, the descriptive and prescriptive roles of the foundation of science, in most of the authors here analysed, coexist: the establishment, clarification, description, and correction of a methodology are provided in the light of a reflection on the limits of scientific knowledge and go along with the changes in actual scientific practices. The development of a philosophy of science out of its foundation, accordingly, serves to unravel the gradual change in function of some academic disciplines (logic and metaphysics) towards this twofold role. This double characterisation, in fact, can also be found in the second notion of foundation, since the unravelling of the ontological premises of natural philosophy (and of the other branches of philosophical investigations, as well as of related disciplines such as medicine) can be intended both as a prescription and a description of those entities and features which science is about. In the third case, foundation as the introduction of students to new (i.e. Cartesian or Newtonian) ways of thinking is more prescriptive, given its intrinsic, didactic character. The fourth sense of foundation, in turn, designates the specific use of a discipline – especially metaphysics or rational theology – as an actual part of another, namely, natural philosophy: this is the case of Descartes's 'metaphysical physics' as this is set out in the first and second part of his *Principia philosophiae*, although Descartes himself assigned to metaphysics (as this is expounded in his *Discours de la méthode, Meditationes* and also in the first part of his *Principia*) the function of reflecting on the possibilities, limits, and concepts of philosophy. As all these notions of foundation were often linked in early modern philosophy, I will from time to time consider all the notions of foundation. Foundation or 'fundamentum' is indeed a metaphor: the use of this concept allowed philosophers to convey the rough idea of a 'construction' of disciplines provided with a basis. The very terminology adopted by early modern philosophers did not often discriminate between metaphysics, rational theology, and logic. Moreover, there was no agreement on the possibility of providing philosophy with a foundation by rational means. Therefore, rather than imposing strict distinctions between the concepts and methods of the foundation, I will focus on the ways in which the meanings of the term and the different kinds of solutions – logical, metaphysical, and rational-theological – were dealt with by theories, in which one of these alternatives played a leading role. This does not mean that one cannot ascertain an evolution in foundationalism: on the contrary, this book is dedicated to unravelling the reasons for the emergence of a foundation intended mainly in the first two senses, that is, as a reflection (both descriptive and prescriptive) on the conceptual-ontological apparatus, method and limits of scientific knowledge. As to the parts of philosophy provided with a foundation I will consider, first and foremost, physics or natural philosophy, broadly conceived as the study of the natural and material world. However, I will also take into account the possible relations of natural philosophy with other disciplines, such as ethics, medicine, and law. Accordingly, this study will make it possible to define the assumptions and the scope of philosophy, the relations between disciplines, their epistemic status, i.e. their being capable of reaching different kinds of certainty, and their ends. Finally, an analysis of the problem of the foundation will make it possible to answer the problem of the functions of philosophy in early modernity, assessing the specificity of Cartesianism and Newtonianism with respect to Aristotelianism.

2 The 'crisis' of foundationalism: Regius and Descartes

2.1 Regius and the Utrecht Crisis (1641)

For a history of the Cartesian foundation of science, two events are to be credited with prompting the early discussion over the methods and concepts of natural philosophy. These events are well known to intellectual historians: the Utrecht Crisis (1641–1642) and the Leiden Crisis (1647) (Verbeek 1992a, Verbeek 1994). These crises had a direct impact on the development of foundational strategies by early Dutch Cartesians, as they touched upon the topics of the nature and functioning of mind, the method of philosophy, and the relations between philosophy and the higher arts of theology and medicine. The protagonists of these debates were, on the one hand, declared adversaries of Descartes such as the theologians Gysbertus Voetius, Martin Schoock, Jacob Revius, and Cyriacus Lentulus, and on the other, philosophers who expressly aimed to provide a defence of Descartes's philosophy: as to the 'first generation', Henricus Regius and Adriaan Heereboord; as to the second, Johannes Clauberg and Johannes de Raey. Yet, the development of a foundationalism internal to Cartesianism is not explained only by considering the clash between the Cartesians and the Aristotelians. Since its early phase, indeed, a main factor of development had been the use Regius made of Descartes's philosophy. This caused the Utrecht Crisis and led Regius to clash with Descartes himself in 1645.

Regius was the first to teach Descartes's natural philosophy at a Dutch university.[1] Yet, he was not educated as a Cartesian.[2] After having obtained his MA in Franeker in 1616, he matriculated at the medical faculties of the Universities of Groningen (1617) and Leiden (1618). In 1623, after his grand tour of France and Italy, he graduated from the University of Padua (Farina 1975, Bos 2002). When he came back to Utrecht in 1634, he started to lecture privately on Cartesian natural philosophy, to which he had been introduced by Reneri. Apparently, he was able to obtain a position at the University of Utrecht in 1638 – at first as extraordinary professor of theoretical medicine – thanks to such private teaching and Reneri's recommendation.[3] In his *Epistola ad Patrem Dinet*, moreover, Descartes reports

---

1 Regius's friend Henricus Reneri did not provide systematic teaching of his philosophy in the university: see Sassen 1941, Buning 2013.

2 On Regius's education, see De Vrijer 1917, Dechange 1966, Rothschuh 1968, Farina 1975, Bos 2002, Strazzoni forthcoming b.

3 See the official report of the University on the Utrecht crisis, *Testimonium Academiae Ultraiectinae, et Narratio historica* (Voetius *et al.* 1643, also in Verbeek 1988): p. 9.

https://doi.org/10.1515/9783110569698-003

that Regius was appointed at the medical faculty because he wrote a comprehensive textbook on Cartesian physiology, which he presented to the friends supporting him before the city authorities.[4] This textbook is important because it is an original development of Descartes's physical principles, which Regius was able to derive from Descartes's *Essais* and *Discours de la méthode*, having read a draft of Descartes's *Le monde* only after April 1641 (Verbeek 1994). This early acquaintance with Descartes's natural philosophy explains in part the rejection by Regius of Cartesian metaphysics, which was fully developed only in Descartes's *Meditationes de prima philosophia* (1641–1643) and *Principia philosophiae* (1644). Besides this early contact with Cartesianism, indeed, it was his medical approach and interests that shaped his use of Descartes's philosophy. After defending Descartes's model of blood circulation in a disputation (Regius 1640), his developments of a Cartesian philosophy resulted in two series of disputations on medical matters he held in 1641: his *Physiologia sive cognitio sanitatis* and the shorter *De illustribus aliquot quaestionibus physiologicis*. The *casus belli* of the Utrecht Crisis was his infamous characterisation of man as *ens per accidens* in the latter series. In discussing the notion of form, Regius distinguishes between a general form, which pertains only to matter and consists of the *comprehensio* of movement, rest, position, and figure of its parts (Regius 1641b, disputation II, thesis 16), and a special form, which is human mind, whose nature cannot be accounted for by the features of the general form. For Regius this position has two consequences: first, no entities such as substantial forms can be recognised in matter; second, as it is a complete substance, mind does constitute a *unum per se* with the body: therefore, man is an *ens per accidens* (Regius 1641b, disputation III, theses 8 to 10). This expression, which Regius borrowed from the Dutch atomist David Gorlaeus (Verbeek 1992b), aroused harsh criticism from the Reformed theologian Gysbertus Voetius, who took occasion to attack the whole *nova philosophia* of Descartes in his *Appendix ad Corollaria theologico-philosophica nuperae disputationi de Iubileo romano: De rerum naturis et formis substantialibus*, discussed in December 1641. Besides rebuking Regius's idea of man as *unum*

---

4 "Doctor quidam medicinae [...] legit Dioptricam meam et Meteora, cum primum edita sunt in lucem, ac statim aliqua in iis verioris philosophiae principia contineri iudicavit. Quae colligendo diligentius, et alia ex iis deducendo, ea fuit sagacitate, ut intra paucos menses integram inde Physiologiam concinnarit, quae, cum privatim a nonnullis visa esset, eis sic placuit, ut professionem medicinae, ibi tunc forte vacantem, pro illo, qui antea ipsam non ambiebat, a magistratu petierint et impetrarint," AT VII, pp. 582–583. Regius's *Compendium physicum*, revised through the years, is mentioned in the correspondence of Regius and Descartes, in their *Responsio*, in Descartes's *Epistola ad Voetium*, and in the *Narratio historica*, under various titles: *Physiologia*, *Compendium physices*, *Prodromus novae philosophiae*, *Physica fundamenta*: see Bos 2002, p. 40.

per accidens as endangering belief in the resuscitation of bodies (Voetius *et al.* 1643, 48), Voetius was mostly concerned with the negation of the existence of substantial forms. This had pernicious consequences: first, it entailed the denial of the existence of any individual things in nature. Accordingly, Cartesianism would make it impossible to use philosophy for the explanation of Bible, in books such as *Genesis* and *Proverbs*, where species and natures are overtly mentioned.[5] Second, it would equate (divine) creatures with (human) artefacts; again, this is something decidedly unwarranted according to *Psalms*, *Numbers*, and *Hebrews*.[6] Even though denouncing Regius's use of the formula 'unum per accidens' as offensive to theologians and his overt rejecton of substantial forms,[7] Descartes backed the answer of Regius to Voetius which appeared in February 1642, *Responsio sive Notae in Appendicem*. This piece, however, made the situation worse, as Regius quoted and explained the very passages referred to by Voetius, claiming that these do not mention any substantial forms; rather, natures, faculties, and species – which can be accounted for on mechanical principles – nor do they contradict the possibility of a creation of the world according to mechanical principles (Regius 1642, pp. 14, 18–19). For Voetius, this would amount to Regius venturing into biblical interpretation. As he presented an official complaint, together with other professors, to the Utrecht town municipality,[8] the Senate of the University of Utrecht officially condemned Regius for having entered into a quarrel with theologians, and his new philosophy as prejudicial to the comprehension of the concepts used in the higher arts.[9]

---

5 "Considerent, an sibi satisfaciant in conciliatione huius opinionis cum Sacra Scriptura [...] Vide Gen. 1.11.21.22.24.25. Proverb. 30.24.25 26. 26 25. Ubi permanentes naturas, facilitates, et species rerum distinctas innui putamus," Voetius *et al.* 1643, p. 39.

6 "Sequeretur facultates proprias, et intrinsecas earumque principia in animalibus, alterius generis nulla esse, quam in automatis [...] et consequenter opificia Dei et naturae per creationem aut generationem producta, essentialiter et univoce eadem esse cum operibus artis. Quod quomodo cum Psal. 104.29 et 7.14.15. Numer. 16.22 et 27.16 Hebr. 11.9.10 satis conveniat, fateor me nondum videre," Voetius *et al.* 1643, p. 41.

7 See the letters of Descartes to Regius of the second half of December 1641 and of late January 1642: AT III, pp. 460–461, 491–492; also in Bos 2002, pp. 90–91, 98–99.

8 "Quoniam collega noster, propugnator novae philosophiae, non parum abuti potest, atque etiamnum, ad ea, quae intendit, efficienda, abutitur lectionibus illis philosophicis, quae certis de causa a vestr. ampl. ei sunt concessae, ut de illis tale quid a vobis decernatur, ut tota res eo melius extra omne periculi discrimen constituatur. Et quandoquidem edito isti libello responsum videtur opponendum, rogamus vestr. ampl. ut consideret, qua ratione, modove hoc potissimum fieri possit," Voetius *et al.* 1643, p. 61.

9 "Displicere sibi eum agendi modum, quo collega alius in alium libros aut libellos publice edat, praesertim expresso nomine, [...] pugnantes cum caeteris disciplinis et facultatibus atque inprimis cum orthodoxa theologia," Voetius *et al.* 1643, p. 66.

Although in his subsequent answer to the Utrecht municipality and to Voetius himself, conveyed in the *Epistola ad Patrem Dinet* (published with the second edition of the *Meditationes metaphysicae*, 1642) Descartes claimed that his philosophy was a rediscovery of the most ancient truths and could not contradict theological truths,[10] the Utrecht Crisis made clear the problems of the use of the new philosophy as part of the university curriculum. These problems emerged as a consequence of the peculiar character of Regius's Cartesianism, which would lead him to clash with Descartes himself in 1645, and to which the following generation of Dutch Cartesians reacted in order to secure the place of Cartesian philosophy in the university. Such a peculiar character consists in the impossibility, for Regius, of solving metaphysical problems by rational means, and was ultimately determined by the medical orientation of his natural philosophy. Though aimed at improving the scientific status of medicine, Regius's use of Descartes's natural philosophy undermined its very metaphysical basis and exposed the new philosophy to the issue of scepticism.

The erosion of Descartes's metaphysics was twofold: on the one hand, by assuming an empirical standpoint on the sources of knowledge, and adding medical evidence to corroborate his claims, Regius could not prove the immateriality and immortality of the soul. On the other, the method he used to expound his natural philosophy, based on that of the physicians, prevented the formulation of any demonstratively certain physical models. *A posteriori*, the medical orientation of Regius also explains his lack of acumen in dealing with the notion of *unum per accidens*, i.e. the foremost early signal of Regius's difficulties in dealing with metaphysical matters.

2.2 A medical standpoint on philosophy

The contents of Regius's first, unpublished treatise were probably used for his first comprehensive work, i.e., his series of disputations *Physiologia* (1641).[11] Since

---

10 "Quantum ad theologiam, cum una veritas alteri adversari nunquam possit, esset impietas timere," AT VII, p. 581.

11 Regius 1641a (also in Bos 2002, pp. 195–248: for bibliographical details, see pp. 195–196). The two first disputations are *De sanitate*; the next four are *De actionibus animalibus*. Later, other disputations completed Regius's *Physiologia*: they took place in 1641 and 1643 and were on medical topics only: *De morbis, De symtomatis specialibus, De morborum signis* (1641), *De diagnosi et prognosi morborum, De hygienia, De therapeutica* (1643). They cover the topics later addressed in his *Fundamenta medica* (Regius 1647a). On Regius's early physiology, see Dechange 1966, Rothschuh 1968, Strazzoni forthcoming b.

the official chair of natural philosophy belonged to Arnold Senguerd (1610–1668), the disputations were not in physics but in physiology, that is, the theoretical explanation of bodily functions, or the physical premises of medicine (Voetius *et al.* 1643, p. 18). Actually, this was not only a measure to avoid conflicts with the academic authorities: Regius's philosophy, indeed, was structured on physics as the condition of a better understanding of diseases, resulting in a large treatise on medicine published in 1647, his *Fundamenta medica*, which focused on the causes, symptoms and healing of diseases. The relations between philosophy and medicine are outlined by Regius in this 1647 treatise. Following the traditional structure of medical textbooks, he divides medicine into *cognitio* and *curatio*. *Cognitio*, or theoretical medicine, is divided into *physiologia*, or knowledge of health (concerning *bona temperies* and *apta conformatio* of bodily parts), and *cognitio pathologica*, or knowledge of diseases (Regius 1647a, pp. 2–3). However, at the beginning of the treatise Regius points out the continuity between *cognitio* and *curatio*, or between theoretical and practical medicine, as explanations of bodily functions are aimed at healing.[12] Moreover, he rectifies the vulgar definition of physiology as the study of the natural things concerning the human body.[13] This is, indeed, only an imprecise definition of physiology, which more properly concerns human health. He distinguishes two kinds of physiology: a 'general' physiology, or the study of natural things in the human body (which is a branch of physics), and a medical physiology, concerning human health. Such a distinction prevents the repetition of the same notions in medical and physical treatises.[14]

---

12 "Medicina inter artes numero: quia omnia eius praecepta ad aliquid agendum sunt delineata. Atque hinc constat illa vulgo male in theoretica et practicam dividi: cum omnes artes [...] doctrinae sint practicae [...]. Neque his adversatur prior medicinae pars, quae cognitio a nobis appellatur: uti nec sanitatis, remedii, et multorum aliorum in medicina tradendorum, definitiones. Nam haec omnia revera sunt practica, cum ad actionem medicam, sive medendum, cuncta dirigantur," Regius 1647a, p. 1. On Regius's medical theories, see Verbeek 1989, Gariepy 1990, Bitbol-Hespériès 1993, Kolesnik-Antoine 2013b, Strazzoni forthcoming c. On Descartes's medical theories and their reception, see Lindeboom 1978, Aucante 2006, Caps 2010, Mahut-Gaukroger 2016.
13 Regius seems to refer to the (standard) definition of Jean Fernel: Fernel 1567, *Praefatio*, p. IV (unnumbered): "φυσιολογικὴ, quae hominis integre sani naturam, omnes illius vires functionesque persequitur."
14 "Cognitio physiologica est de cognoscenda sanitate. Haec vulgo appellatur physiologia, definiturque pars medicinae, quae agit de rebus naturalibus, seu talibus, quae corpus humanum constituunt. [...] Sed meo iudicio, non satis bene: cum talis physiologia sit pars physicae, cuius munus est de rebus naturalibus, quales hae sunt, agere. Nec iuvat, quod physici alia ratione et alio respectu haec tractent, quam medici. Si enim artes et scientiae, pro diversitate tractationis usus et respectus, diversae essent, diversisque locis deberent tradi, una eademque doctrina infinitis pene locis esset repetenda. Haec autem cognitio recte physiologica vel physiologia dicitur, quia illud, quod homini, secundum naturam inest, nempe sanitatem, cognoscere docet," Regius 1647a, p. 3. See also the

Physics, thus, is to be considered a necessary premise for medicine, from which it is detached only for pragmatic reasons. Regius's study of the nature and functioning of the mind, in fact, will be dominated throughout his career by a medical approach. In his 1641 disputations, furthermore, the two kinds of physiology are mixed: his *Physiologia* concerns both human health and animal actions, namely, the 'medical' physiology and the study of the main functions of the human body. Given such topics, Regius's disputations include some remarks on the different kinds of knowledge embodying the premises of his mature epistemology, focused on the sensory origin of knowledge. Providing a detailed classification of *actiones animales* in the third disputation, Regius distinguishes between cognitative and automatic actions, both involving the human mind, whereas *actiones naturales* concern only the body.[15] Cognitative actions are performed by mind and are divided into intellect and will. On the other hand, automatic actions are those of which the soul is not aware: sense reception, natural appetite, and spontaneous motion.[16] As these are performed by the body only, they are similar to *actiones naturales*. However, whereas natural actions are mere organic processes such as generation and *alitura*, animal automatic actions can be turned into cogitative ones, if the mind pays attention to them (Regius 1641a, p. 17). The background for this explanation is Cartesian: Regius provides a mechanistic account of sense perception distinguishing it from the purely mental acknowledgment of immaterial entities, such as God or mind.[17] Perception or intellect can be either organic or

---

second edition of Regius's main work in natural philosophy, his *Fundamenta physices* (Regius 1646), published as *Philosophia naturalis* (Regius 1654): "philosophia naturalis, quae vulgo physica et physiologia dicitur, est rerum naturalium scientia," p. 1. Regius's *Philosophia naturalis* had another edition in 1661 (Regius 1661). In the second edition of *Fundamenta medica*, which appeared in 1657 as *Medicinae libri quatuor*, Regius added some lines to the paragraph here quoted, underlining that he had to repeat some notions already developed in physics in order to make the treatise more understandable: "itaque, si quicquam istarum rerum naturalium dictarum, in cognitione hac physiologica, a me tradatur, vel designetur, id non ut ad medicinam proprie pertinens, sed, tanquam per necessarium istius cognitionis commentarium hic repetitum, tantum est habendum," Regius 1657 a, p. 3.

15 "Absolutis actionibus naturalibus sequuntur animales, quae non tantum a natura partium, seu naturali temperie et conformatione fiunt, sed etiam vi animae seu mentis perficiuntur," Regius 1641a, p. 33.

16 "Expositis actionibus cogitativis aggrediamur automaticas, quae anima seu mente ad rem non attendente per solum organorum animalium, nempe spirituum, nervorum, cerebri, aut musculorum, ab obiecto externo vel interno agitatorum motum ab homine tanquam aliquo automato peraguntur. [...] Actiones automaticae sunt receptio, appetitus sensitivus simplex, et motus spontaneus. Receptio est actio (vel potius passio) animalis automatica, qua motus rerum recipimus. [...] Haec triplex est sensus simplex, reminiscentia simplex, imaginatio simplex," Regius 1641a, p. 46.

17 "Inorganica perceptio est, qua mens nostra sine organo ullo percipit res imagine corporea carentes, ut Deum, animam rationales, etc." Regius 1641a, p. 33.

inorganic, that is, working with or without the body. However, a short remark on the difference between organic and inorganic perception is the only concession to Descartes's theory of pure understanding, as Regius's account of perception concerns only the sensory acquaintance of movements through *sensus reflexus*, *reminiscentia*, and *imaginatio*.[18] Even the perception of universals is organic, as these are gained through the imagination.[19] The point had been discussed by Regius in his correspondence with Descartes, who found it questionable.[20] Few variations will then be necessary to transform Regius's physiology into a fully empiricist account of human knowledge.

2.3 Regius's clash with Descartes

These points came into full light in Regius's *Fundamenta physices* (1646), where, in the chapter *De homine*, he makes explicit his refusal of any innate ideas by identifying intellect with sensory perception, therefore called *sensus cogitativus*, and omitting any reference to inorganic perception (Regius 1646, p. 252). In this way, he retains the arguments of his earlier *Physiologia* as the basis for an explanation of every mental activity depending on sense perceptions. When a first draft of the text was submitted to Descartes before publication, indeed, the Frenchman distanced himself from its contents in a letter of July 1645. In his missive, Descartes objects to two of Regius's main points. The first objection is that Regius has not provided adequate proofs for his physics, as he displays definitions and divisions, going from the general to the particular, without grounding them on adequate *probationes*.[21] Moreover, Descartes criticises Regius's consideration of

---

18 "Intellectus est rerum obiectarum cognitio. Estque perceptio et iudicium. Perceptio est intellectus, quo res mente percipimus. Estque inorganica et organica. Inorganica perceptio est, qua mens nostra sine organo ullo percipit res imagine corporea carentes, ut Deum, animam rationalem, etc. Perceptio organica est, qua mens nostra instrumento corporeo percipit res imaginationem corpoream habentes. Haec triplex est, sensus reflexus, reminiscentia, imaginatio," Regius 1641a, p. 33.

19 "Receptio universalium ad imaginationem pertinet. Universalia enim sunt singularia in abstracto considerata sine notis individuationis hoc, hic, nunc, ut loquuntur scholastici. Itaque haec fiunt per imaginationem, quae detrahit," Regius 1641a, p. 42. The same point is recalled in Regius 1646, p. 285. For a full account of Regius's theory of understanding, see Bellis 2013.

20 See the letter of Descartes to Regius of July 1641: AT III, p. 66 (attributing Descartes's remarks on universals to a letter of 24 May 1641), also in Bos 2002, p. 76 (dating it after the disputation).

21 "Fateor quidem eas per definitiones et divisiones, a generalibus ad particularia procedendo, recte tradi posse, atqui nego probationes debere tunc omitti. [...] Alii autem legentes assertiones sine probationibus, variasque definitiones plane paradoxas, in quibus globulorum aethereorum,

the soul as a bodily modification, objecting to his lack of acumen in dealing with metaphysics and theology.[22] These criticisms are fundamental to understanding the distance between Regius and Descartes. Let me focus, first, on the latter.

In his *Explicatio mentis humanae* (1647), as anticipated in Descartes's letter, Regius will indeed admit that the human soul can be an accident of the body, or *modum corporis*,[23] while in his *Physiologia* and *De illustribus quaestionibus physiologicis* he maintained that mind is a special form provided with its own substantiality.[24] Eventually, in the attempt to make the (possible) material nature of the soul consistent with its immortality, he would admit in the second edition of his *Fundamenta physices* that mind could be an indestructible atom.[25] How did he come to support such a tenet? Regius's rejection of Descartes's metaphysics was fully stated, for the first time, in his answer to the letter by Descartes (23 July 1645). Since Descartes, according to Regius, supported his positions in metaphysics the way any enthusiast would do with his fantasies, i.e. just claiming

---

aliarumque similium rerum, nullibi a te explicatarum, mentionem facis, eas irridebunt et contemnent, sicque tuum scriptum nocere saepius poterit, prodesse nunquam," Regius 1646, p. 249. On Regius's natural philosophy, see Farina 1977, Verbeek 1994, Verbeek 2000, Bellis 2013.

22 "Nuncque omnino subscribo illorum sententiae, qui voluerunt, ut te intra medicinae terminos contineres. Quid enim tanti opus est, ut ea quae ad metaphysicam vel theologiam spectant scriptis tuis immisceas, cum ea non possis attingere, quin statim in alterutram partem aberres? Prius, mentem, ut substantiam a corpore distinctam, considerando, scripseras hominem esse *ens per accidens*. Nunc autem e contra, considerando mentem et corpus in eodem homine arcte uniri, vis illam tantum esse *modum corporis*," Regius 1646, p. 250.

23 The *Explicatio mentis humanae* was published as corollaries attached to a 1647 disputation presided over by Regius: Regius 1647b, corollary 2: "quantum ad naturam rerum attinet, ea videtur pati, ut mens possit esse vel substantia, vel quidam substantiae corporeae modus. Vel, si nonnullos alios philosophantes sequamur, qui statuunt extensionem et cogitationem esse attributa, quae certis substantiis, tanquam subiectis, insunt, cum ea attributa non sint opposita, sed diversa, nihil obstat, quo minus mens possit esse attributum quoddam, eidem subiecto cum extensione conveniens, quamvis unum in alterius conceptu non comprehendatur. Quicquid enim possumus concipere, id potest esse. Atqui, ut mens aliquid horum sit, concipi potest, nam nullum horum implicat contradictionem. Ergo ea aliquid horum esse potest."

24 "Nos substantiam corpoream esse unicam omnium corporum materiam agnoscimus, nullasque isti materiae substantiales formas realiter ab illa distinctas adiungimus (excepta sola anima rationali)," Regius 1641a, p. 5. "Forma specialis est mens humana, quia per eam cum forma generali in materia corporea homo est, id quod est. Haec ad formam generalem seu materialem nullo modo potest referri: quoniam ipsa (utpote substantia incorporea) nec est corpus, nec ex motu aut quiete, magnitudine, situ aut figura partium oriri potest," Regius 1641b, disputation III, thesis 8.

25 "Illa tum in minima sensorii communi atomo, sive corpusculo propter parvitatem et soliditatem suam naturaliter indivisibili, posset existere," Regius 1654, pp. 345–346.

their evidence, he was subjected to the fanciful imaginations caused by bodily
constitutions:

> Vous ne serez pas surpris de ma conduite, lorsque vous saurez que beaucoup de gens d'es-
> prit et d'honneur m'ont souvent témoigné qu'ils avaient trop bonne opinion de l'excellence
> de votre esprit, pour croire que vous n'eussiez pas, dans le fonds de l'âme, des sentiments
> contraires à ceux qui paraissent en public sous votre nom. Pour ne vous en rien dissimuler,
> plusieurs se persuadent ici que vous avez beaucoup décrédité votre philosophie, en pub-
> liant votre Métaphysique. Vous ne promettiez rien que de clair, de certain et d'évident; mais,
> à en juger par ces commencements, ils prétendent qu'il n'y a rien que d'obscur et d'incer-
> tain, et les disputes que vous avez eues avec les habiles gens à l'occasion de ces commence-
> ments, ne servent qu'a multiplier les doutes et les ténèbres. Il est inutile de leur alléguer que
> vos raisonnements se trouvent enfin tels que vous les avez promis. Car ils vous répliquent
> qu'il n'y a point d'enthousiaste, point d'impie, point de bouffon qui ne pût dire la même
> chose de ses extravagances et de ses folies. (AT IV, p. 255)

Regius's accusation of enthusiasm was not a novelty in the early history of
Cartesianism, as it is a restatement of the main critique advanced by Voetius's
pupil, Martin Schoock, in the *Admiranda methodus* (1643), which was the answer
devised by Voetius to Descartes's *Epistola ad Patrem Dinet*, as a continuation of
the *querelle d'Utrecht*. This book is notorious for its slanderous overtones against
Descartes (Verbeek 1992a, p. 21), but it also provides an argument against Descartes's reliance on the principles of pure reason in setting the basis of his philosophy. For Schoock, pure rational principles have to be tested by the senses to
conform to others's testimony, even logical principles. The use of the method of
Descartes, based on radical doubt and on the rejection of the use of the senses,
would lead us to accept fantasies as evident truths. Schoock, in fact, compared
Descartes to the autodidacts and to religious enthusiasts, who avoided the use of
the senses or Revelation.[26] The critiques by Regius of Descartes, therefore, echoed

---

26 "Eadem methodus recta ad enthusiasmum ducit. [...] Periculosae vero aleae plenissima haec
methodus est. Mens enim sive intellectus ob caligantes, quos habet oculos, sensibus externi-
bus ut ducibus, haut aliter indiget ac coecus suo ductore. Ipsa axiomata solis radiis clariora,
ut indubitata non amplectitur, nisi sensuum ministerio eorum instituerit examen illorumque
certitudinem in praxi manibus quasi palpaverit. [...] Quando primo enim a sensibus abducitur
ad contemplationem eorum axiomatum, quae ei insculpta videntur, exuitur ammussi ac norma
sua, sibique relicta facile axioma quod fingere potest, quod si normae exhibeatur, postea falsum
ac sublestae fidei deprehendatur: quoniam nihilominus, prerogativa ei defertur, (quasi eiusmodi
iudice cum regula non indigeret) pertinaciae callo obducitur, audetque se munire per axiomata
contemplando adinventa contra quascunque rationes et apertissimam etiam veritatem. Quod
αὐτοδιδάκτοις accidere solet, qui opinionum Helenas, a se inventas, prae amore comprimendo
enecare quae derelinquere malunt, contemplantibus talibus accidit. [...] Antiqui et recentiores
enthusiastae [...] Scripturam contemnere inceperunt, et vice divinorum oraculorum ea obtrudere
quae mens dictabat. Fateor, quidem, hos inter enthusiastas antesignanos, pro mente iactasse

those of the expounders of Aristotelian philosophy: whereas his rejection of substantial forms endangered the use of philosophy for theology, his theory of knowledge was somehow more traditional than that of Descartes. Which kind of theory of knowledge was Regius embracing? As it has been labelled elsewhere (Bellis 2013, pp. 169–172), it was an original form of 'radical empiricism', or 'Cartesian empiricism'. This fully emerged with the appearance of his *Fundamenta physices* and of the texts prompted by his clash with Descartes, namely, Regius's *Explicatio* and *Brevis explicatio mentis humanae* (1648) – which was his reply to Descartes's answer, i.e. the *Notae in Programma quoddam* (1647).[27] According to Regius, mind is *organica* insofar it relies on the empirical data provided by the body in the formulation of the first principles (Regius 1646, pp. 247, 251–253) and even in knowing immaterial things.[28] This was the outcome of a double move by Regius. On the one hand, he rejected the traditional idea that knowledge is the result of an abstraction of *species* by understanding: in accordance with Descartes's physiology, he proposes a purely mechanical account of sense perception.[29] On the other hand, Regius does not accept Descartes's theory of knowledge. If sensory knowledge is explained in the same terms as Descartes's,[30] once Regius comes to the perception of purely immaterial entities he denies that their knowledge can be otherwise based than on sensory data. This has a medical reason, as conditions like sickness or variations in temperaments, for Regius, show that the mind does not always think and that knowledge is hindered or favoured by the constitution of the body.[31] In his 1654 *Philosophia naturalis* Regius goes further in his rejection of Descartes's supposed enthusiasm, as he stresses that the apprehension of immaterial things, i.e. the object of special metaphysics as it deals

---

internum hominem, spiritum, Deum loquentem, somnia, et quae alia fanaticorum vocabula esse solent, sed unius rei, mentis nempe variae tantum denominationes fuerunt," Schoock 1643, pp. 255–257. On the critiques of enthusiasm, see Heyd 1995.

27 For a full account of their debate, see Rodis-Lewis 1993, Verbeek 1993a, Bos 2002, Strazzoni 2014a.

28 "*Mens* [...] *est organica*, ita ut actiones suas sine corporeis organis perficere non possit, eaque utatur corpore, corpus vero non utatur mente. In omnibus enim actionibus saltem cerebro satis sano et satis recte disposito indiget; ut passim in pueris, senibus, deliris, sanis, aliisque quotidiana docet experientia. Idque non tantum in rebus corporeis, sed etiam spiritualibus et divinis, considerandis," Regius 1648, p. 10. Words in italics are from the original text of his *Explicatio mentis humanae*. On Regius's theory of soul, see Verbeek 1992b, Olivo 1993, Rodis-Lewis 1993, Alexandrescu 2013, Strazzoni 2014a.

29 See Regius 1641a, p. 34: "itaque ad sensus movendos nullae species intentionales, vel qualitates spirituales, requiruntur, sed solus motus eiusque varietates sufficiunt."

30 See Descartes's *Dioptrique*, AT VI, p. 85.

31 Regius 1646, pp. 246–247. Regius will explicitly deny that we are constantly thinking in his 1654 *Philosophia naturalis*: Regius 1654, p. 344.

with immaterial substances, is particularly influenced by diseases such as apoplexy and epilepsy.[32] In the edition of 1661, eventually, he appeals to Aristotle's authority to support the sensory origins of all knowledge (Regius 1661, p. 419). What are the consequences of this approach? First, he negates the validity of the *cogito*. In Regius's 1654 and 1661 *Philosophia naturalis* this is defined just as a general concept, which has its origin in the senses and from which it is not possible to deduce the existence of any innate ideas.[33] Accordingly, it is not possible to demonstrate the immateriality of the soul, which is granted only by Revelation.[34] Second, since the idea of God is shaped by the mind according to ordinary experiences and imagination, a rational theology based on Descartes's proofs of the existence of God is no longer tenable.[35] Rejecting Descartes's proofs of God's existence and goodness, in his *Fundamenta physices* Regius appeals to Holy Writ as the only means to ground our knowledge of external reality, which is however subjected to sense deception:

> Cum itaque sic a natura mens sit comparata, ut a variis motibus variae perceptiones et iudicia ipsi possint excitari, cumque illi motus non tantum a corporibus veris, sed etiam a

---

32 "Cum itaque mens a rebus mundanis abstracta, eaque his instrumentis purioribus est instructa. Tum nihil est mirandum aegros illos, quamvis reliquum corpus sit debile, sapientiorum cogitationum proferre indicia," Regius 1654, p. 343. This can be read, actually, as a crypto-accusation of enthusiasm against Descartes.

33 "Patet sensum aliquem omnis cognitionis, reliquarumque actionum cogitativarum esse principium, ac proinde non esse omnis cognitionis principium, sive primum cognitum, *cogito*: nedum, *cogito, ergo sum*. Hi enim sunt conceptus generales, qui ex speciali aliquo sensu primam originem duxerunt," Regius 1661, p. 399.

34 "Quod autem mens revera nihil aliud sit quam substantia, sive ens realiter a corpore distinctum et actu ab eo separabile et quod seorsim per se subsistere potest, id in Sacris literis nobis clarissime est revelatum," Regius 1646, p. 246. Regius refers then to *Ecclesiastes*, 12.9, *Matthew* 17.3, *Luke* 16.9, 16.22, 24.43, *Second Epistle to the Corinthians*, 5.8, 12.3–4, *Apocalypse* 6.9–10. In the corollaries 2, 3, and 5 of his subsequent *Explicatio* (1647) Regius manifests his doubts on the possibility of defining the nature of the soul, as its belonging to the body implies no contradiction, nor does our doubt of the existence of the body prove that it is a substance different from mind. Therefore, it is not possible to clearly conceive mind as distinguished from body: see Regius 1647b.

35 "Imo ipsa idea Dei, quae scilicet non est ex revelatione vel inspiratione divina, non videtur nobis innata, sed vel ex rerum observatione in nobis primum producta [...]. Nam in ente summo, quod Deum appellamus, humanum ingenium nihil quicquam considerat, quam bonum aliquod, quod quotidie in homine observatur," Regius 1646, p. 252. "Conceptus noster de Deo, sive idea Dei, in mente nostra existens, non est satis validum argumentum ad existentiam Dei probandam: cum non omnia existant, quorum conceptus in nobis observantur; atque haec idea, utpote a nobis concepta, idque imperfecte, non magis quam cuiusvis alius rei conceptus, vires nostras cogitandi proprias superet," Regius 1647b, corollary 15. See also Regius 1648, pp. 12–13, 15.

> causis imaginariis, et a potentissimo directore tantum imaginariis productis, animae offerri queant: hinc sequitur per naturam dubium esse, vera an falsa, seu imaginaria, mente percipiamus et diiudicemus. Verum hoc dubium nobis tollit divina in Sacris revelatio [...]. Unde patet ea quae recte percipimus, esse res veras, et non imaginarias [...]. Atque ita magna illa dubitatio, quae in animis recte philosophantium per naturam necessario utramque paginam etiam in evidentissimis faceret, per Verbum Dei penitus evertitur. Unde recte quilibet verus philosophus, iam cum propheta canit: *verbum Dei est lucerna pedibus meis.* (Regius 1646, pp. 249–250; see *Psalms* 119.105)

The reliability of knowledge, to the extent that it is provided by the senses only and not by God Himself as Descartes's innate notions are, can be subjected to the power of God, "a potentissimo directore," who can deceive us for our own good – again – like a physician.[36] In *Philosophia naturalis* (1654) Regius concludes that we cannot avoid a natural scepticism.[37] Although Revelation ensures us that our knowledge is not illusory, the Aristotelian ideal of *scientia* is attainable neither in metaphysics – i.e. with regard to the nature of mind – nor in natural philosophy, whose method does not provide incontrovertible conclusions.

2.4 Medicine and the method of natural philosophy

As shown in section 2.2, Regius oriented his natural philosophy towards physiology and carried out his 'metaphysical' considerations in physiology itself: first, in his *Physiologia*, where he provides, *in nuce*, an empiricist account of knowledge, and then (section 2.3) in the chapter *De homine* of his *Fundamenta physices*. This did not go unnoticed by Descartes, who complained of Regius's reversal of the order of presentation of his physics, and also of the omission of adequate proofs in a letter to Elizabeth of March 1647.[38] This was another outcome of his medical

---

36 "Nec obstat, si quis dicat per naturam constare Deum esse, eumque non posse fallere [...]. Respondeo enim primo, Deum pro summa sua [...] potestate, fallacia posse uti, primo innocua et sapienti, quali medici et prudentes patres familias utuntur, et deinde paenali [...], quod testatur Scriptura, cum dicit: *et tradidit ipsos in sensum perversum,*" Regius 1648, p. 11; see *Epistle to the Romans*, 1.18. In his 1661 *Philosophia naturalis* a new quotation from the Bible is added to confirm that God can be a deceiver: see Regius 1661, p. 414, *Ezekiel*, 14.9. Regius remarks, however, that God is not responsible for human faults as men are free in suspending judgment: Regius 1648, p. 11.

37 "Quicunque [...] omnipotentem et liberrimum cognoscit Deum [...] talem omnium rerum verisimilitudinem, vel scepticismum naturalem [...], qualem proposuimus, negari non potest," Regius 1661, pp. 350–351.

38 "Il ne contient rien, touchant la physique, sinon mes assertions mises en mauvais ordre et sans leurs vraies preuves, en sorte qu'elles paraissent paradoxes, et que ce qui est mis au commencement ne peut être prouvé que par ce qui est vers la fin," Descartes to Elizabeth, March 1647, AT IV, p. 625.

orientation. First, Regius could not grant the metaphysical certainty Descartes aimed at providing his physics with, so that he concludes his *Fundamenta physices* by claiming that his arguments are not intended to be compulsory for everyone because human temperaments are various and no argument can convince everyone in the same way.[39] So that when he describes his method of discovery, he proposes a problem-solving method, which consists of posing a problem, imagining an intelligible cause from which effects can flow, and then looking for other causes, until a better one is found.[40] This method had a mathematical origin: it was the same analytical method of discovery that Descartes himself appropriated from Pappus. As shown by Stephen Gaukroger, it amounts to Pappus's 'problematical analysis', in which one (1) poses a problem, (2) proceeds by unfolding its sub-problems, then (3) finds a question whose solution is clear, according to the accepted criteria of truth, such as that which is evident according to experience. In the hands of Descartes, this method amounts to (1) observing a phenomenon, (2) formulating a hypothesis which is consistent with the first principles of physics, and by which one can derive different effects related to the same phenomenon (3) testing by experience this hypothesis. This is the case, for instance, in Descartes's law of refraction, which is consistent with his principle of the conservation of the quantity of motion: so that its conformity with metaphysical truths is granted,[41]

---

39 "Atque ita universae Physicae fundamenta, brevi, quantum potui, systemate comprehensa [...] absolvimus. Sicut autem, nullius consentientis vel dissentientis habita ratione, libere id proposui, quod mihi rationi maxime consentaneum fuit visum ita hic nemini assentiendi vel dissentiendi legem praefigo. [...] Imo, ex Terentiano proverbio iam olim puer didici tot esse sententias, quot sunt homines. Neque hoc mirum. Cum enim infinita pene temperamentorum sint discrimina, quae iudiciorum producunt diversitatem, innumerae etiam de rebus humanis iudiciorum debent esse dissimilitudines. [...] Dissentiat igitur quilibet," Regius 1646, pp. 305–306.
40 Regius expresses what is to be a better explanation using adverbs such as 'commode', 'probabiliter', 'intelligibiliter': "cum enim problema aliquot in physicis proponitur solvendum, primo excogitanda est causa intelligibilis, qua effectum, in problemate proposito observatum commode et intelligibiliter peragi possit. Deinde circumspiciendum an non alia commodior vel aeque commoda queat inveniri. Quae si inveniatur, commodior priori est praeferenda aequalis vero ipsi aequiparanda. Sin alia commodior vel aeque commoda excogitari nequeat, solutioni inventae tamdiu acquiescendum, donec melior vel aequalis alia fuerit inventa," Regius 1654, p. 441.
41 See Gaukroger 1989, pp. 73–88, 110–114. It is worth quoting some of Gaukroger's words: "the approach, as Descartes outlines it, in the case of the discovery of the sine law, the calculation of the angles of the bows of a rainbow, and the solution of Pappus's locus-problem, is the same, and in each case it consists purely in analysis. In each case we take a specific problem bequeathed by antiquity and solve it using procedures compatible with the basic precepts of Cartesian science. We then try to incorporate the solution within a general system which has as its foundations those truths which we cannot doubt because we have a clear and distinct grasp of them (and because God guarantees those truths of which we have such a grasp)," p. 114; see Descartes's *Dioptrique*, AT VI, pp. 97–100.

and the whole 'chain of reasoning' is – for him – provided with metaphysical certainty.[42] In the case of Regius, it amounts to being just an inference to the best explanation.[43] The cognitive process involved in the formulation of hypotheses is imagination alone: since in *Fundamenta physices* intellective processes are reduced to *sensus cogitativus*, *reminiscentia*, and *imaginatio*,[44] the operations of the intellect are performed through imagination or the manipulation of sense data. Making the point more explicit, Regius adds a reference to 'imaginative deduction' in the 1654 edition, whereas in the first edition he attributed his hypotheses to the intellect only.[45] Insofar as he cannot base all his explanations on metaphysical principles, Regius adopts a method of explanation borrowed from medicine and the practical arts in general. As signalled by Descartes, it consists of going from general to particular notions, displayed by definitions, divisions and explanations. For Regius, a natural-philosophical treatise should consist of the arrangement of propositions in a clear order. Such order, however, is not observed in long treatises such as his own. The method he follows, rejected in Descartes's 1645 letter, consists in proposing a series of definitions and explaining them in order to provide an ideal model for natural phenomena. As he puts it in 1661:

> Methodus sive ordinatio est, qua mens, per plures e notionibus compositas sententias discurrens, eas sibi mutuo homogeneas, pro naturae suae claritate, praeponit et in ordinem redigit, unde ordinis et confusionis iudicium, in rerum examine, consequitur. Talis, in rerum longioribus tractatibus, passim observatur. Optima artium inventarum traditio fit per definitiones, distributiones, et additas dilucidationes, analytica methodo procedentes.

---

42 See Descartes's *Principia philosophiae*: "praeterea quaedam sunt, etiam in rebus naturalibus, quae absolute ac plusquam moraliter certa existimamus, hoc scilicet innixi metaphysico fundamento, quod Deus sit summe bonus et minime fallax, atque ideo facultas quam nobis dedit ad verum a falso diiudicandum, quoties ea recte utimur, et quid eius ope distincte percipimus, errare non possit. Tales sunt mathematicae demonstrationes, talis est cognitio quod res materiales exsistant, et talia sunt evidentia omnia ratiocinia, quae de ipsis fiunt. In quorum numerum fortassis etiam haec nostra recipientur ab iis, qui considerabunt, quo pacto ex primis et maxime simplicibus cognitionis humanae principiis, continua serie deducta sint," AT VIII/1, pp. 328–329. See Gaukroger 1989, pp. 73–88, 110–114, Strazzoni forthcoming c.
43 "An autem satis clare et distincte rem perceperimus et examinaverimus, mens secundum apparentiam tantum diiudicat. Illique tamdiu acquiescendum, donec contrarium vel aliud per experientiam vel alia ratione fuerit probatum," Regius 1646, p. 287.
44 See *supra*, section 2.2.
45 "Per manifestam [...] imaginationis demonstrationem," Regius 1654, p. 8. See Regius 1646, p. 3: "insensibiles sunt, quae, propter exiguitatem [...] sensus fugientes, solo intellectu [...] observantur." Regius 1654, p. 6: "insensibiles sunt, quae, propter exiguitatem [...] sensus fugientes, solo imaginationis et iudicii intellectu [...] observantur." On the evolution of Regius's views on the sources of knowledge, see Bellis 2013, Bos 2013. On Regius's theory of matter, see Rothschuh 1968, Farina 1977, Strazzoni forthcoming b.

Haec enim est clarissima et brevissima. Atque in his tota logica, eiusque rectus usus consistit. (Regius 1661, pp. 476–477)

This method was the one usually prescribed for medical treatises. For instance, in his *Institutiones logicae* (1626) Franco Burgersdijk distinguishes between a natural and an artificial method. The natural one proceeds from universal to particular notions, insofar as universal concepts are better known than particular ones. The artificial one proceeds in the other direction. In turn, the progression from general to particular notions is made by divisions. Hence, this method can be divided into synthetic or analytic. The synthetic proceeds from first, simple principles to those notions which are composed by these – and it is used in speculative disciplines, physics, metaphysics, and mathematics. The analytical method, on the other hand, proceeds from the definition of the goals of an art, to that of their means, until the "prima ac simplicissima" of this art are found (Burgersdijk 1626, pp. 289–292). Regius, *contra* Burgersdijk – who avails himself of the views of Jacopo Zabarella (Burgersdijk 1626, pp. 380–381; see Zabarella 1597, pp. 181–187, 193–198, 222) – extends this method to natural philosophy. As he claims that in long treatises no order can be kept, he instead applies an analytical method, working by definitions, divisions, and explanations, to the singular parts both of natural philosophy and medicine. This is the so-called *methodus partialis* that Burgersdijk allows to convey sub-parts of the arts.[46] In the hands of Regius, the method of analysis is applied to all the parts of all disciplines, which, deprived of their Cartesian, metaphysical foundation, cannot be granted that internal unity that Descartes attempted to provide them with by making them consistent with metaphysical principles. So that Regius can claim, in the *Dedicatio* of his *Fundamenta medica*, to have exposed Descartes's natural philosophy – before his medicine – following an "expeditam et compendiariam viam" (Regius 1647a, *Dedicatio*, p. IV (unnumbered)), i.e. getting rid of his metaphysical unity. In short, Regius discarded Descartes's own project in two ways: first, he adopted a medical standpoint towards Descartes's own proceeding in metaphysics. Second, he applied a method typical of medical treatises to the exposition of his physics.

----

46 "Saepe evenire, ut in aliqua disciplinae parte, quae tota disponitur methodo synthetica, servetur ordo analyticus, aut contra, ut in partibus methodi analyticae, servetur ordo syntheticus. Ex. gr. physica disponitur ordine synthetico, si tota consideretur: attamen in ea parte, ubi agitur de corpore animato, apte servari potest ordo resolutivus, facto initio a vita, quae est corporis animati finis, indeque progrediendo ad vitae causas et principia, quae sunt animae facultates, temperamentum, partes corporis animalis organicae et c." Burgersdijk 1626, p. 383. The resolutive method had a Ramist origin, and had trespassed in the 'semi-Ramist' logic of Burgersdijk, which was based on that of Keckermann: see Van Rijen 1993, pp. 9–28. See also Ashworth 1974, Ashworth 1988, Verbeek 2001, Hotson 2007, Strazzoni forthcoming a.

2.5 The necessity of a foundation?

Prima facie, the use by Regius of Cartesian physics shows that this did not need a metaphysical, nor any kind of philosophical foundation in order to be accepted in the university. In fact, Regius developed and taught his natural philosophy and medicine without the support of Descartes's theory of knowledge. Insofar as he did not assume Descartes's revolutionary point of view on the sources of philosophical knowledge, he did not need to defend such a standpoint in the university. Accordingly, one can assume that Regius was not interested in metaphysical and foundational issues as such, since he did not need a metaphysics to develop his natural philosophy. Moreover, Regius's solution of metaphysical problems through an appeal to Revelation is to be interpreted as the result of his medical orientation, which led him to dismiss Descartes's foundation. Since his theory of philosophical knowledge could not avoid sceptical arguments, Revelation turned out to be the only way to ensure the reliability of such knowledge in providing a hypothetical, consistent explanatory model for phenomena – even if it is not *scientia*. Regius's foundation turned out, therefore, to be unfeasible in the view of the other supporters of Cartesian philosophy in the Netherlands, who distanced themselves from his solution and attempted to provide philosophy with the status of *scientia*, confronting the natural scepticism which Regius set out. The evolution of Dutch philosophy can be interpreted as a reaction to Regius's approach, as well as to the problems he raised. First, Regius put into question the reliability of evidence as a criterion of judging the truth of concepts and natural-philosophical theories, aiming to show that what is 'evident' is convincing only according to some physiological (i.e. temperamental) conditions. Secondly, Regius's development of a physics without a definition of its metaphysical assumptions would endanger the introduction of Cartesianism in the university as a whole corpus of disciplines, including ethics, logic, metaphysics as the science of being, and as the basis for law and theology, requiring a comprehensive conceptual and methodological apparatus. Thirdly, Regius raised the problem of the use of experience in natural philosophy, that is, as a source of 'scientific' knowledge. He showed, in fact, that Descartes's physics could be developed on the basis of *ad hoc* hypotheses on the causes of phenomena, rather than proven by metaphysical principles – such as the constancy of God. The debates brought about by Regius, as well as his own positions, would force other Cartesians to develop reflections on how natural philosophy provides its theories, both in prescriptive and descriptive manners.

3 Cartesianism as the Philosophy of the School: Logic, metaphysics, and rational theology

3.1 Critiques and replies

The Utrecht Crisis and the subsequent quarrel between Descartes and Regius were just the beginning of the debates over Cartesianism in the Netherlands. A third event shaped the introduction and uses of Descartes by Dutch and Dutch-related philosophers, prompting new approaches to the use of metaphysics. This event is known as the Leiden Crisis (1647), and the long-standing debate which followed it, saw the intervention of the two foremost figures in Dutch Cartesianism: Johannes Clauberg (1622–1665) and Johannes de Raey (1620–1702). Their role in the ongoing quarrels over the new philosophy is revelatory of different attitudes towards the use of Cartesian philosophy, and testifies to different ways of promoting its acceptance into the curriculum in the 1650s. On the one hand, Clauberg provided a Cartesian scholasticism, encompassing all the academic disciplines. On the other hand, De Raey – himself a student of Regius – taught Cartesian physics by omitting its metaphysics and offering a Cartesian interpretation of Aristotelianism. This was nothing but concealment: it served, for De Raey, to convey Cartesian ideas by keeping himself within the boundaries of the *philosophia recepta*. In fact, De Raey played an active role in organising the strategy of the defence of Cartesian metaphysics and logic undertaken by Clauberg. Moreover, the very contents of Clauberg's logic and metaphysics would pull De Raey, a decade later, to reconsider the functions of these disciplines. Considering the enduring criticisms of Cartesianism, and the ways in which Dutch Cartesians reacted, is crucial to understanding the whole of the subsequent evolution of Cartesian foundationalism.

At the University of Leiden, Cartesian ideas were first introduced by Adriaan Heereboord (1614–1659), who had previously studied under Franco Burgersdijk at the same university. Heereboord had been extraordinary professor of logic from 1641 and ordinary professor of ethics from 1644. His Cartesian sympathies were signalled for the first time in a letter from Descartes to Pollot of 8 January 1644, where Descartes remarks how Heereboord – who praised him more than Regius ever did – did not rely on any support from him, nor had he straightforwardly attacked the Aristotelians in the disputations he had recently held.[1] If in Utrecht

---

1 "Je viens de lire les Theses d'un professeur en philosophie de Leyde, qui s'y declare plus ouvertement pour moy, et me cite avec beaucoup plus d'eloges, que n'a iamais fat Mr De Roy. Il a fait cella sans mon conseil et sans mon sceu; car mesme il ya trois semaines qu'elles sont imprimées, et ie ne les receus que hier. Mais elles sascheront fort mes ennemis; car il ya quelques temps que

https://doi.org/10.1515/9783110569698-004

Regius broke the *pax academica* by attacking Voetius, in Leiden Heereboord limited himself to expounding both old and new ideas in corollaries and theses,[2] in accordance with an ideal of the *libertas philosophandi* which, for him, was the necessary outcome of religious Reformation.[3] As early as March 1643, indeed, Heereboord had presided over a disputation *De notitia Dei naturalis*, which had an evident Cartesian ring, as he claims that the knowledge of God can be acquired "ex libro naturae interno," which is nothing but the "notitia Dei naturalis [...] insitam vel innatam" (Heereboord 1654, pp. 22–23). The disputation seems not to have caused him any problems. In fact, the discussion over Cartesian theses, insofar as these were compared with those of other anti-Aristotelians, became a standard in the teaching of Heereboord in subsequent years.[4] Between February

---

ce mesme, en ayant fait d'autres, *de formis substantialibus*, ou il sembloit estre pour Aristote, et toutefois en effet il estoit pour mou, a ce qu'on m'a dit, car ie ne les ay point veues, Voëtius luy escrivit aussytost, pour luy gratuler et l'exhorter a continuer," Descartes to Pollot, 8 January 1644, AT IV, pp. 76–78. Descartes seems to refer to a disputation presided over by Heereboord on 18 July, now lost, defending the recourse to substantial forms against Regius. Fragments are reported in Revius 1650, pp. 87–89, 173–174. As to the more recent theses, they were probably defended in a (lost) disputation presided over by Heereboord on 19 December 1643, where Cartesian arguments are given in support of the immortality of the soul: see Revius 1650, pp. 38–39. For a full account, see AT IV, pp. 656–657.

2 "Quarum ipse Disputationum exstiti author, in iis Aristotelis principia fui secutus, quod Physicarum, Ethicarum, Selectarum disputationum docent curricula, in corollariis, ut vocant, et thesibus studiosorum proprio Marte et arte confectis, aliorum etiam philosophorum placita et principia ventilari fui passus, ut meum simul et illorum exerceretur ingenium, ac quo ratio nos ducere valeret, palam fieret. Nulli haec res fuit unquam obnoxia culpae apud [...] Curatores: nihil incommodi creavit ulli mortalium, nullam pacem turbavit aut animorum concordiam rupit," Heereboord 1654, *Epistola ad Curatores*, p. 9.

3 "Densissimis istis tenebris nova lux affulsit [...] Dante Aligerio et Francisco Petrarcha, primis philosophiae, bonarum artium, et omnis eruditionis restauratoribus [...]. Impetiit Germaniam hoc lumine primus Rodolphus Agricola, aeternum Belgii decus, qui acrius quaedam, adversus receptum philosophandi modum, socratica dixit libertate. Exinde plures purgando Augiae stabulo manus auxiliares admovere: prae caeteris, Hollandiae nostrae ac totius orbis miraculum, Desiderius Erasmus, Martinus Lutherus, Philippus Melanchton, primi apud nos religionis simul et philosophiae restauratores. [...] Intravit academiarum apud reformatos limen religio purior, impurior tamen remansit philosophia," Heereboord 1654, *Epistola ad Curatores*, pp. 5–6.

4 "Dixi tum inter sermocinandum, eum facile visurum ex thesibus, liberum hoc nobis esse disputandi modum, et me cum esse, qui multum indulgere studiosis, in corollariis et annexis, ut meum atque illorum ingenium experiar, me pro et contra, ut loquuntur, seu in utramvis multa disputare partem, quod sic maxime argumentorum robur ponderetur et expendatur, ita a me factum fuisse semper, a quo professioni fuissem admotus: inter quas invenit Theses de principiis cognoscendi, [...] contra Cartesium: et De primo [...] pro Cartesio, a doctissimis iuvenibus, utraque eodem anno 1644 me praeside disputatas," Heereboord 1654, *Epistola ad Curatores*, p. 11. The disputations are now lost. See the corollaries to two disputations Heereboord presided over on 14

and March 1647, however, Jacob Revius – at that time Regent of the theological college of the University of Leiden (*Statencollege*) – presided over a disputation *De cognitione* accusing Descartes of Pelagianism, as the Cartesian account of free will (more extensive than the powers of intellect) would lead to assume that man is capable of saving himself by the sole force of his will (Revius 1647, article 13). In March of the same year, the theologian Jacob Triglandius accused Descartes of blasphemy, as Descartes used the hypothesis of a deceiving God.[5] Kept informed by Heereboord – who later wrote an *Epistola* to the Curators of Leiden University, giving a report of the whole quarrel (1648)[6] – Descartes himself addressed a defensive letter to the Curators on 4 May 1647, after which a ban on the discussion of Cartesian ideas followed in the same month (Molhuysen 1913–1924, vol. III, pp. 109–112). Yet the situation worsened at the end of the year as De Raey, who had recently obtained his MA under Heereboord (15 July 1647), intervened in a disputation *De deo* presided over by the Aristotelian professor of physics and metaphysics Adam Stuart. In the disputation, Stuart had implicitly mentioned the Cartesians as "neoterici nonnulli philosophi, qui certam omnem fidem sensibus abrogant, et philosophos Deum negare, et de eius existentia dubitare posse contendunt," so that De Raey publicly accused him of having broken the ban on Cartesian ideas. Besides causing uproar in the audience (Heereboord 1654, *Epistola ad Curatores*, pp. 18–19), it also prompted the intervention of Heereboord, who in the subsequent year wrote a short *Praefatio* to Descartes's *Notae in programma quoddam* (1648), attacking Revius as calumniating Descartes, and Stuart as attributing to him absurd theses, as revealed by his own concealment of the name of Descartes by using the term "neoterici."[7] Eventually, in February 1648, the University Curators, summoning Heereboord, Revius and De Raey, proscribed for the second time discussion of the opinions of Descartes (Verbeek 1992a, pp. 34–51, Van Bunge 2001, chapter 2).

These events prompted the publication of texts embodying the first, extended discussion of Descartes's metaphysics and methodology. Such discussion was substantiated in a body of critiques, replies, and counter-replies – mainly with an eristic character – involving Dutch and German philosophers and theologians. The debate took place from 1648, when Revius published his *Methodi cartesianae*

December 1644 and January 1645, *De angelis*: Heereboord 1654, *Epistola ad Curatores*, pp. 10–11. See also Verbeek 1992a, p. 37.

5 "Internum S. Spiritus testimonium de certitudine salutis ad tempus negare sub quocunque praetextu non licet, multo minus ipsum Sp. S. (seu ipsum Deum ut male Carthesius,) pro impostore ac deceptore habere seu fingere, quod plane blasphemum est," Triglandius 1647, corollary 7.

6 Later published in his *Meletemata*: see Heereboord 1654, pp. 1–20.

7 Descartes 1648, *Praefatio*; also in AT VIII/2, pp. 347–352.

consideratio theologica, followed in 1650 by his *Statera philosophiae cartesianae* and in 1651 by the *Nova Renati Descartes sapientia* of the German philosopher Cyriacus Lentulus (1620–1678), professor of history and politics at the Academy of Herborn. One year later, Clauberg replied to these attacks in his *Defensio cartesiana* (1652), setting out the program of his logic, followed by the publication of Lentulus's *Cartesius triumphatus* and Revius's *Thekel, hoc est levitas Defensionis cartesianae* (both of 1653). After these interventions, Clauberg's *Logica vetus et nova* saw the light in 1654, followed by his *Initiatio philosophi, sive Dubitatio cartesiana* (1655), and his *Exercitationes de cognitione Dei et nostri* in 1656. In the same years, De Raey published his *Clavis philosophiae naturalis* (1654), which was part of a coordinated strategy of defence of Cartesianism. Which kind of critiques did Cartesian philosophy undergo? And how did the Cartesians react?

3.1.1 The critiques of Descartes

By attacking primarily the *methodus cartesiana,* Revius and Lentulus moved the attack to the metaphysical basis of Descartes's philosophy. In his *Methodi cartesianae consideratio theologica,* for instance, Revius takes Descartes's method into account in a broad sense. Considering the historical narration of his own proceeding into philosophy given in Descartes's *Discours de la méthode,* Revius outlines eight stages of the method, namely, the arguments starting from radical doubt up to the demonstrations of the existence of God, thus covering the main steps of Descartes's metaphysics and reflections over his methodology. The first two stages of the method concern, respectively, Descartes's learning and examination of Scholastic knowledge (Revius 1648, p. 14). Through them, Revius focuses on Descartes's analysis of Aristotelian philosophy and on his rejection of Scholastic logic. Whereas the Frenchman rejected this logic as a mere expository means in the second part of his *Discours,* according to Revius no conclusion can be argued without any formal argumentation and logical notions. The rejection of logic, therefore, is inconsistent with the very first rule of Descartes's method, as clarity and distinction is provided only by well-formed arguments. On the other hand, the other rules of the method cannot provide such order: as the second increases the difficulties in understanding, the third presupposes that what is simpler is easier to be understood, and the last requires an infinite ability in revising all the factors involved in a problem. Moreover, because the model of Descartes's method is that of mathematicians, Descartes's spurning of syllogisms turns out to be contradictory, as mathematics is syllogistic (Revius 1648, pp. 27–31). The very demonstration of the existence of the *ego* is also syllogistic: like Gassendi, Revius considers the argument of the *cogito* an enthymeme

(Revius 1648, pp. 31–33; see Gassendi 1644, pp. 38–39). From the third stage of the method, Revius focuses on more theological topics, reverting to those already discussed during the Utrecht and Leiden crises. In fact, this stage concerns the relinquishing of all bookish knowledge and of Revelation, opening the way to enthusiasm, as it was for Voetius and Regius (Revius 1648, p. 35). The fourth step consists in doubting every kind of knowledge, including mathematical truths and those concerning God, through the hypothesis of a deceiver genius.[8] In this way, Descartes introduced a provisional, atheistic hypothesis in order to refute atheism itself (Revius 1648, pp. 47–51).[9] Likewise, Revius's consideration of the fifth and sixth stages is devoted to theological problems, as these concern Descartes's rejection of the truths of faith and of the use of the senses, in virtue of his radical doubt.[10] The *cogito* being the only first principle in his philosophy, no place is left for any truth of faith as a first principle, whereas he admitted that they were beyond any doubt (Revius 1648, pp. 59–71). The last two stages of Descartes's method, or his search for something certain, namely, the *ego* (in the seventh stage) and his demonstrations of the existence of God (in the eighth), are rejected in the same way. The argument of the *cogito* was in fact borrowed by Descartes from Augustine; however, the Frenchman impiously negated the existence of everything else besides the *ego*.[11] Moreover, Revius rejects Descartes's proofs of the existence of God as they are based on an ambiguous account of 'idea'. The theologian addresses Descartes's misuse of the term, since he replaced its original, Scholastic meaning with an unclear one (Revius 1648, p. III (unnumbered)). According to Revius, indeed, Descartes ascribed to the term 'idea' eight different meanings, making its use inconsistent (Revius 1648, pp. 86–87). Consequently he rejects Descartes's argument for the existence of an extra-mental entity on the basis of the properties of mental content (Revius 1648, p. 119).

Such critiques were furthered in other texts. In his *Statera philosophiae cartesianae* (1650), namely, his answer to Heereboord's *Epistola ad Curatores*, Revius carries on the critiques expounded in the *Consideratio*, presenting Cartesian philosophy as striving for an unreachable degree of certitude in every kind of knowledge. The Dutch theologian focuses first on Descartes's supposed application of the geometrical method to metaphysics or rational theology, assuming the deductive rearrangement of Descartes's metaphysics – displayed in his *Responsio*

---

8 To such genius, according to Revius, Descartes ascribed divine features, without distinguishing him from the true God: Revius 1648, p. 48. Revius rehashes the accusations of Triglandius.

9 On the difference between Revius's and Schoock's accusation of indirect, speculative atheism, see Goudriaan 2002, p. 35.

10 "Ventilantur haec, et tum contradictionem, tum impietatem continere," Revius 1648, p. 53.

11 Revius 1648, pp. 75–77; see *Romans* 1.19–20; Augustine, *Civitas Dei*, book XI, chapter 26.

to Mersenne – as representative of all his arguments.[12] Roughly following Heereboord's *Epistola*, Revius esteems Descartes as a mathematician – thanks to Frans van Schooten's Latin edition of Descartes's *Géométrie* (1649) – rejecting, however, his geometrical or synthetic metaphysics. In fact, one cannot have any clear and distinct notion of God serving as premise in geometrical demonstrations (Revius 1650, pp. 7–10). In his *Statera*, Revius also takes into account Descartes's natural philosophy: through a comparison of relevant passages from Descartes's *Principia*, he shows that the Frenchman was concerned with false, imaginary, or merely probable principles, though he promised geometrical certainty for his conclusions. Accordingly, he made his whole philosophy a sophism (Revius 1650, pp. 15–20), and his method ultimately *vitiosus*, as it presupposes the existence of a deceitful God. Revius extensively quotes his own *Brevis explicatio mentis humanae* with regard to the considerations on God being a deceiver, like a good father or a physician may be. Instead of clear and distinct conclusions, then, Descartes's philosophy leads to scepticism and an appeal to Revelation as the only means to guarantee the truth of our statements, as Regius had shown.[13] The appeal to the interpreters of Descartes, which is aimed at showing the contradictory consequences of his thought, is actually a frequent strategy in Revius's *Statera*. Indeed, roughly following the progression of Descartes's *Principia*, after having criticised the Cartesian principles of motion (Revius 1650, pp. 108–124), Revius focuses on the problems of the soul, and thus on metaphysics again. Confronting the questions of the soul of beasts and of the immateriality of the human mind (Revius 1650, p. 144), he refers to the critiques of Kenelm Digby (Revius 1650, pp. 151–154, 160–161, 164),[14] Regius (Revius 1650, pp. 24–28, 170–175),[15] and Gassendi in his *Obiectiones* and *Disquisitio metaphysica* (1644) (Revius 1650, pp. 184–186), which he regards as drawing the necessary consequences of Descartes's philosophy.

Revius's criticisms are then continued in Lentulus's *Nova Renati Descartes sapientia* (1651), a commentary on Descartes's *Discours* and *Principia* straightforwardly confronting Descartes's rejection of Scholastic logic. Following Revius's arguments, Lentulus describes Descartes as a good mathematician who, however, wrongly applied a method inspired by mathematics to every discipline (Lentulus 1651, pp. 57–58). Thus, he made a small set of notions the basis of an inquiry into more difficult topics, being unable, however, to reach any evident

---

12 Revius 1650, pp. 9–11; see AT VII, pp. 160–170.

13 Revius 1650, pp. 39–48; see Regius 1648, pp. 10–11.

14 Revius extensively quotes from Digby 1645, chapters 26, 32, 35.

15 He is referring to Regius's positions on man as *ens per accidens*, mainly through Descartes's *Epistola ad Dinet* and *Epistola ad Voetium*.

conclusion. In fact, he did not clarify the first notions on which philosophy is to be based, as these are in one place conceived as those of mathematics, and in another place as the notions of mind and thought. Besides the application of Descartes's method to every field, Lentulus rejects clarity and distinction as truth criteria, since this coincides with Descartes's personal perception of things and paves the way, once again, for enthusiasm in philosophy. Doubt, therefore, turns out to have no role in Descartes's metaphysics except in eradicating all previous knowledge from the minds of his followers in order to supplant it with Descartes's convictions (Lentulus 1651, pp. 55, 57, 79). On this critique, Lentulus grounds his refutation of the argument outlined in Descartes's *Discours* against the old logic, rejected as merely expository or a means of disputing (AT VI, p. 549). Lentulus replies to Descartes's argument underlining that the Frenchman made no distinctions between *syllogistica* and *topica*, that is, between formal reasoning and the discovery of the topics of argumentation, and that syllogisms are the only means of arriving at conclusions. Moreover, even the expository role of logic should not be condemned, since it has a didactic value (Lentulus 1651, pp. 50–51). Lentulus, in fact, traces it back to the *Organon*, to Petrus Ramus's dialectics and to Bartholomäus Keckermann's systematisation, being consistently represented as a developed theory of reasoning,[16] whereas Descartes's proposal is dismissed as an enthusiastic, solipsistic, and reckless attempt to replace a well-ordered system of sciences (Lentulus 1651, p. 16).

These critiques move, mostly, on a logical and metaphysical field. The Cartesian reaction, i.e. the development of a Cartesian foundationalism, will therefore take the form of a metaphysical and logical theory, which came to be entangled in the works of Dutch Cartesians. How did they come to react at first?

3.1.2 The co-ordinated strategy of defence of Cartesianism

The Cartesian reaction was co-ordinated across the Netherlands and Germany. As far as the development of a full-blown logical and metaphysical defence is concerned, this was deployed by Clauberg.[17] Born in Westphalia in 1622, Clauberg

---

16 "Audeat Cartesius ieiunitatem suam cum una pagina Keekermanni nostri, vel etiam Rami comparare, audeat cum Organo Aristotelis contemnere," Lentulus 1651, pp. 223–224; see also pp. 30–31.

17 The philosophy of Johannes Clauberg is the object of increasing interest: after Francesco Trevisani has reconstructed Clauberg's role within the dissemination of Cartesianism in German universities (Trevisani 1992, Trevisani 2012), Massimiliano Savini more recently provided us with a comprehensive survey of Clauberg's views on the relations among ontology, metaphysics,

studied philosophy, theology, and Hebrew philology in Bremen under the guidance of Gerard de Neufville, who first prepared him for the reception of a new philosophy,[18] as his teaching brought the influences of Bacon and Comenius.[19] The crucial part of Clauberg's studies, however, took place in the Netherlands after having moved to Groningen in 1644, where he studied with the Tobias Andreae, professor of Greek and history and sympathiser with Cartesianism. There Clauberg published the first edition of his *Elementa philosophiae sive Ontosophia* (1647). In 1648, after his *grand tour* of England and France, he headed to Leiden in order to deepen his knowledge of Cartesian philosophy with De Raey, who at that time was giving private lectures on Cartesian philosophy. Eventually, he moved to Herborn Academy to start his teaching activity in 1649,[20] which he continued from 1651 at the University of Duisburg. During his years in Duisburg he published most of his works. Besides the *Defensio* and *Logica*, his *Initiatio philosophi, sive Dubitatio cartesiana* and *Exercitationes de cognitione Dei et nostri* (both published in Leiden, 1655 and 1656), his *Physica* (Amsterdam, 1664), and the second and third edition of his *Ontosophia* (Amsterdam, 1661 and 1664), testified to intense activities as a supporter of Cartesian philosophy.

His publishing activities in the early 1650s entailed a division of labour with the Cartesian network. In Holland, De Raey provided a defence of Descartes's physics by showing its concordance with the original thought of Aristotle, and purportedly avoided any metaphysical or theological issue in his *Clavis*. From Germany, Clauberg defended Cartesian logic and metaphysics. This strategy was established in two letters from De Raey to Clauberg of 1651 and 1652, which is worth briefly commenting on here (for a full account, see Strazzoni 2014b) as they outline the extent of the Cartesian network in the early 1650s.[21] Sent to Herborn, which Clauberg left in December 1651, the first letter can be dated back to August 1651.[22] The letter focuses on the answer to be given to Revius's *Statera*

---

and logic, and of their development with respect to the parallel debates over Cartesianism in the Netherlands and to the internal progress of Clauberg's positions (Savini 2004, Savini 2006, Savini 2011a, Savini 2011b). For a bibliographical overview, see Verbeek 1999.

18 Clauberg 1658, *Tobiae Andreae epistola*, pp. II–III (unnumbered); also in Clauberg 1691, p. 767.

19 See Leinsle 1999, Savini 2006, pp. 85–88, Savini 2011a, pp. 25–33, Strazzoni 2012, pp. 258–261, 267–270.

20 Clauberg 1691, *Vita per Henninium descripta*, p. IV (unnumbered).

21 A full account and version of these letters is provided in Strazzoni 2014b. The letters are preserved at the Leiden University Library (Special Collections (KL), BPL 293: B)

22 It refers to a letter from Clauberg received on 31 July: "pertactis gratissimis tuis literis quas prid[ie] cal[endas] aug[usti] ad nos dabas" (lines 3–4). Moreover, it mentions three disputations on Aristotle's *Problemata*: "ter iam disputavi: quatuor adminimum restant theses de dictis praecognitis, quas ubi omnes absolvere non exiguo nec inutili labore defunctus mihi videbor"

philosophiae cartesianae, (published in late 1650).[23] At first sight, it is likely that in summer 1651 Clauberg had not written his *Defensio* yet, since, according to De Raey's letter, Clauberg expressed to him his intention to refute Revius's *Statera*. This forthcoming answer is urged on by De Raey and by the Cartesian theologian Abraham Heidanus:[24] hence, De Raey asks for further information about the method Clauberg wants to follow for such confutation.[25] Whether by following the arguments of Revius with a commentary ("per notulas vaga hominis vestigia premendo"), or by discussing the main topics at stake with a more fruitful and brilliant discourse and method ("uberiori sermone et viva magis methodo praecipuas materias [...] vindicando"), the refutation of Revius's *Statera* would be a confrontation with obtuse adversaries,[26] there being no need of a too detailed method to carry on such a confutation.[27] Still, De Raey reminds Clauberg that he would tell him many things about the refutation of Revius, if only they could talk in person.[28] In fact, as De Raey himself started to think about a reply to Revius, being however prevented from carry out this purpose by some private reasons concerning the University of Leiden.[29] Most probably, he refers to the 1647 prohibition of overtly treating Cartesian philosophy, and to the need to avoid any conflict with the academic authorities by a direct attack on Revius. Such circumstances, on the other hand, did not affect Clauberg.[30] The central lines of the letter reveal the actual cooperation between Clauberg and De Raey in drawing up the *Defensio*, since De Raey asks his friend to be kept informed about the method he would choose and, suggesting secrecy, to read the proofs in advance. Such a communication, actually, was required by De Raey as in such a field extraordinary

---

(lines 27–29): De Raey's first three disputations took place on 3 and 17 of May, and on 14 June 1651, according to the front pages of De Raey's disputations: see De Raey 1651–1652.

23 See Revius 1650, *Jacobo Triglandio epistola*, p. II (unnumbered): the letter is dated "V kal. octob.".

24 "Dici non potest quanta laetitia ego pariter atque Dominus Heidanus affectus fuerim hisce diebus [...]. Probamus consilium vestrum de refutatione Staterae, idq[ue] ut quamprimum exequamini suademus," lines 4–5.

25 "Vellem mihi significasses qua methodo id efficiendum censeas," lines 5–6.

26 "quodcumque fiat, margaritae proiciendae erunt porcis, aut permiscendae saltem illorum sterquiliniis," lines 7–8.

27 "Methodo quidem aliqua sed non nimis accurate opus est," line 13.

28 "Nosti quam multa conferenda tecum haberem super hoc negotio, siquidem coram agere inter nos liceret," lines 8–9.

29 "Cum Statera primum edita esset, levi cum attentione sed maiori fastidio eam perenni [?], visusque mihi tunc fui methodum aliquam concipere foeliciter ipsam refutandi. Idque tunc temporis effectum etiam dedissem, nisi privata rationes quae me et hanc Academiam spectant praecaeteris, obstitissent," lines 9–12.

30 "Te vero nihil tale movere potest," line 12.

prudence was required.[31] Not willing to be lengthy, De Raey finally recommends Clauberg to keep him informed about his decisions in his next letters, with the due precautions.[32]

De Raey's concerns about secrecy are clearer in his second letter, dated 2 and 12 November 1652. This letter is in reply to two lost letters of Clauberg of September and October 1652[33] and was written after the publication of the second main text addressed in Clauberg's *Defensio*, the *Nova Renati Descartes sapientia* of Cyriacus Lentulus, published in Herborn in 1651, as well as after Clauberg's *Defensio*.[34] In this letter De Raey addresses some circumstances related to the publication of Lentulus's book. After declaring his acquaintance with the book,[35] De Raey declares that his own correspondence had been violated. Presumably referring to a previous letter from Clauberg, De Raey declares that he had never sealed his letters with a coin: this has been done by someone who broke the original seal.[36] This detail might reveal the occasion of the publication of Lentulus's *Sapientia*, if compared with the very text of his *Dedicatio*. In this text, Lentulus mentions the spread of the 'Cartesian poison' in Germany.[37] As Clauberg was teaching in Herborn from the end of 1649, apparently without any problems, Lentulus seems to refer to a forthcoming over-run of the controversy about Cartesianism to Herborn, which would explode with the publication of Clauberg's *Defensio*. This forthcoming publication, in fact, seems to have been foreseen by Lentulus through the violation of De Raey's correspondence. This may explain the urgency of Lentulus's writing, assessed in his *Dedicatio*, as well as the encouragement

---

31 "Vellem methodum quam praeconcepisti mihi indicares si liceret per otium. Et, si nemine conscio fieri posset, non inconsultum foret scriptum tuum a me perlegi priusquam typo mandaretur, non quod tua diffidam scientia, sed quod maiori quam in aliis solet opus sit prudentia," lines 19–22.

32 "Sed nimis longum foret in praesentiarum me dimittere in hunc campum. Si grave tibi non fuerit, quam primum certior fieri velim consiliorum tuorum, atque una scrupulos aliquos, qui occurrent forte in istohoc labore, in literis proximis consignatis videre," lines 32–35.

33 "Binas a te accepi literas, unam mense sept[embris] alteram octobr[is] datam," line 2.

34 The *Praefatio* of Clauberg's *Defensio* is dated February 1652: Clauberg 1691, *Defensio*, p. 941.

35 "Vidimus et obiter inspeximus librum Lentuli," lines 2–3.

36 "Quod sciam multas unquam literas nummo obsignavi, sed id factum proculdubio fuit a scelestis manibus posteaquam sigillum meum perfregerant," lines 4–6.

37 "Renatum dico Des Cartes [...] Sententiarum eius peregrinantem adeo contempseram, apud Batavos cum degerem, ut ne paginam quidem liberorum contra eum scriptorum legere dignarer. Peregre deinde agens, de fama eius nihil audivi, nec inquirere curavi. Postquam vero e Gallia Narbonensi ab Illustrissimo Comite Nassoviae ad docendam politicam et historiarum usum Herbornam evocatus, Lugduni etiam turbatum esse cum indignatione percepi, et virus illud ulterius serpere, et venas Germaniae tentare animadverti," Lentulus 1651, pp. 7–8.

of his friends (probably, those from Leiden).[38] Indeed, when Lentulus wrote his *Sapientia*, a large number of them commented on quotations from Descartes's works (whereas Revius's *Statera* has a more consistent structure) in a few weeks. De Raey himself, moreover, describes its appearance as something unexpected, "ante eventum" (line 9), precipitating the clash over Cartesian philosophy. As can be presumed from De Raey's words, the violation of his correspondence took place in Leiden, where he had some ambiguous friends, even if the University Curators and city authorities (to which he would dedicate his *Clavis*) were supporting him.[39] Clauberg wrote his *Defensio*, as declared in the *Praefatio*, in order to respond to the accusations of Lentulus, who attacked him apparently without any reason,[40] and on the suggestion of some friend of his.[41] On the other hand, in his *Thekel hoc est levitas Defensionis cartesianae* (1653) Revius will accuse Clauberg of having been pushed by Leiden Cartesians to write his *Defensio*, since they could not attack Revius directly. This had been communicated to him in a letter of October 1651, before the appearance of Clauberg's *Defensio*.[42] The contents of the two letters by De Raey demonstrate that Revius was somehow right because it is true that De Raey urged Clauberg to answer Revius since he could not personally do so. Aware of this fact, Revius and Lentulus anticipated the publication of

---

38 "Seposita meliorum meditationum cura, in castra Cartesii speculatum transii, animumque simul cepi ac operae iudicavi, cursoria functione quid sentirem de famosi authoris opinionibus, chartae illinere. Quod paucarum septimanarum praecipitata scriptitandi opera factum. Cum vero moecenatibus meis et amicis visa essent non indigna luce, quanquam apud Batavos meliora aut edita esse aut edi posse credebam, ut qui et scribendi acumine pollerent, et hunc conflictum iam non novum haberent. Cessi tamen in eorum, quibus negare nihil poteram, benevolam coactionem et pro veritate protegenda vel periculum famae subii. Famae vel a scribendi festinatione, vel ab adversariorum solita invehendi petulantia et obtrectandi libidine denigrandae," Lentulus 1651, pp. 8–9.

39 "Pro certo tibi affirmare possum omnes Curatores causae nostrae et philosophiae favere, ne ipsis quidem Consulibus, qui hisce diebus electi et inaugurati sunt, exceptis," lines 6–8.

40 "Herbornae philosophiam tranquille docebam, cum Cyriacus Lentulus successibus invidens, professoribus plerisque insciis, librum prelo daret, hunc prae se ferentem titulum: Nova Renati des Cartes sapientia [...]. Me vero iam sub discipuli cartesiani, iam sub sectatoris Cartesii aliisque appellationibus, iam sub initialibus nominis mei et cognominis literis I. CL. immerito vellicatum esse deprehendo," Clauberg 1691, *Defensio*, p. 939.

41 "Opportune igitur hortabantur amici atque instabant, ut eadem opera ab insultibus Iacobii Revii [...] philosophiam nostram liberarem," Clauberg 1691, *Defensio*, p. 939.

42 "Id fortasse coniiciet e verbis ad me Herborna prescriptis kal. nov. 1651 ita autem habent: *philosophi cartesiani* SCENAE AD TEMPUS INSERVIRE *(haec enim sunt illorum verba)* LUGDUNI *decreverunt. Claubergium nostrum,* VESTRI *rogant, ut vel refutationem Revii, vel Lentuli urgeat:* AB IPSIS ENIM HOC NON DEBERE," Revius 1653, *In praefationem Claubergii*, pp. II–III (unnumbered). One can find an account of these reciprocal accusations, among the reasons of the publication of Lentulus's *Sapientia* and Clauberg's *Defensio*, in Savini 2011a, pp. 117–119.

Clauberg's *Defensio* with the *Sapientia*. Whether De Raey's suspicions were valid or not, they testify to the acrimony characterising the debates on Cartesianism in the early 1650s. In any case, in his letter De Raey suggests Clauberg keep on writing the esoteric part of his *Defensio*, to be edited along with the first part.[43] The esoteric part, however, would never been published under the title of 'Defensio'. In 1652 only the first part of Clauberg's *Defensio* saw the light. According to Clauberg his text had to be edited in two parts: the first, *exoterica maius*, concerning the Cartesian method or logic, which had to serve as an introduction to the more complex topics of Cartesian philosophy, such as those treated in metaphysics. This had to be the object of Clauberg's planned *Defensio acroamatica*, which he never published: still, he devoted to Descartes's metaphysics his *Initiatio philosophi* (1655), likely to be considered as the continuation of his *Defensio*, and his *Exercitationes de cognitione Dei et nostri* (1656).[44] Being a commentary on Descartes's *Discours de la méthode*, or the exoteric introduction to his philosophy, Clauberg's *Defensio* mainly focuses on sections I to III of the *Discours*, avoiding any close enquiry into the metaphysical problems treated in section IV,[45] and on Descartes's natural philosophy considered in section V and VI.[46] In fact, the *Defensio cartesiana* sets the ground for the comprehensive development of Cartesian logic expounded by Clauberg in his *Logica vetus et nova*, where the four rules of the method are integrated in syllogistic reasoning. At the same time, it anticipates the metaphysical topics – first of all, the use of doubt as the preparation for philosophy and the demonstration of the existence of God – at stake in the metaphysical treatises of Clauberg.

---

43 "Auctores tibi sumus ut pergas hieme in acroamaticis: probo consilium de edendo textu cum defensione," lines 20–21.

44 Clauberg 1655, Clauberg 1656 (also in Clauberg 1691).

45 Whilst not openly aiming to refute Regius's metaphysics, Clauberg separated Regius's thought from Descartes's original philosophy in his *Logica*: see Clauberg 1691, *Logica*, p. 859. Clauberg rejects Jacob Revius's assessment of Regius as drawing the necessary consequences from Descartes's metaphysics, i.e., that Descartes's philosophy leads to scepticism and to an appeal to Revelation as the only means to guarantee the truth of our statements. See Strazzoni 2013, pp. 130, 143.

46 See chapter 23: "ad sectionis quintae initium. I. Cum in hac sectione physica tractentur, similes ob causas ad tres Principiorum libros physicos a nobis reservabuntur, ob quas ea quae in antecedente sectione proponuntur, ad Meditationes metaphysicas retulimus," Clauberg 1652, p. 251; also in Clauberg 1691, *Defensio*, p. 1013. Chapters 1 to 18 concern the method, or the first three sections of Descartes's *Discours*; chapters 19 to 21 concern Descartes's provisional ethics (section IV), chapter 22 concerns metaphysics (section V), and chapters 23 to 30 concern some paragraphs of sections V and VI, on physics and its method. The other chapters (31 to 37) are about various arguments. On Clauberg's *Defensio*, see Savini 2011a, pp. 117–139.

3.2 Logic as introduction to Cartesian philosophy: Clauberg's *Defensio cartesiana* and *Logica vetus et nova*

The criticisms moved by Revius and Lentulus brought Clauberg to develop a comprehensive foundation of Cartesian philosophy, as well as the development of a new 'Scholasticism', in which Cartesianism could serve as the theoretical basis for any philosophical discipline, and for the higher arts: medicine, law, and theology (Viola 1975). The starting point of Clauberg's construction of philosophy is logic, insofar as the quarrels over Cartesianism primarily involved the problem of method. Moreover, in order to provide an introduction for students to a new way of philosophy and an instrument to conduct reason within it, logic became the first discipline to be developed in a Cartesian scholasticism, as it had to replace Aristotelian logic as the instrument of philosophy. In his *Defensio*, the development of a Cartesian logic takes the form of a vindication of Descartes.

Clauberg points out that insofar as Descartes was concerned with the old logic he did not completely reject it; he underlined, however, its merely expository value and erroneous precepts. In fact, Descartes just left syllogistic theory out of his arguments without rejecting it.[47] In addition, his apparent rejection of the old logic appears to concern dialectics more than syllogistic theory; that is, only the arguments contained in the *Topica*, as stated in Descartes's *Epistola ad Voetium*. Such a distinction is borrowed from the late Scholastic tradition, as Clauberg quotes Burgersdijk's *Institutiones logicae* to support it.[48] In sum, Descartes's purpose was not to reform logic: Clauberg had such a purpose and developed a new logic embodying the four rules of the method with a consistent

---

[47] "Cartesius non oppugnat logicam, sed eam, quam in scholis didicerat, sibi minus prodesse asserit ad suum propositum, interim aliis quibus prodest eam relinquit," Clauberg 1691, *Defensio*, p. 973. See also p. 972.

[48] "Operae pretium fuerit etiam illum locum expendere, quandoquidem vix alibi occasio dabitur. Exstat autem in Epistola ad Voetium pag. 26. 27. [...] *Artes quibus te uti ex scriptis tuis deprehendo, et* [...] *tales esse mihi videntur, ut eas vilissima multa ingenia perfacile possint addiscere* [...]. *Prima* [...] *est puerilis illa dialectica, cuius ope olim sophistae, nullam solidam scientiam habentes, de qualibet re copiose disserebant ac disputabant.* [...] Notandum est, Cartesium in citato Epistolae loco de logica universa non agere, sed tantum de dialectica. [...] Notandum est, maximam esse differentiam inter logicam analyticam, demonstrativam, scientificam, de qua Aristoteles in Posterioribus analyticis, et dialecticam, popularem, disputatricem, de qua Aristoteles agitur in Topicis [...] Hinc Franco Burgerdicius in praefat. Logicae: *in Topicis traduntur praecepta disputandi ex iis, quae revera sunt probabilia. In Analyticis traditur ratio disputandi seria veramque scientiam adipiscendi,*" Clauberg 1691, *Defensio*, pp. 973–974; see AT VIII/2, p. 50, Burgersdijk 1660, *Praefatio*, pp. V–VI (unnumbered).

syllogistic apparatus, grounded on Descartes's four rules, which are the core of the first part of logic:

> totam logicam, quam vocavi analyticam, tanquam a suo instituto alienam aliis reliquit. Cum etiam docere alios eo tempore non institueret, sed tantum mentem propriam vellet cognitione informare. [...] Mero meridie clarius est, Cartesius non velle quatuor illa praecepta toti reipublicae literariae ad quoscunque logicae universae fines assequendos sufficere, sed sibi duntaxat ad scopum quem animo destinaverat. Observandum tamen est, me in explanatione istorum praeceptorum (ad priorem logicae geneticae partem [...] proprie spectantium) ut usum eorum uberiorem patefacerem, etiam ad reliquas Logicae partes multis in locis respexisse. Illius enim quadripartitae prima portio simul est fundamentum sequentium. (Clauberg 1691, *Defensio*, p. 998)

The first task of logic, according to Clauberg, is to help man avoid error. This purpose is first stated in the *Prolegomena* to the second edition of his *Logica*, whose first chapter sounds as "futuro logico et philosopho errorum et imperfectionum humanae mentis in rebus cognoscendis orginem et causas investigandas esse" (Clauberg 1691, *Logica*, p. 769). In fact, logic is conceived as *medicina mentis*, that is, something new (but not unheard of) in the history of philosophy.[49] Like his *Defensio cartesiana* (Clauberg 1691, *Defensio*, pp. 1050–1097), Clauberg's *Logica* focuses on the causes of error. Borrowing several arguments from Bacon, Clauberg lists the causes of error by paying attention to the ages of men and to the social context (Clauberg 1691, *Logica*, pp. 770–778). In fact, even if doubt is regarded as the main remedy against error, and the very first step in philosophy, in his *Logica* more specific tools are set forth against it, making logic more of a practical art than a theoretical science (Clauberg 1691, *Logica*, pp. 778, 800). For Clauberg, logic is a set of rules meant to avoid error not only in philosophy but also in everyday life. It is a *didactica*, serving to teach how to avoid error.[50] The functions of logic determine its structure, including, like Ramus's dialectic, *genetica* and *analytica* (Savini 2006, p. 75, Savini 2011a, pp. 197–208, Strazzoni

---

49 "Novum hoc esse et insolitum in logicae vestibulum," Clauberg 1691, *Logica*, p. 769. On the *medicina mentis* tradition, see Corneanu 2011, Savini 2012.

50 "Iam quia in vitae societate alii saepe homines veritatis praeceptis imbuendi sunt, certe nulli unquam fallendi, nec voce, nec scripto, nec alio signo. Inde hinc oritur secunda logicae necessitas, quae respicit eos, quos docemus, aut quibuscum disserimus. Quis enim neget, singulari nobis arte opus esse, ut homines a praeiudicatis infantiae opinionibus atque inde descendente erroneo iudicandi modo liberatos paulatim ad rectiorem rationis usum ducamus?" Clauberg 1691, *Logica*, p. 779. Logic is defined as dialectic and didactic in Clauberg's *Logica contracta*, a compendium of some sections of *Logica vetus et nova* appeared in Clauberg 1660 (also in Clauberg 1691, p. 913).

forthcoming a).[51] Maintaining its practical role, Clauberg's logic loses, however, the main rhetorical features that Ramistic dialectics had. The structure of Clauberg's logic, expounded in his *Logica vetus et nova*, is as follows: 1) *genetica* concerns, in its first part, the clear and distinct formulation of concepts, made possible by attention and diligence; their ordering by different kinds of method, the formulation of judgments and syllogisms, and the aids to the memorisation of concepts. Such operations are mostly based on Descartes's rules of the method expressed in the *Discours*: for instance, the formulation of concepts has to be guided by diligence, which is embodied by the fourth rule of Descartes's method, while in judgments one has to proceed by a careful application of doubt. In turn, the clarity and distinction acquired by these 'Cartesian' means make the memorisation of concepts easier (Clauberg 1691, *Logica*, pp. 786–787, 797–814). 2) The second part of logic (*genetica hermeneutica*) concerns the instruments by which one can express concepts and judgments through brevity and perspicuity: by using simple words, examples, similitudes, or by repeating definitions (Clauberg 1691, *Logica*, pp. 819–830). The other two sections of *Logica* (i.e. the *analytica*) concern the interpretation or analysis of the sentences of other men. So, 3) the third one explains how to understand their meanings by the application of attention, diligence and memory (Clauberg 1691, *Logica*, pp. 843–845), and by using the conceptual tools of *lexica* and *rhetorica* (Clauberg 1691, *Logica*, pp. 849–850). 4) Eventually, the fourth part (*hermenutica analytica*) teaches how to assess the truth of others's texts, and consists of their analysis in the light both of the principles of formation of concepts expressed in the first part – by perception and judgment – and of their communication, dealt with in the second (Clauberg 1691, *Logica*, p. 866).[52]

Actually, it is in the first and in the fourth section that metaphysical topics can be found. As it treats the "inveniendi veri methodum" (Clauberg 1691, *Logica*, p. 780), in the first section Clauberg put forward three main questions: "quid sit cognoscendum," "quis ipse sit, qui vult cognoscere," "quomodo possit cognoscere, ubi de methodo" (Clauberg 1691, *Logica*, p. 783). Stating the basics of Cartesian metaphysics, some words are devoted to the objects of knowledge, matching Descartes's hierarchy of knowledge:

> quod omnibus necessario cognoscendum est, ante omnia cognoscamus, v.g. Deum et nos ipsos, in caeteris vero eorum, quae potioris sunt dignitatis et usus, potiorem rationem

---

51 See also Ong 1958, pp. 363–367, Petrus 1997.

52 On Clauberg's logic as hermeneutics, see Savini 2011a, Strazzoni 2013. On the history of philosophical hermeneutics, see Hasso Jæger 1974, Alexander 1993, Danneberg 1998, Danneberg 2001, Danneberg 2005, Danneberg 2006.

> habeamus, non necessariis atque inutilibus omissis, cum sapientia non paretur ex quarumvis rerum notitia, sed ex earum duntaxat quae maioris sunt momenti. (Clauberg 1691, *Logica*, p. 784)

The subsequent considerations are on the division of knowledge into the practical and theoretical: according to Clauberg, indeed, even the knowledge of God is ultimately practical, as it grounds disciplines such as medicine or law.[53] As he would state in his *Initiatio philosophi* (1655), indeed, logic and metaphysics come to be entangled as they are at the beginning of philosophy, and serve to ground all knowledge.[54] Foundational arguments are then set forth in the fourth part of logic, as he considers the epistemic status of sentences expressing eternal truths, such as those embodying the demonstration of God's existence. In the first part, he had outlined two degrees of certitude and truth: contingent (or moral) certitude, and necessary certitude, embracing in turn three further degrees: *certitudo physica sive de omni*, *certitudo metaphysica per se* and *certitudo metaphysica universaliter prima* (Clauberg 1691, *Logica*, p. 801). Metaphysical or eternal truths are grounded on the clear and distinct perception of the connection of subject and predicate.[55] A difference, however, is made between those metaphysical truths, whose utmost certitude relies only on the definition of the subject, as in the sentences "omnis homo est animal rationale"[56] or "Deus necessario existit," analysed in the last part of the *Logica*:

> examinantur veritas et falsitas, et gradus utriusque in enunciationibus [...] ubi illa [...] Deus necessario existit, habet certitudinem [...] metaphysicam, estque per se et universaliter primum, ideoque magis necessaria hac, binarius est par. (Clauberg 1691, *Logica*, p. 891)

Because God is defined as *ens summe perfectum*, this definition implies a necessary existence. Therefore, "Deus necessario existit" is to be considered even more necessary than "binarium esse parem," because the truth of the former sentence depends on the definition of the subject, whereas that of the latter relies on the notion of the predicate (Clauberg 1691, *Logica*, pp. 892–893).

---

53 "At nunquid caeli notitia ad Creatorem agnoscendum ac celebrandum adducimur, et nunquid sequitur, omnem cognitionem quodammodo practicam esse oportere, nullam otiosam aut sterilem," Clauberg 1691, *Logica*, p. 784.

54 "Nimirum, tantum abest ut scepticis faveamus, ut eos non solum initio philosophiae, verum etiam in logica refellamus," Clauberg 1691, *Logica*, p. 784.

55 "Unde vero existit summa illa seu metaphysica de axiomatibus nonnullis in animo nostro certitudo? Resp. Certitudo axiomatis affirmantis proficiscitur e subiecti et predicati nexu insolubili a mente clare et distincte percepto," Clauberg 1691, *Logica*, p. 802.

56 Clauberg 1691, *Logica*, p. 803. This, in fact, is the Aristotelian definition of man: regarded, however, as absolutely necessary in the light of its clear and distinct perception.

Ultimately, Clauberg finds in logic the proper place for an analysis of the degrees of certitude of the bases of Cartesian metaphysics, which is treated according to the Cartesian criterion of truth perception and by paying attention to the kinds of subject-predicate connection. This analysis is applied not only to the *a priori* proof, but also to Descartes's *a posteriori* argument. In the last section of *Logica*, indeed, the principle of causality is examined from an analytical point of view,[57] involving, in fact, metaphysical considerations. It is considered in the light of the concepts of efficient and exemplary cause, which in the case of the idea of God must be identified as it contains perfection in its objective being (Clauberg 1691, *Logica*, p. 897). This argument will be properly developed in the *Initiatio philosophi* and in the *Exercitationes de cognitione Dei et nostri*, Clauberg's main treatise on metaphysics along with *Ontosophia*.

3.3 Metaphysics and natural theology in the foundation of philosophy and arts

Clauberg's *Initiatio philosophi* offers an outlook on his metaphysical foundation of philosophy. As he has to confront the problem of the introduction of the student to a new paradigm, together with a demonstration of its reliability, in this treatise Clauberg maintains that the initiation of scholars into the new philosophy through doubt is to be identified with the theoretical justification of philosophical knowledge itself. This introduction and justification is provided, first of all, by means of doubt. In accordance with a Baconian *expurgatio intellectus*,[58] doubt serves as an emending instrument through which it is possible to reach a metaphysical or absolute certainty on the notions of self, God, and matter, and to become acquainted with a new way in reasoning. It is the very first step into the new philosophy,[59] or the initiation into Cartesianism for everyone who has

---

57 "Examini analytico subiecimus veritatis certitudinisque gradus ac differentias, iam etiam, Logicae nostrae ordinem secuti, gradus universalitatis in axiomatibus, et quae quibus superiora, quibus inferiora sint, expendamus: hoc enim multum prodesse ad iudicium formandum ipsa nos docuit. Sumamus vulgatissimum illud: quod quid non habet, id alteri dare non potest," Clauberg 1691, *Logica*, p. 894.

58 Clauberg 1691, *Initiatio philosophi*, pp. 1125–1126, see § 7. On the use of Bacon by Clauberg, see Savini 2006, pp. 73–88, Strazzoni 2012, pp. 258–261, 267–270.

59 "Dubitatio nostra, quae aliis debito generalior esse videtur non spectet ad eum qui firma iam philosophicae cognitionis fundamenta iecit, quasi ea deberet in dubium revocare ac reiicere. Verum ad illum duntaxat, qui fundamenta eiusmodi nondum posuit quique non aliter consideratur, quam ut vulgaris homo, nihil adhuc scientifice demonstratum habens, nihil clare distincteque perceptum, cui iudicium superstrui queat indubitatum," Clauberg 1691, *Initiatio philosophi*, p. 1138.

not been acquainted with clear and distinct perceptions, allowing no further doubting or suspension of judgment. Such a first, emending step in philosophy belongs to metaphysics, *philosophia prima* (Clauberg 1691, *Initiatio philosophi*, pp. 1142, 1144, 1208–1209). The function of metaphysics is outlined by Clauberg against its Aristotelian definition as the discipline coming after physics: as he writes in his *Differentia inter cartesianam et in scholis vulgo usitatam philosophiam* (1657, 1680),[60] the very name of 'metaphysica' suggests that this discipline was definitively misplaced in the traditional order of the sciences.[61] This order is reversed by Clauberg by embracing Descartes's plan of disciplines, according to which the tree of philosophy is composed of the roots of metaphysics, the trunk of physics, and the boughs of mechanics, medicine and moral philosophy.[62] Hence, Clauberg underlines the absence of a foundation in Aristotelian philosophy. Describing several differences between Scholastic and Cartesian philosophy, Clauberg counts among them a different introduction to philosophy:

> vulgaris philosophiae sententia est, omnem scientiam ex praecognitis oriri debere [...]: quamobrem illa multas res, praesertim illas, quas ab infantia ipsi vidimus, audivimus et sensimus, praesupponit tanquam certissima fundamenta quae nulla demonstratione indigeant [...]. Atque hoc modo illa introitum suum facit. (Clauberg 1691, *Differentia*, p. 1225)

Clauberg opposes to such a state of philosophy a consistent architectonic as a guarantee for the validity of the new philosophy. Descartes's architectonic metaphor, to which Clauberg also refers in his *Defensio cartesiana* when he considers a provisional ethics (Clauberg 1691, *Defensio*, p. 1002), is the starting point of the outline of a philosophy which has its first *introitum* into doubt as the 'door' to metaphysics. The development of an architectonic structure of philosophy is made possible by Descartes's personal renovation of philosophy:

> cartesiana philosophia similis est alii cuidam urbi, quam architectus unicus eodem tempore iuxta regulas artis suae dimensus est, et e fundamento exaedificavit. [...] Et ut breviter [...]

---

60 The German first edition appeared in 1657, as *Unterschied zwischen der cartesianischer und der sonst in Schulen gebraeuchlicher Philosophie* (Clauberg 1657), followed by the Latin translation in 1680 (Clauberg 1680, also in Clauberg 1691).

61 "Notabilis differentia inter cartesianam et aristoteleam [...] siquidem illa a rebus spiritualibus aut intellectualibus et ratione utentibus, haec autem a corporalibus initium sui scrutinii atque doctrinae primum facit, atque ita in scholis prima philosophia dicitur, dignitatis et naturae, non cognitionis nostrae ordine, contra quam fit in philosophia cartesiana. Atque etiam haec primum est inventa, illa autem postremo, indeque nominata metaphysica, quasi post-physica," Clauberg 1691, *Differentia*, p. 1226.

62 Clauberg 1691, *Differentia*, p. 1157, quoting Descartes's words on the tree of knowledge: AT IX/2, p. 14.

> dicamus quod res est, vulgaris philosophia farrago est ex opinionibus variorum hominum [...]. Cartesiana philosophia solum unius viri opiniones comprehendit. (Clauberg 1691, *Differentia*, p. 1220)

The peculiar character of Descartes's philosophy is to have been developed by a single man. Its theoretical foundation coincides with the very initiation of the vulgar man into new thought, since it begins as an introspective endeavour based on suspension of judgment. For Descartes, as he is referred to by Clauberg, metaphysics concerns the principles of human knowledge, among which the divine attributes are to be counted[63] as from them it is possible to deduce natural laws.[64] In fact, God is required both to ensure the reliability of our faculties, and to explain the ultimate cause of natural laws.

The proper justification of philosophical knowledge – *scientia* – on a rational-theological basis is then developed in Clauberg's *Exercitationes de cognitione Dei et nostri*.[65] As stated in his *Defensio cartesiana*, metaphysics coincides with natural theology[66] and proceeds from the acknowledgment of the notions of self and God to that of bodily reality. In fact, such concepts lead to the demonstration of the reliability of the human faculties, as they include the acknowledgment of the goodness of God and are, at once, the basis of natural-philosophical explanations, such as the notions of physical modes and natural laws (Clauberg 1691, *Differentia*, p. 1233). Yet, the rational theology developed in his *Exercitationes* is not only meant to ground physics. Since Clauberg conceives of all disciplines as relying on philosophical knowledge, even disciplines like law find their foundation in *philosophia prima*, in accordance with an attempt to integrate Cartesianism into the academic curriculum.[67] God, as the first cause – whose acknowledgment

---

63 "In praefat. editionis Princip. gallicae, ubi explicaturus ordinem, quem quis tenere debet in se instruendo hac philosophia, *cum iam*, inquit, *acquisivit habitum quendam inveniendae veritatis in his quaestionibus* (nempe mathematicis) *debet serio incipere se applicare verae philosophiae, cuius prima pars est metaphysica, quae continet principia cognitionis, inter quae est explicatio praecipuorum Dei attributorum, immaterialitatis animarum nostrarum et omnium notionum clararum et simplicium, quae sunt in nobis. Secunda est physica*," Clauberg 1691, *Initiatio philosophi*, p. 1154; see AT IX/2, p. 14.

64 "Vide qui Cartesius regulas de motu et corporum existentiam ex Dei natura et existentia derivet," Clauberg 1691, *Initiatio philosophi*, p. 1155.

65 "Cartesiana philosophia seponit in principio omnia visibilia et corporalia, quo unusquisque philosophiae studiosus ante omnia in propriam suam mentem descendat. [...] Porro menti nostrae nihil propius arctiorisque cognatione iunctum est, quam ipse Deus. [...] De hac materia latius egi in tractatu De cognitione Dei et nostri," Clauberg 1691, *Initiatio philosophi*, p. 1226.

66 "Theologiam seu metaphysicam," Clauberg 1691, *Defensio*, p. 1011.

67 "Utilis [...] est naturalis Dei cognitio propter alias omnes humanas disciplinas, quarum firma et evidens notitia expetitur. Non enim possunt satis refutari sceptici neque conclusionis ullius

allows for the attainment of *scientia* as the knowledge of first causes – is to be taken into consideration in all disciplines, not only in the deduction of physical laws.[68] Natural theology is thus present from the first to the last step of philosophy,[69] which ends in physics, ethics, and politics.[70]

The *Exercitationes* contain the presentation and clarification of Descartes's proofs, enriched with corollary considerations borrowed mainly from Scholastic and Renaissance authors. Actually, Clauberg does not add any new points to such proofs: with his considerations, however, he discloses some points implied in Descartes's arguments. His first focus is on the imitative nature of ideas, which may thus be conceived as images, in accordance with the views of Bartholomäus Keckermann and Rudolph Goclenius. Presenting the first proof, in addition to the twofold nature of ideas – formal and objective – Clauberg highlights the relation between human and divine ideas, envisaged as ectypes and archetypes.[71] God,

---

vera certitudo haberi, nisi ante probetur, Deum summe veracem et causam omnis veri et boni necessario existere, a quo proinde accipiamus omnem intellectum, quo si recte utamur, hoc est, si non nisi de clare distincteque perceptis iudicemus, fallere aut falli nequeamus. Et constat inter cunctos logicos et philosophos, non posse obtineri ullius rei creatae veram scientiam, nisi perspectis causis. Causas autem non posse perfecte cognosci, nisi ad primam et supremam causam, quae Deus est, recurratur," Clauberg 1691, *Exercitationes*, p. 594.

68 "Addo peculiari de causa tractationi de Deo locum esse dandum in primis philosophiae principiis, quoniam perfecta rerum scientia, quam philosophando acquirere laboramus, ex causarum praecipue notitia resultat. At prima rerum omnium causa, et sine qua reliquarum causalitates nec sunt, nec accurate cognosci possunt ullae, est Deus," Clauberg 1691, *Exercitationes*, p. 596.

69 "Nam initio philosophiae non ulterius agitur de Deo, quam quatenus eius cognitio ad iacenda omnis scientiae humanae fundamenta desideratur. Sed in fine absoluta de Deo tractatio instituitur, omniaque eius attributa, quae ex naturae lumine cognosci queunt, expenduntur, quod initio necessarium non erat, quoniam non omnia Dei attributa se habent ut principia rerum creatarum, et quae huiusmodi relationem possunt recipere, non tamen absolute ideo aut planius, quam originis illa relatio postulat, opus est explicare. Nec possunt sane attributa Dei absolute et plene satis explanari, antequam rerum ab eo creatarum tractatio praecesserit," Clauberg 1691, *Exercitationes*, p. 596.

70 "Dico per universam philosophiam diffusam esse theologiam naturalem, quia dum in operibus Dei rite contemplandis occupamur, fieri nequit, quin ipsius opificis potentiam, bonitatem, sapientiam passim admirando, in eius notitiam magis magisque assurgamus. Quaecunque enim sunt in rerum natura creata et ordinata, ad ipsum tanquam suum principium et originem sunt referenda. Quod respiciens S. Augustinus in Epist. ad Volusianum, ipsam quoque physicam, ethicam, politicam aliasque disciplinas theologiae terminis contineri asseruit," Clauberg 1691, *Exercitationes*, p. 596.

71 "Observo 1. conceptum seu ideam omnem habere duplicem dependentiam, unam a concipiente sive cogitante intellectu [...] altera, a re concepta aut simili, cuius scilicet repraesentatio sive imago est, sive unde per imitationem expressa est. [...] Observo 2. intellectum esse causam conceptus efficientem, [...] rem vero conceptam [...] esse causam conceptus exemplarem (quae quidem etiam ad efficientem reducitur) atque eo modo ad conceptum referri, quo archetypon ad ectypon," Clauberg 1691, *Exercitationes*, p. 606. See p. 618: "ex ideis aliae sunt ectypae, qualis est

therefore, is at the same time the efficient and the exemplar cause of our innate ideas. As these are conceived as images or imitations, they cannot be more perfect than what they are images of, as stated by Aristotle and Keckermann:

> *imago est*, inquit Aristoteles lib. 6. Top. cap. 2, *quae per imitationem efficitur*, sive cuius generatio est per imitationem. Imitatio autem, veram eius naturam si intueamur, per se nihil aliud est, quam *conformatio imperfectioris ad perfectius*, ut bene inter alios definit Keckerm. Syst. phys. lib. 4 cap. 8. (Clauberg 1691, *Exercitationes*, p. 606; see *Topica*, 140a 14–15, Keckermann 1623, p. 564)

Assuming the existence of an imitation or an *esse obiectivum seu vicarium* (a term borrowed from Goclenius (Goclenius 1613, p. 1047)), an archetype is thus required.[72] Besides Aristotle, Keckermann, and Descartes,[73] the position of Clauberg relies on the works of Eustache de Saint-Paul and Goclenius.[74] Clauberg's strategy is to refer the basics of Descartes's arguments to the Scholastic theories. However, it is not clear to what extent Clauberg supports the view that ideas are truly mental images or visual contents: "imago quaedam," "tanquam imago," or "per modum imaginis" (Clauberg 1691, *Exercitatones*, p. 617) suggest that Clauberg is only using a comparison with images more than identifying ideas with them,[75] following the philosophical terminology adopted by Descartes himself.[76] In fact, the status of ideas remains ultimately unexplained, even if Clauberg maintains Descartes's classification of ideas into fictitious, innate, and sensory.

---

idea Dei et aliarum rerum ab homine non factibilium, aliae archetypae, quae rerum faciendarum formulae et exemplaria sunt et a philosophis ad causam efficientem referuntur."

72 "Idea secundum esse vicarium spectata non potest esse perfectior sua causa, hoc est, suo exemplari, imperfectior esse potest. Imo nulla imago plus realitatis et perfectionis repraesentare, quam reperitur in ea re, unde talis imago desumta sive expressa est. [...] Probatum [...] tum ex natura imaginis atque imitationis, tum ex illo axiomate, quod effectus non possit esse nobilior causa. Et sane, quam necessarium est, ut omnis idea habeat causam exemplarem, tam necessarium est, ut omne praeclarum quod habet idea, procedat ac derivetur ab exemplari illa causa. [...] Ut ex nihilo nihil fit a natura: ita nec potest mens nostra ullum realem conceptum formare, nisi rem aliquam imitata, et cuius totum esse in imitatione consistit, id non potest plus continere, quam est in imitabili sive exemplari. [...] Si ergo [...] summae perfectiones non sunt in mente [...] sequitur eas esse extra mentem nostram [...] hoc est, in Deo," Clauberg 1691, *Exercitationes*, p. 609.

73 Clauberg 1691, *Exercitationes*, pp. 608, 609–610, quoting Descartes's *Meditatio prima* and *tertia*: see AT VII, pp. 19–20, 40, 51–52.

74 Eustache de Saint-Paul 1620, pp. 54–55, Goclenius 1613, pp. 208–209.

75 See *Exercitatio* VIII: "cogitationis et picturae comparatio, ad melius intelligendum pro Dei existentia allatum argumentum utilis," Clauberg 1691, *Exercitationes*, p. 609.

76 See Descartes's *Meditatio tertia*, AT VII, p. 37.

Clauberg's alternative in characterising ideas is to describe them as definitions. The propositional nature of ideas is supported by Clauberg in reporting Descartes's argument for the existence of the idea of God. It can be acknowledged, indeed, by understanding the very definition of 'God', which is not just an idea.[77] Actually, such a characterisation does not add anything to our comprehension of the nature of ideas, stating only the linguistic meanings of words and sentences. This is confirmed by the identification of ideas with *themata*, or with whatever can be conceived by the mind, following Descartes and traditional logic. In *Logica vetus et nova* Clauberg states that the difference between the first and the second logical 'degree', or perception and judgment, matches the difference between simple and complex *themata* (Clauberg 1691, *Logica*, pp. 799–800). Complex *themata*, actually, are propositions (Clauberg 1691, *Logica*, p. 829). Therefore, insofar as every simple *thema* can be rendered into a complex one, every idea is expressed by a definition. Such intersections of logic and metaphysics, again, do not shed light on how ideas represent things. This is also the case with the epistemological considerations developed in physics. Clauberg's *Theoria corporum viventium*[78] contains an overview of mental faculties. He defines the functions of the soul as thoughts (*cogitationes*), divided into actions and passions. Passions are perceptions or *conversiones mentis ab obiecto*, that is, modifications of the soul determined by a form or figure. Actions are wills, or *lationes animi ad obiectum*: "adeo ut voluntas latio quaedam animi esse videatur, tendens ad obiectum in idea propositum; perceptio autem quaedam eius figuratio vel in varias formas conversio, veniens ab obiecto" (Clauberg 1691, *Physica*, p. 190). 'Obiectum', 'idea', 'figuratio' and 'forma' are the terms used by Clauberg to express what is involved in mental activities. Thus, the conceptual apparatus of Descartes's theory of knowledge is rendered by Clauberg into Scholastic terms. Eventually, these metaphysical insertions into logic and physics show that the justification of philosophical knowledge is provided by an appeal to the veracity of God rather than by a consideration of the actual ways in which ideas match things. Ultimately, the nature of ideas is considered insofar as it serves the demonstration of the existence of God.

---

77 "Addo, quod definitio rei nihil aliud sit, quam clara et distincta rei idea, ita ut, si omnia vocabula in definitione Dei adhibita sint intelligibilia, necessum sit, quid Deus sit, intelligi, Deus, aiunt, est maximum id, quod cogitari potest. Inde sic infero: ergo Deus cogitari potest, hoc est, Dei idea, sensu cartesiano, haberi potest," Clauberg 1691, *Exercitationes*, p. 604.

78 This treatise is contained in Clauberg's *Physica*, together with his *Physica contracta, Disputationes physicae, Theoria corporum viventium* and *Corporis et animae in homine coniunctio*: Clauberg 1664a. Also in Clauberg 1691. On Clauberg's theory of animated body, see Smith 2013. On his theory of knowledge, see Mueller 1891, Spruit 1999.

In the same manner, the second demonstration of the existence of God is borrowed from Descartes's writings and is clarified through references to Scholastic philosophy. The proof is based on the experienced continuity of our existence, due to Divine conservation action.[79] According to Clauberg, since it is not possible to infer our persistence in being from our past existence, a conserving cause is to be postulated. The demonstration relies on the principle of the successive nature of time and is supported by quotations from Samuel Desmarets's *Systema theologicum* (1645),[80] used to prove that time is experienced in the same way by men and angels. The reference confirms Clauberg's theological interests: he also demonstrates that we cannot be conserved by angels.[81]

Finally, the third proof is explained by Clauberg in the light of his logic, stating that Descartes did not develop his *a priori* argument according to the canons of the *Topica* but from the intuition of the idea of God:

> ille canon logicus: quod convenit definitioni [...] etiam convenit definito (v.g. Deo) [...]. Quaeris, si canon ille definitionis [...] cur eum non retinuit, cur alio potius loquendi modo, quam vulgato et communi usus fuit? Responsionem pete ex Logicae nostrae part. 2 quaest. 134. Voluit potius a notione prima naturae atque ideae, quam a notione secundae definitionis argumentum ducere, [...] cartesiana maior clarius exponit quam definitionis canon [...]. Hae et similes rationes fuerunt Cartesio, cur non uteretur topico isto canone. (Clauberg 1691, *Exercitationes*, p. 648)

The introduction of a logic guided by the criterion of clarity and distinction supersedes the use of dialectical canons in philosophy. Indeed, in his *Logica vetus et nova* only the syllogistic theory of Aristotle's *Analytica* is accepted, whereas the dialectics of the *Topica* is not considered as being admitted by Descartes. This acceptance is ultimately allowed by the propositional nature of ideas, which enables the inclusion of a proceeding based on the intuition of clear and distinct ideas in a demonstrative, syllogistic system. The use of the notion of *thema complexum*, in fact, is what allows such insertions of ideas into syllogisms. Since *themata* or ideas have a propositional nature, they can be combined

---

79 As *duratio* is existence, conservation and creation are the same thing: see Clauberg 1691, *Exercitationes*, pp. 645–646.

80 Clauberg 1691, *Exercitationes*, pp. 636–637; see Desmarets 1645, section 5, § 34; cf. 2nd ed. (Desmarets 1649), pp. 97–98.

81 "Quod difficilius sit aliena curare et conservare, quam propria et sua, unde sequitur, si anima mea non possit suas proprias cogitationes [...] nec suum corpus [...] conservare, tum nulla probabili ratione posse angelo tribui potentiam conservandi animam meam," Clauberg 1691, *Exercitationes*, p. 639.

in demonstrations. Descartes's third proof, indeed, is presented in a syllogistic form by Clauberg.[82]

The demonstrations of the existence of God open two ways of ensuring our knowledge. First of all, Clauberg supports an 'ontological' criterion of truth, according to which an idea is more true insofar as it represents something more real than other beings, such as God is. If truth is a matter of correspondence between model and imitation, or between thing and idea, it is still maintained by Clauberg that truth is first of all in the model and by consequence in its ectype. *Veritas rei*, thus, is the condition of truth as correspondence.[83] Since it contains more perfection, the idea of God is the truest: moreover, the ideas of eternal essences are intrinsically truer than all the others. A traditional point that has its counterpart in Clauberg's theory of transcendentals, which is maintained by him in a Cartesian context.[84] The ultimate argument in the foundation, however, is that of *veracitas Dei*, confirming the validity of Descartes's criterion of truth (or evidence in perception) and to be defined as *veritas ethica*:

> quoniam philosophaturus ante omnia certam habere debet hanc regulam: quicquid clare et distincte percipio, verum est. Haec autem e veracitate Dei eruitur et a priori demonstratur in metaphysica, licet etiam propriam mentis attendentis conscientiam testem suae certitudinis habeat. [...] Quid intelligitur per Dei veracitatem? Resp. illa quae in scholis veritas ethica dicitur, et a logica nec non metaphysica et physica veritate distinguitur. Consistit autem in dictis, factis, promissis, signis aliis. (Clauberg 1691, *Exercitationes*, p. 651)

This statement of God's truthfulness is to be related to the other proof of divine veracity: as *summum ens*, God is the most true being[85] and cannot deceive us. The

---

82 "Sic proponi potest: quod continetur in idea seu conceptu, id ipsum de ea verum est. Atqui existentia necessaria continetur in idea seu conceptu Dei, seu, necessitas existendi in Dei idea continetur. Ergo verum est Deum necessario existit. Maior probatur inductione [...]," Clauberg 1691, *Exercitationes*, p. 647.

83 In fact, *veritas rei* is the very correspondence of something with its own idea or definition: "quam ad rem observa, quod alii veritatem rei censent consistere in conformitate eiusdem cum sua idea, alii in convenientia cum sua definitione, ubi res eadem diversis tantum modis effertur," Clauberg 1691, *Exercitationes*, p. 648.

84 "Per se esse manifestissimum, quod idea Dei mihi exhibeat omnem realitatem, est enim idea Dei, hoc est, entis perfectissimi sive realissimi [...] exhibitio sive repraesentatio. [...] Et hinc sequitur, ideam Dei esse maxime vera, id est, maioris perfectionis, realitatis, veritatis, bonitatis repraesentatricem, quam ulla alia in mente nostra idea, cum nulla alia omnimodam nobis perfectionem repraesentet. [...] Habent etiam realitatem aliae aliis maiorem: veritas enim, realitas, entitas, perfectio hoc loco idem revera sunt," Clauberg 1691, *Exercitationes*, p. 616.

85 This is suggested by Clauberg in proving the ethical truthfulness of God, through a quotation from Descartes's *Secundae responsiones*: "probatur [...] a summi entis et non-entis oppositione. Resp. 2. p. 76. *Qui est summum ens, non potest non esse etiam summum bonum et verum, atque*

ontological *veritas* of God leads to His *veracitas*. That is, the divine attribute of perfection – or His very reality, goodness, and unity – is the ultimate guarantee of the truth of our judgments. God is regarded as the most perfect being, thus as the first cause of things, of truth (as He is truth itself) and therefore as ethically trustworthy: these points are the very ground of every science.

A further point is to be stressed in Clauberg's *Exercitationes*. Even if primarily intended to ground physics, they are about topics belonging also to the other parts of philosophy. They focus on a broader scope of subjects: some considerations concern the demonstration of the immortality of the soul (Clauberg 1691, *Exercitationes*, pp. 675–684), the ethical problems related to Descartes's theory of passions (focusing mainly on wonder) (Clauberg 1691, *Exercitationes*, pp. 722–735) as well as the topic of body-mind relation (Clauberg 1691, *Exercitationes*, pp. 752–755). His *Exercitationes* are functional, ultimately, to the development of a Cartesian scholasticism, or a comprehensive system designed to replace the whole philosophical *curriculum* as the foundation of superior studies. Moreover, his *Exercitationes* reveal some intersections with the last part of philosophy. According to him, rational theology has to be developed as the concluding part of the system:

> nam initio philosophiae non ulterius agitur de Deo, quam quatenus eius cognitio ad iacienda omnis scientiae humanae fundamenta desideratur. Sed in fine absoluta de Deo tractatio instituitur, omniaque eius attributa, quae ex naturae lumine cognosci queunt, expenduntur, quod initio necessarium non erat, quoniam non omnia Dei attributa se habent ut principia rerum creatarum, et quae huiusmodi relationem possunt recipere, non tamen absolute ideo aut plenius, quam originis illa relatio postulat, opus est explicare. (Clauberg 1691, *Exercitationes*, p. 596)

Clauberg will not develop such a complete rational theology. In fact, he will develop only a Cartesian ontology, or a branch of philosophy that replaces the old discipline μετά τα φυσικά. Such a discipline finds its systematisation in Clauberg's renowned *ontosophia*, or the first attempt to develop a theory of being in a Cartesian context. *Ontosophia* is the crown of the system, or a metaphysics that can only be developed after the other disciplines have been established. It can attain the original place of metaphysics and, insofar as it is not designed to ground the knowledge of things as they are, it can deal with mere concepts besides the actual features of substances. It is a replacement, thus, of the traditional *metaphysica*, and can be legitimately developed after physics. However, first philosophy, logic, and ontology are ultimately interconnected.

---

idcirco repugnat, ut quid ab eo sit quod positive tendat in falsum," Clauberg 1691, *Exercitationes*, p. 652; see AT VII, p. 144.

3.4 The 're-duplication' of metaphysics and the birth of ontology

Clauberg's metaphysics is to be evaluated in the light of the end of his system, that is, the study of *ens quatenus ens*, or a metaphysics that goes beyond Descartes's foundation of philosophy. 'Metaphysica' has in Clauberg's philosophy a double meaning. Besides being a *philosophia prima*, it is also a *philosophia universalis, ontosophia,* or the discipline concerning all the attributes of being, no matter if they are only our modes of thinking. It comes after first philosophy: however, like logic, it has some intersections with foundational theory. An examination of it can highlight the whole structure of Clauberg's system.

Ontosophia had three main editions (Clauberg 1647, Clauberg 1660, Clauberg 1664b (also in Clauberg 1691)). Whereas the 1647 version precedes Clauberg's adoption of Cartesianism, the other editions contain Cartesian notions and omit some parts of the first edition (*Prolegomena, Didactica* and *Diacritica*), retaining only *Primae philosophiae elementa* (Clauberg 1648, pp. 37–102). Cartesian insertions can be noticed, for instance, in the definition of being as extended or immaterial substance, or in the note on the distinction between first philosophy, based on *cogito*, and *ontosophia*, based on the non-contradiction principle (Clauberg 1691, *Metaphysica*, pp. 283, 286). Even if Clauberg rejects Aristotle's ten categories as the principles of being, he still finds in the Scholastic tradition those concepts allowing the development of a science of being.

Clauberg proposes, in all the editions of *Ontosophia*, a threefold distinction of the meaning of 'ens': *intelligibilis, realis,* and *res*. His 1664 *Ontosophia* is mainly devoted to the properties of *ens* in the third meaning. However, as *philosophia prima* begins with the consideration of the mind, *ontosophia* begins with that of intelligible being (Clauberg 1691, *Metaphysica*, p. 283).[86] *Ens*, in this meaning, cannot be opposed to anything: indeed, if intelligible being is opposed to a non-intelligible entity, this, in turn, will become intelligible.[87] The second meaning is

---

86 According to Carraud, the Cartesian foundation of ontology relies on the identification of being with *ens cogitabile*: the Cartesian concept of *mens*, indeed, becomes central in the 1660 and 1664 editions: see Carraud 1999. On Clauberg's ontology see also Brosch 1926, Mancini 1960, Viola 1975, and the mentioned studies of Savini.

87 "Non posse quicquam opponi enti sive intelligibili, de quo in praesentia agimus, ne per mentis quidem fictionem. Nam si quid proprie ei opponi posset, id utique foret non ens sive non intelligibile. At eo ipso quo non ens sive non intelligibile opponimus, hoc intelligimus, quia per intellectum ista fit oppositio. Ergo quod non intelligibile tunc dicitur in oratione, fit intelligibile in ratione, unde rationis ens nominatur," Clauberg 1691, *Metaphysica*, pp. 283–284.

aliquid, or whatever may have a formal being.[88] *Aliquid*, thus, can be opposed to *non ens* as whatever has no formal being.[89] *Non ens* can be, therefore, a sort of *ens* according to the first meaning: even if in this case it is only an *ens rationis*.[90] The third meaning of 'ens' is a sub-class of the second one: it is substance as opposed to modes.[91] However, 'ens' in its third signification does not only mean mind and extension: it can also mean modes modified by other modes: that is, *modi mediati* and *immediati*.[92] Therefore, rather than being substances in a strict sense, *res* are substances or modes (*sensu cartesiano*) conceived as subjects of other modes. Rather than referring to Cartesian real substances, Clauberg seems to refer to the notions of *subiectum* and *adiunctum* as described by Franco Burgersdijk (Karsksen 1993), which Clauberg himself counts among the relative attributes of being in his *Logica contracta* and in *Ontosophia*.[93] This categorisation can be explained

---

88 "Aliquid igitur est, quod non tantum mente cogitatur vel cogitari potest, sed alio praeterea modo est aut certe esse potest: sive in mente, ut omnes cogitationes nostrae, sive in mundo," Clauberg 1691, *Metaphysica*, p. 285.

89 "Nihilum, quod alicui generatim opponitur [...] non ens appellatum, est quicquid nullum esse reale habet, hinc dicitur aliquid negativum et sua natura, hoc est, cum nulla accedit fictio, tantum negative, id est, per remotionem et absentiam entis animo concipitur, et negativo solum nomine dignum est," Clauberg 1691, *Metaphysica*, p. 286.

90 "Haec dicta sunt de nihilo sive non ente, quod enti in secunda significatione accepto contradictorie vel privative opponitur. Hoc vero non obstat, quo minus ipsum quoque nihilum in prima et generalissima significatione ens dici queat. Nempe omnis privatio et negatio, dum rationis nostrae obiectum est, utcunque proprium, hoc est, negativum tantum, de ea conceptum ratio formet, ens rationis dici potest," Clauberg 1691, *Metaphysica*, pp. 288–289. *Entia rationis*, in fact, have different kinds, as the fiction of a golden mountain does not imply contradiction, whereas that of a square circle does: see Clauberg 1691, *Metaphysica*, p. 289.

91 "Ens in significatione tertia acceptum propriissime quoque res dicitur [...]. Vulgo quidem substantia, id est, rei quae ita existit, ut aliquo ad existendum subiecto non indigeat, opponitur accidens, quod in alio existit, tanquam in subiecto, sive, cuius esse est inesse. At non omnia, quae in substantia considerantur, accidentia [...] dici debent: cum plurima sint entis attributa essentialia et inseparabilia, a quibus distinguuntur accidentalia [...]. Et haec proprie modi appellantur, nempe modi rerum ipsarum, a quibus illae afficiuntur et variantur, ut pilei a suis formis," Clauberg 1691, *Metaphysica*, p. 290.

92 "Porro res cum opponitur modo [...] non perpetuo significat substantiam, sed interdum etiam accidens, cui alius modus specialior additur, cuius intuitu prior modus tunc res appellatur. Hinc modi alii mediati, alii immediati perhibentur," Clauberg 1691, *Metaphysica*, p. 290. As Clauberg's *Exercitationes de cognitione Dei et nostri* were published in 1656, well before the circulation of Spinoza's works, this expression cannot have been borrowed from them. An influence of Clauberg on Spinoza, on the other hand, is discussed in Lagrée 2002. It concerns, however, biblical hermeneutics rather than ontology.

93 "Essentiae nomine non intelligimus omnia quae rei insunt vel adsunt, sed primum et praecipuum aliquid in ea [...]. Et quicquid praeter illam in re consideramus ut additum et, vel accedens vel accidens (quod neque constituit neque consequitur necessario essentiam, utpote

by recalling the proper place of Clauberg's *Ontosophia* in the system: its concepts are not designed to be employed in other disciplines but are the result of a speculation on being in its most abstract meanings. Such meanings, therefore, are not regarded as matching the actual features of extended and spiritual substances.

In his *Exercitationes*, moreover, Clauberg states that even if it is possible to find some attributes common to God and creatures,[94] this does not justify *ontosophia*'s status as *scientia*. In other words, even if *ontosophia* is the crown of his system, it is not grounded on first philosophy, since it does not deal with clear and distinct concepts:

> hactenus dicta eo faciunt, ut rerum omnium similitudo et convenientia quaedam agnoscatur: at si quis putet me existimare, illis ipsis satis esse probatum tradendam esse ontosophiam seu universalem [...] scientiam, is a mente mea aberrat. [...] Nam si conceptus illi quos habere potest mens nostra, a Deo et creatura quodammodo abstracti et utriusque conceptui communes, non sint satis clari et distincti, sed confusi nimis, et quae mentem veritatis studiosam non satis afficiant, multo minus impleant, dubitari sane cum ratione poterit, an pertineant ad scientiam, utpote quae obiectum requirit quod clare distincteque percipiatur. (Clauberg 1691, *Exercitationes*, p. 703)

Clauberg, however, sets aside the deeper consequences of Descartes's metaphysics for the theory of being. His *Ontosophia* has a heterogeneous composition, according to which a Cartesian distinction of being in extended and spiritual substance[95] is followed by a survey of its attributes given in traditional terms. As *ens* is first of all a concept or a second notion, *ontosophia* is first of all a study of concepts or *modi considerandi*. However, because it is not a foundational discipline, according to him it is still possible to pursue it as a branch of philosophy, or the 'science' dealing only with abstract concepts. An evaluation of Clauberg's use of ontological concepts and of his definition of their status, however, reveals some other ambiguities in this architectonic of philosophy.

An emphasis on attributes of being reduced to mere concepts or *modi considerandi* can be found in different places of the treatise. Discussing real, modal, and rational distinctions and the notions of identity and difference, Clauberg admits that

---

inseparabilem cum ea nexum non habens) adiunctum vocamus," Clauberg 1691, *Metaphysica*, pp. 334–335; Clauberg 1691, *Logica contracta*, p. 918.

94 See *Exercitatio* LX: "Deum et creaturam habere aliquam in re similitudinem et convenientiam," Clauberg 1691, *Exercitationes*, p. 694.

95 "Primaria igitur entis realis divisio est illa sine dubio, quae maxime opposita et contraria attributa (intellige positiva) in rebus divisis menti nostrae consideranda exhibet. Nulla autem realia attributa magis opponi sibi queant, quam ex una parte longum, latum et profundum esse [...], ex altera parte intelligere, velle, nolle et c.," Clauberg 1691, *Metaphysica*, p. 291.

> tota haec disputatio de eodem ac diverso potius ad modum cogitandi et loquendi pertinet, quam ad res ipsas in se spectatas. Quod nihil hic novi videri debet, cum similis aliorum generalium entis attributorum sit ratio. (Clauberg 1691, *Metaphysica*, p. 331)

Moreover, in the dedicatory letter he states that he is speaking only about our ways of considering things, without clarifying, however, to what extent our abstraction of their attributes is legitimate.[96] Despite these remarks, there is a foundational reason for treating transcendentals as real attributes, in Clauberg's perspective: that is, to ground truth in the notion of God as the utmost being, whose archetypes are more real, true, and good than any other created thing. The definition of God as ethically veracious, provided on the ground of divine perfection and goodness, is laid down in the light of the doctrine of transcendentals. The ontological proof of the existence of God has its counterpart in the consideration of being as perfect, true, and good. Our thoughts, moreover, are true insofar as they imitate divine archetypes. The correspondence truth is based on the ontological truth, since our ideas of things are truer to the extent that they imitate the models present in the divine mind.[97]

Clauberg's *Ontosophia* reveals, in sum, a problem intrinsic to Cartesianism: that of the adherence to classical metaphysics within a philosophy based on the *cogito*. A tension, therefore, is to be noticed in his metaphysics: between *philosophia prima* and *ontosophia*. Indeed, his metaphysics implies an overestimation of the ontological value of the attributes of being, accordingly of a foundation of truth on a theory deploying the notions of transcendentals. At the same time, it is stated that they are mere ways of considering substances. In any case, insofar as *unum*, *verum*, and *bonum* are deemed as actual attributes of things, *ontosophia* seems to have a foundational value more consistent than that admitted by Clauberg himself.

In conclusion, some words are to be devoted to the relations between logic, first philosophy, and *ontosophia*, or the study of being.[98] In the first edition of his *Ontosophia*, before his acceptance of Cartesianism, Clauberg states that logic has priority in a didactic order, as it teaches how to use the intellect, whereas metaphysics (still identified with *ontosophia*, as Descartes's *philosophia prima* has not

---

96 "Generalissimos istos conceptus et terminos, uti vocant, ad certum prorsus numerum atque ordinem reduci non posse experiendo didici. Adeo transcendentia illa non solum connexa, verum etiam innexa sibi sunt. Quin imo nihil aliud sunt, quam diversi de re eadem cogitandi modi, qui, animo iam huc iam illuc se convertente, mille formis variari solent et possunt. Id quod hac editione tertia vel inprimis demonstrare studui," Clauberg 1691, *Metaphysica*, *Lectori salutem*, p. 279 (unnumbered).

97 See *supra*, on *veritas ethica*.

98 This topic is well considered in Savini 2011a, pp. 44–69 (*Le rôle de la logique dans l'instauration de la métaphysique, La configuration du rapport entre logique et* ontosophia *dans la fondation de la métaphysique*) and pp. 184–193 (*Philosophie première et ontologie*).

been yet received by him[99]) comes first in the natural order of the sciences, since it deals with the first genres of things.[100] Therefore, students are introduced to logic through some 'anticipations' of metaphysical concepts.[101] In the following editions of Clauberg's *Ontosophia*, and along with the development of his more mature views, the plan of the disciplines changes. As a Cartesian *philosophia prima* is introduced, logic and *ontosophia* come after it. Logic is based on the evidence criterion prescribed by the method. It maintains, however, its instrumental role (Clauberg 1691, *Exercitationes*, p. 592): it teaches how to organise and interpret speech in the light of an adequate formation of concepts. Such a logic is implied by Descartes's metaphysics because it makes explicit the rules of reasoning underlying that part of philosophy.[102] Moreover, it shares with first philosophy its starting points, or the assumption of the evidence criterion and doubt. If first philosophy discovers the first notions and truths, according to the evidence criterion, logic teaches, at least in principle, its proper method. It is, somehow, a corollary discipline of first philosophy. The natural order of disciplines outlined by Clauberg prescribes starting with first philosophy and ending, with the help of logic, in *ontosophia*, after all the other disciplines have been established: physics (the trunk of philosophy, also embodying foundational arguments), moral philosophy, medicine, politics, law, mechanics. In fact, their development is interconnected, since metaphysical considerations are implied both by logic and ontology. Eventually, such interconnectedness of logical, metaphysical, and ontological problems will be addressed by Johannes de Raey, on the one hand, who proposed a simplification of the system of knowledge. In doing this, he aimed at defining the proper scope and the boundaries of Cartesian philosophy. On the other, by Arnout Geulincx, who stressed the purely instrumental role of logic and finalised metaphysics to the development of a rational ethics. Their theories, however, were not a mere – or direct – reaction to Clauberg's, but were developed through the ongoing philosophical debates taking place in the Netherlands in the 1650s and 1660s, namely, the debates over the use of Cartesianism in metaphysics, on the one hand, and in practical matters, including theology, on the other.

---

99 Clauberg 1647, p. 2; see Savini 2011a, pp. 25–27, 44.

100 Clauberg 1647, pp. 33–34; see Savini 2011a, pp. 64–65.

101 Clauberg 1647, pp. 291, 309; see Savini 2011a, pp. 54–55, 61–63.

102 "Nuspiam apertius Cartesius est logicus, quam in libello de Passionibus animae, sed maxime etiam logicus est, ubi artem celat, ut in Meditationibus metaphysicis. Confer. Log. II. 14. [...] Ad recte definiendum opus esse praemittere divisiones, sancit Logica I. 103. Id quod videmus factum esse ab auctore," Clauberg 1691, *Exercitationes*, p. 723.

4 Dutch Cartesianism in the 1650s and 1660s: Philosophy, theology, and ethics

4.1 Cartesianism in Leiden in the 1650s: Physics without metaphysics

As mentioned above, the debates over Cartesian philosophy continued after the publication of Clauberg's *Defensio cartesiana* in 1652, as in 1653 and 1654 new anti-Cartesian texts appeared, such as the *Cartesius triumphatus* (1653) of Lentulus and the *Thekel* (1653) of Revius, which have mainly an eristic character and focus on the *stylus scripturae* of Descartes.[1] Other texts were issued in these years and constitute a body of replies and counter-replies: Revius's *Statera* finds a reply in the *Appendix* of Andreae's *Brevis replicatio* (1653). In this text, Andreae aims to defend, against Regius, Descartes's theory of the soul as *res cogitans*, thereby vindicating Descartes against Revius's reading of Descartes through his 'radical' interpreter (see Fowler 1999, pp. 42–43). In turn, Revius replied to Andreae in his booklet *Psychotheomachia* (1654), in which topics that characterised the quarrel between Descartes and Regius recur, such as the immortality of the soul and the veracity of God. Moreover, Andreae addressed to Revius's *Consideratio theologica* (1648) his *Methodi cartesianae assertio*, in two parts (1653–1654), which follows the structure of Revius's book and adds a defence, in the second part, of Descartes's first philosophy. Andreae's text was replied to by Revius in his *Kartēsiomania* (1654–1655), in which Revius also criticised another ally of Descartes, namely, Christopher Wittich, colleague of Clauberg at Herborn and Duisburg (1651–1652) and then professor of theology at Nijmegen from 1656. Wittich published his *Dissertationes duae* in 1653, and his *Consideratio theologica de stylo Scripturae* in 1656, mainly addressing the problem of the use of Cartesian philosophy in biblical hermeneutics rather than the philosophical use of Descartes's method. Revius would reply to the former text in his booklet *Anti-Wittichius* (1655), focusing on the topic of the infinity of the world. The analysis of these texts is beyond the scope of the present survey.[2] What is of interest to this study, in fact, is that such debates over Cartesianism shaped a change in function of philosophy in the academic

---

1 "Quod ut in doctrinae errore demonstrando et defendendae doctrinae modo recensendo fecimus, iam denudando stylo et scribendi genere faciamus," Lentulus 1653, p. 37. The hermeneutics contained in Clauberg's *Logica vetus et nova*, in fact, can also be considered as a means against the misinterpretation of philosophical texts – especially Cartesian ones – brought forth by Revius and Lentulus: see Strazzoni 2013.

2 For a more detailed account of such a body of texts, see Goudriaan 2002, *Introduction*.

https://doi.org/10.1515/9783110569698-005

curriculum. We have seen that in the hands of Clauberg, Cartesianism became the basis of all academic disciplines and shaped a new function of metaphysics as the introduction and foundation of a new way to do philosophy. In Dutch universities, on the other hand, the academic teaching of Cartesianism – in accordance with the coordinated strategy of defence of the new philosophy revealed above – took a different form. First, in the early 1650s Cartesian philosophy was restricted, in Leiden, to natural philosophy, as a consequence of the quarrels and bans over Cartesianism. Secondly, Cartesian philosophers and theologians embraced a separation thesis, according to which philosophy and theology have different methods and ends, as the former is devoted to the discovery of truth (i.e. it has a theoretical function), while theology has a practical aim and is devoted to salvation. The foremost character in the shaping of Cartesian philosophy according to these strategies was Johannes De Raey: while Clauberg was supporting a full-blown version of Cartesianism with his *Defensio, Logica*, and *Exercitationes*, De Raey disseminated Cartesian physics in Leiden in a somehow concealed form, after the bans on Cartesianism issued in May 1647 and February 1648.

Born in Wageningen in 1620, De Raey first studied at the University of Utrecht under the guidance of Henricus Regius, being the *respondens* of some of his theses on *Physiologia* in 1641. He then studied at the University of Leiden, where he graduated in arts and medicine with Adriaan Heereboord and Adolf Vorstius in 1647. In 1648 and 1649 he gave private lectures on Cartesian philosophy, which were attended by Johannes Clauberg, while in 1651 and 1652 he held a series of disputations *Ad Problemata Aristotelis*, before being appointed as extraordinary professor of philosophy in 1653.[3] This series of disputations, later published as his *Clavis philosophiae naturalis* (1654), served to prevent any further attack by anti-Cartesian theologians such as Revius, who intervened with the University Curators to forbid De Raey to lecture on the new philosophy (Molhuysen 1913–1924, vol. III, p. 11). According to the *Epistola dedicatoria* of the *Clavis*, indeed, De Raey states that the Curators themselves bestowed on him the task of teaching the new philosophy by showing its agreement with the 'original' thought of Aristotle.[4] Accordingly, from 1651 he devoted his disputations on the pseudo-Aristotelian *Problemata* to show, from a Cartesian standpoint, how the main tenets of Descartes's physics were already present in the Aristotelian corpus,

---

3 On De Raey's early thought, see Ruestow 1973, pp. 61–72, Bodeüs 1991, Schuurman 2001, Schuurman 2003c, Verbeek 1993c, Strazzoni 2011.

4 "Vos estis, qui me ex doctore privato publicum professorem creastis, et ut eam philosophandi rationem, quam pluris a me fieri atque etiam Aristoteli valde adversam ab aliquibus censeri notum erat, cum Aristotele componerem, haud obscure imposuistis," De Raey 1654, *Ad Curatores epistola*, p. XXIV (unnumbered). On De Raey's *Clavis*, see Strazzoni 2011.

and were then neglected and corrupted by the Aristotelians, first and foremost by Averroes. The 'concordance' shown by De Raey, in fact, cannot be labelled as part of the 'novantique' philosophy given in Holwarda's *Philosophia naturalis, seu Physica vetus-nova*, published in Franeker in 1651, by Clauberg himself in his logic, and by Du Hamel in his *De consensu veteris et novae philosophiae*, which appeared in Paris in 1663. De Raey, indeed, offered a specimen of the new physics only concealed as commentary on Aristotle. He provides his concordance on the basis of four 'praecognita' i.e. first principles or axioms grounding Descartes's physics: (1) the existence of an extended matter which is the substance of bodies;[5] (2) the principle of the extrinsic origin of motion with respect to bodies, whose essence does not include movement (whose ultimate cause is God);[6] (3) the three principles of motion and the rules of impact set by Descartes – and proven by means of Aristotelian, textual evidences by De Raey;[7] (4) the existence of a subtle matter, i.e. Descartes's third matter, which allows the explanation of the apparent autonomous movement of bodies and the existence of a void.[8] In fact, De Raey is careful to avoid the ontological claims set forth, on the other hand, by Regius, such as the rejection of substantial forms. Moreover, he justifies the validity of such axioms or *praecognita* by claiming their being evident to anyone provided with a healthy mind, in accordance with Aristotle's *Analytica posteriora* and *Topica*. Thus, De Raey's Cartesian *praecognita* – objects of noetic knowledge – may become in the section *De praecognitis* of the *Clavis* the principles for demonstrative i.e. dianoetic knowledge;[9] moreover, they are evident to intellect alone (*intelligentia*), in the same way to mathematical axioms, since matter can be described by means of the notions of geometry.[10] Accordingly, De Raey maintains a Cartesian theory of scientific knowledge based on purely intellectual

---

5 De Raey 1654, *Clavis*, pp. 50–51, 53–54, quoting from *Metaphysica*, VII, 3, 1029a 10–12, 20–21, *Physica*, I, 9, 192a 31–32; II, 3, 194b 24–25; IV, 8, 216b 4–15.

6 De Raey 1654, *Clavis*, pp. 68–71; see *Metaphysica*, XII, 6, 1071b 29–30, and *De motu animalium*, IV, 700a 16.

7 De Raey 1654, *Clavis*, pp. 106–108; see *Physica*, IV, 8 19–22.

8 De Raey 1654, *Clavis*, p. 127; see *Meteorologica*, I, 3, 339b 25–26.

9 De Raey 1654, *Clavis*, pp. 37–38; see *Analytica posteriora*, I, 10, 76b 10–14, and *Topica*, VI, 4, 141b 7–13.

10 These are evident as the truths of mathematics, the principles of metaphysics (as "nihili nullae sunt affectiones"), and those concerning immaterial entities: De Raey 1654, *Clavis*, pp. 36–37. See also p. 41: "quantum denique ad corporum naturalium, quae physica considerat, scientiam, ad eam imprimis opus est praecognitis [...]. At vera et prima axiomatum seu notionum communium, unde solidior, certior ac profundior naturae contemplatio pendet, evidentia ac demonstratio, non ab externo sensuum sed ab interno solius mentis lumine est petenda, quia, haud secus atque axiomata multa mathematica, [...] fugiunt omnem sensum."

notions, without expressly rejecting the Aristotelian theory of knowledge and by no recourse to Descartes's path of the *cogito* so harshly debated by Revius and Lentulus. Yet, in his disputations and *Clavis* De Raey did not merely present an account of Cartesian physics in order to disseminate Cartesianism in Leiden, but set the basis for the further development of foundational arguments, as he had already devised in 1651 a fundamental distinction between practical and theoretical, i.e. truly philosophical, knowledge. In the same year, similar positions were formulated by Wittich and led to the development of the 'separation thesis' between philosophy and the higher arts.

4.2 Philosophy, theology, and ethics (and the separation thesis)

The disputations of De Raey on the *Problemata* were preceded by his *Dissertatio de cognitione vulgari et philosophica* (1651, included in the *Clavis* itself), in which he distinguishes between philosophical and commonsensical or vulgar knowledge, that is, between the Cartesian and the Aristotelian approaches to philosophy. This text is devoted to showing the errors characterising the commonsensical way of understanding nature, based on sensory experience, memory, imagination.[11] According to De Raey, Scholastic philosophy reflects such 'understanding' of natural phenomena as it ascribes the visible effects of bodies to occult qualities,[12] whereas philosophical knowledge (*scientia*) is based on concepts graspable by intellect alone, namely, the *praecognita* or common notions and axioms matching the first causes of observable phenomena. In fact, in his *Dissertatio* De Raey does not reject, *in toto*, the Aristotelian approach but rather restricts it: for instance, the acknowledgment of sensory qualities and substantial forms can still be useful to the exploitation of natural powers in the practical arts.[13] From this kind of knowledge, anatomy, medicine – including surgery – and all

---

11 "Plurima eorum quae cognoscit sapiens cunctis mortalibus communia sunt, ac iis etiam obvia, qui vel mente capti, vel barbari, vel [...] in sapientiae studio exercitati non sunt [...]. Talisque maximam partem est omnis ea vulgi notitia, quae sensuum experimentis primam originem, memoriae conservationem, imaginationi [...] perfectionem atque incrementum debet," De Raey 1654, *Dissertatio de subsidiis, gradibus ac vitiis notitiae vulgaris*, p. 8. First edition De Raey 1651.

12 De Raey 1654, *Dissertatio de subsidiis, gradibus ac vitiis notitiae vulgaris*, pp. 18–19.

13 "Illi vero qui posita barbarie in magnas coiere societates, vitamque ducunt tranquillam ac mansuetam [...] in hac vitae tranquillitate et otii abundantia magis circumspecti et ad ea quae quotidie in vita occurrunt attenti esse solent homines. Ubi observant quippiam, quod usum, vel commoditatem in vita allaturum sperant, id non contemnunt, vel negligunt, verum diligenter eius qualitates, formam, operationes, vires ususque notant," De Raey 1654, *Dissertatio de subsidiis, gradibus ac vitiis notitiae vulgaris*, p. 18.

the mechanical arts, based on experience, derive.[14] The distinction of De Raey between philosophical and commonsensical knowledge, so far, was not aimed against the Aristotelians but it rather served to show the proper place of the new philosophy, which had a theoretical function. Yet, De Raey in 1651 still did not address the relation of philosophy and theology, which would on the other hand be confronted by Christopher Wittich in his *Dissertationes duae* and *Consideratio theologica de stylo Scripturae*, then enlarged and published together as *Consensus veritatis* (1659), recently analysed by Antonella Del Prete and Rienk Vermij (Del Prete 2001, Del Prete 2002, Del Prete 2013, Vermij 2002). Addressing the reconciliation of the heliocentric hypothesis and the Bible, Wittich maintains – in accordance with the hermeneutic principle of *accomodatio* – that the language of the Bible, when it refers to natural things, reflects a *cognitio vulgaris* matching the common experience of men and describes things *relate ad hominem*, conveying, in any case, some kind of truth essential to salvation. The ascertainment of such truth, in turn, is not guaranteed by philosophy but by Scripture itself, as it gives us the means to understand the aim of its own contents. Accordingly, if philosophy can decide upon the kind of knowledge involved in biblical passages – contrary to Gysbertus Voetius and his followers, considering philosophy as the handmaid of theology, and defending the reliability of the *physica sacra* – it does not provide hermeneutic criteria. Moreover, theology is essentially aimed at practice, whereas philosophy is a theoretical discipline. Accordingly, philosophy is to be based on the epistemic principles set out by Descartes, such as clarity and distinction as the mark of truth, while theology can rely on the appearance of things for the sake of Salvation (Wittich 1659; see Del Prete 2002, Pesce 1992).

The distinction of philosophy and theology became a main tenet of the so-called 'Cartesio-Cocceians' faction in Dutch universities, which included Wittich and De Raey's friend Heidanus, but also Balthasar Bekker, Salomon van Til, Petrus Allinga, Campegius Vitringa the elder, Frans Burman (see Van der Wall 1996). In particular, Wittich and Heidanus – although theologians – are to be regarded as shaping the development of Cartesian philosophy itself, as it was taught in the faculties of arts. If Wittich set the standard for the separation of theology and philosophy – to the extent that this does not serve as a hermeneutic criterion – Heidanus, minister and then professor of theology in Leiden from 1648, used Cartesianism in the early 1640s as a source of arguments against Remonstrant theologians such as Simon Episcopius, to whom he directed his

---

14 "Arte naturam perficere, vel superare laboravimus, quae prima mechanices artiumque, quas illiberales vocant, initia fuere," De Raey 1654, *Dissertatio de subsidiis, gradibus ac vitiis notitiae vulgaris*, p. 19.

De causa Dei (1645), and Voetius, who aimed to develop a 'neo-Scholasticism' in support of Reformed theology. In fact, Heidanus defended some principles of Cartesianism in order to rebuke the project of Voetius for using philosophy as the handmaid of theology. Under the pseudonym of Irenaeus Philalethius, Heidanus quarrelled with the faction of Voetius in two pamphlets over Cartesianism which he wrote with the cooperation of De Raey in 1656, namely the *Bedenkingen op den Staat des geschils over de Cartesiaensche philosophie en op de Nader openinghe over eenige stucken de theologie raeckende*, and *De overtuigde quaetwilligheidt van Svetonius Tranquillus*. These were replies to Svetonius Tranquillus – whose identity had not been ascertained, but is certainly a Remonstrant theologian – who had published various anti-Cartesian pamphlets in the same year, attacking Descartes's philosophy as irreconcilable with Christian Faith, forcing Cartesian theologians to adopt the accommodation principle (Svetonius 1656a, Svetonius 1656b, Svetonius 1656c; see Vermij 2002, pp. 305–306). Eventually, the quarrel ended on 30 September 1656, with a resolution by the States of Holland sanctioning the principle of the separation of theological and philosophical discussions. Such a resolution was favourable to the Cartesian faction, as it allowed the teaching of Cartesianism at the faculties of arts.

In these texts, as well as in his earlier *De causa Dei*, the main Cartesian tenet is to be found in the explanation of the interconnected issues of grace and human freedom. Heidanus, as shown by Han van Ruler, used Descartes's explanation of human freedom (which he largely borrowed from Augustine) as it is presented in the fourth *Meditatio*. Descartes conceives freedom not as the *libertas indifferentiae* but rather as the power of acting according to reason, which operates with regard to truth as Grace operates in directing us to good. On this basis, Heidanus makes use of Descartes's account of the interaction of soul and body as it is explained in his *Les Passions de l'âme* (1649), in order to explain the difference between the realm of spirit and grace, and that of flesh and sin, insofar as Grace is compared to reason itself (see Van Ruler 2001, Van Ruler 2003b).[15] In fact, Heidanus was concerned with not reducing theology – and religion – to an ethical system which might become independent from Revelation and Faith, or a 'mera ethica' (see Cramer 1889, p. 92). His main interest, accordingly, was in developing an ethics consistent with religion, on the one hand, and on the other with the principles of Cartesian philosophy, which in his view could serve for a better understanding of the Reformed doctrine of the role of Grace and human

---

15 On the interrelations of Protestant ethics with philosophy in the sixteenth and seventeenth century, see Moltmann 1957, Bizer 1958, Bizer 1965, Menk 1980, Menk 1981, Verbeek 1993b, Van Asselt 2001, Beck 2001, Beck 2007, Goudriaan 2002, Goudriaan 2006, Mulsow/Rohls 2005, Neele 2009, Goudriaan/Van Lieburg 2011.

freedom in Salvation. For this reason, Heidanus was eager to support the assumption by Arnold Geulincx, a scholar from Louvain who came to Leiden in 1658, who was equally interested in developing a rational ethics consistent with Cartesianism and with Augustinian ideas, having been under the influence of Jansenius's *Augustinus* in his Louvain years (see Bizer 1965, Van Asselt 2001, p. 7, Van Ruler 2001, Aalderink 2009, p. 12). Geulincx – as I am going to show in the next section – would strength the Cartesian presence at the University of Leiden; moreover, he would extend its uses from natural philosophy to ethics, thus re-addressing the issue of foundationalism in light of this purpose.

In sum, in the 1650s Dutch Cartesianism was shaped into the form of an academic philosophy with few metaphysical commitments (both for Regius and De Raey, albeit for different reasons), and, for De Raey, with no interference in the higher arts. Yet, as soon as a theology consistent with Cartesian principles came to the fore, from the late 1650s the need for developing an ethics consistent both with theology and the principles of Cartesianism emerged in the Dutch context. Moreover, between the 1650s and the 1660s Dutch Cartesianism had to face the emergence of alternative philosophical standpoints – first of all, the philosophy of Spinoza and of Hobbes – which challenged its metaphysical basis, on the one hand, and its relations with the higher arts, on the other. This led to a change in strategy in providing Cartesian philosophy with a foundation, and to the emergence of a foundationalism in Leiden, aimed at supporting different branches of philosophy, including a rational ethics. Such an evolution of Dutch Cartesianism is embodied in the works of the two main exponents of Cartesian philosophy active in Leiden in 1660s, namely, Geulincx and De Raey, who faced in different ways the problems inherent in the 'improper' uses of Descartes and the relations of Cartesianism and theology: on the one hand Geulincx would develop a metaphysical foundation of physics and of a moral philosophy consistent with Reformed theology. On the other, De Raey limited the application of the Cartesian paradigm to metaphysics (which he identifies with logic) and physics, although he admitted the possibility of developing a philosophical ethics. In what follows, I will provide an account of their foundational theories as these were developed in the 1660s, namely, before Geulincx's death and De Raey's departure for a chair at the Amsterdam *Atheneum Illustre*, both taking place in 1669. Eventually, I will show that their efforts, although disconnected (as Geulincx was not directly supported by De Raey), contributed to the establishment of metaphysics as a foundational discipline in the academic curriculum. Moreover, such foundationalism assumed at the same time a prescriptive and a descriptive role with regard to the limits and uses of academic disciplines, i.e. physics was progressively 'de-metaphysicised' in content, although metaphysics had the function of defining its method and concepts.

4.3 Cartesianism and rational ethics: Geulincx between Reformed theology and Spinozism

Born in Antwerp in 1624, Geulincx obtained a degree in philosophy at the University of Louvain in 1643, and, probably, a degree in theology. In 1646 he became a professor of philosophy at Louvain. In 1652 he was appointed as a *professor primarius* and participated in the final year session of the *Quaestiones quodlibeticae*,[16] whose introductory speech bears witness to Bacon's influence on him (see Aalderink 2009, pp. 52–55). The criticisms of his Aristotelian colleague Vopiscus Plempius (1601–1671) notwithstanding, the speech does not seem to have caused any doctrinal problems for Geulincx. However, six years later he moved to Leiden for having broken the rule of celibacy for professors. There, he obtained a degree in medicine (1658) and started to lecture in philosophy, being nominated lecturer in logic, philosophical exercises, and metaphysics in 1662. Later, he started to teach ethics and was appointed *professor ordinarius* of philosophy in 1665, being allowed to teach moral philosophy only in 1667, two years before his death (which occurred during the Leiden plague) (see Han van Ruler's *Introduction* to Geulincx 2006, pp. XV–XVI).[17] Geulincx's interest in ethics can be seen as the main motive for his original approach to the foundation of the whole 'house' of philosophy, including metaphysics, logic, and physics. The problems he confronted have to be found in the Flemish and Dutch philosophical contexts from which he worked: namely, not only the theological and ethical requirements set forth by Heidanus, but also the issues related to the method of natural philosophy, which, together with metaphysical questions, had been left unaddressed by De Raey in his *Clavis*. In fact, Geulincx's positions on the role of God and on human nature, though primarily of a moral philosophical content, link up with contemporary questions of physics with which they were to be made consistent. Geulincx's physics

---

16 The text of the *Questiones quodlibeticae in utramque partem disputatae* (Geulincx 1653) was published for the second time in 1665, with some variations: Arnold Geulincx, *Saturnalia, seu (ut passim vocantur) Quaestiones quodlibeticae in utramque partem disputatae* (Geulincx 1665a).
17 Several monographs and articles have been devoted to Geulincx in the two centuries since the pioneering studies of J.P.N. Land, editor of a three-volume *Opera omnia* (1891–1893), whose works are the main biographical sources on him (Geulincx 1891–1893, Land 1887, Land 1891, Land 1895). Among the most recent contributions we find those of Mark Aalderink, Bernard Rousset, and Han van Ruler, who have focused on Geulincx's theory of knowledge (Aalderink 2009, part III), on his systematic view of philosophy (Rousset 1999), and on his positions in the debates on causality in the Cartesian context (Van Ruler 1999b, Van Ruler 2000). For a more detailed bibliography, see De Vleeschauwer 1957, Aalderink 2009, pp. 405–423. Other studies on Geulincx's philosophy are Terraillon 1912, Nagel 1930, Dürr 1939–1940, Dürr 1965, Cooney 1978, Nuchelmans 1988, Battail 1993, Verbeek 1998, Nadler 1999, Buys 2011.

had an empirical character, developed within a broader ethical perspective: his physics is to be seen as a systematisation of Descartes's, while making clear the importance of hypotheses or *a posteriori* explanations, based on the experience of natural effects. In Geulincx we find the link between human experience on the one hand and the mechanistic reinterpretation of such experience which is typical of a Cartesian view of philosophical knowledge and offers such a reconstruction of experience as the proper object of physics. In Holland, Geulincx was not supported by his Cartesian colleagues in philosophy but by the theologians: Abraham Heidanus (1597–1678) and Johannes Cocceius (1603–1669), protagonists of Cartesian theology (Bizer 1958, Bizer 1965, Van Asselt 2001, p. 7, Van Ruler 2001). Geulincx's closest supporter was Heidanus, who shared with him an approach deeply influenced by Jansenism (see Van Ruler 2001, pp. 21–28, Aalderink 2009, p. 12). Actually, Geulincx was under the influence of Augustinian ideas since his Louvain years, where the ideas of Jansenius's *Augustinus* (1640) were well known. In Leiden, Cartesianism, Augustinianism, and the theology of Heidanus can be seen as the keystones of Geulincx's philosophy. Whereas De Raey was educated among the first Dutch Cartesians, mainly interested in the introduction of a new physics in the academy (following Descartes's program), Geulincx had a different philosophical agenda, matching the exigences of philosophy and faith by presenting the *vita christiana* as the final stage of ethics.[18] According to Martial Guéroult, Descartes did not develop an ethics because it would deal with the obscure ideas related to the mind-body union (Guéroult 1953, vol. II, pp. 250–259). If Descartes was prevented from developing a rational ethics on account of problems related to his metaphysical dualism, Geulincx tried to develop his ethics while paying attention to the Augustinian positions that enabled him to go beyond Descartes's difficulties. Geulincx's emphasis on the passivity of man with respect to God, which leads to an ethics that concerns our internal attitudes (the cardinal virtues of obedience and humility) more than our habits (Geulincx 1891–1893, vol. III, *Ethica*, section 1, § 3), matches the belief in predestination and in the small value of external acts. Moreover, the need for a philosophical guide in morals was felt in Dutch society, as can be argued from the vulgarisation of his 1665 *Ethica* (Geulincx 1667).

Another possible reason for his philosophical attitude, however, is the spread of Spinoza's ideas around 1660, that is to say, during the years in which

---

18 See his *Ethica*, in Geulincx 1891–1893, vol. III, p. 110. In 1665 the first part of this work was published as *De virtute et primis eius proprietatibus* (Geulincx 1665b). In 1675 a complete version was posthumously published by Cornelius Bontekoe (Geulincx 1675). According to Han van Ruler, "Geulincx was the perfect candidate to fulfil a task Heidanus was eager to support: the invention of a Christian philosophy of morals," Geulincx 2006, *Introduction*, pp. XV–XVI, XXI.

the *Tractatus brevis* and the *Tractatus de intellectus emendatione* were composed (see Mignini 1984, Mignini 1985). Actually, it is still debated whether Geulincx and Spinoza ever had any direct contact (see Rousset 1999, pp. 12–20, Van Ruler 1999a). However, there are several similarities in their approach as well as differences in their solutions. Answering the same demand for a morality based on the new philosophy, both Geulincx and Spinoza were developing a rational ethics. Like Spinoza, moreover, Geulincx had a troubled life: after having been expelled from Louvain University he suffered personal and academic isolation in Leiden, spending his last years in poverty. His ethical system, based on humility as the main virtue and on the acknowledgment of our passivity towards God, can be read as being influenced by his ill-fated life. Similarly, Spinoza developed a philosophy to carry the soul away from the turbulence of the passions. Geulincx's ethics can thus in many ways be seen as a twin to Spinoza's; indeed, Geulincx's philosophy was later portrayed as proto-Spinozistic by Ruardus Andala (Andala 1716). On the other hand, the differences between Geulincx and Spinoza may also justify the hypothesis that Geulincx was reacting to Spinoza's ideas by developing an ethics more consistent with a Christian morality.

All these reasons can make us understand Geulincx's peculiar positions, which I am going to analyse in the rest of this chapter. In fact, Geulincx is not concerned with a 'logical' epistemology, because, in order to present a Cartesian moral philosophy that would meet Christian standards, he fully focuses his system on the relations between God and man and on human passivity with respect to God. The latter ideas are deduced from Cartesian metaphysics, since Geulincx finds in occasionalism a way to explain the interaction of substances in a world deprived of active forms. Moral philosophy is thus at the top of the agenda, provided with the highest degree of certitude and grounded in theology. Physics, on the contrary, though it still has an essential role in the plan of the philosophical disciplines, is a discipline described as the floor of the House of Philosophy.

4.3.1 The architectonic of philosophy

For Clauberg and other Cartesian logicians, such as Arnauld and Malebranche, logic involves epistemological considerations (see Schuurman 2004, pp. 34–50, 63–64; see also Easton 1997). This, however, is not the case for Geulincx. He conceives logic as the science of argumentation, following a traditional approach thoroughly explained in an oration of 1662: the *Oratio de removendis parergis et nitore conciliando disciplinis*. As the title suggests, this oration aims to remove all the introductory questions from logic and to reduce the discipline to a terse body of knowledge that concerned only the forms of demonstration. Logic must avoid proemial questions such as 'what is logic?', which occupy the first pages in the Scholastic manuals, e.g. Burgersdijk's *Institutiones logicae* (Burgersdijk 1626, pp. 1–7), and which involve irrelevant discussions on the whole structure of philosophy.[19] The considerations on the function of logic are described in the *Dictata* to his *Logica restituta* (1662) as belonging to a *scientia scientiarum*, not to logic itself.[20] Indeed, it is metaphysics, or a still unnamed discipline, that must be the science of sciences, later called 'encyclopaedia'.[21] Hence, Geulincx compares the proper content of logic to the cleanliness of the Dutch towns.[22] In his use of this metaphor, besides appealing to the purity of formal reasoning, we can read a criticism of the problem of the improper mingling of disciplines. The development of a logic focused only on the forms of argumentation in fact helps to keep disciplines separated and prevents the object and method of one discipline from

---

19 Geulincx 1891–1893, vol. I, *Oratio de removendis parergis*, p. 153. The text was included in Geulincx 1665a, pp. 350–384.

20 "Hic proprie incipit logica, quae praecesserunt enim ea ex scientia de scientiis mutuati sumus. Nam logicae non est dicendum, quid facere debeat logica, sicut militis non est dicere quid miles facere debeat, sed id facere, nec de armis, sed arma tractare. Haec igitur omnia in prooemio Logices dicta sint, atque ibi maneant, neque se unquam in logicam ipsam exserant," Geulincx 1891–1893, vol. I, *Dictata ad Logicam*, p. 459 (first edition by Land).

21 "Spectat ad disciplinas toto coelo diversas [...] de scientia illa tractare. Logicus aliquis aut physicus tractatus ad logicam pariter aut physicam pertinet. Sed quae sunt de logica vel physica disputationes, non his scientiis, sed metaphysicae, aut, quod malo, disciplinae cuidam, cui nomen non est, quaeque de ipsis disciplinis agit, debentur," Geulincx 1891–1893, vol. I, *Oratio de removendis parergis*, p. 154. See also his *Logica restituta*: "sicut enim datur scientia aliqua de virtute (et haec est ethica), item scientia aliqua de argumentatione (et haec est logica), item scientia aliqua de rebus materialibus (et haec est physica), et sic de caeteris, – sic etiam datur scientia aliqua de ipsis scientiis, ubi desinere, unde incipere debeant, quo tenore, quae tractanda in iis sunt, disponi ordinarique debeant. Proinde haec scientia praescribit etiam logicae," Geulincx 1891–1893, vol. I, *Logica restituta*, p. 455. *Encyclopaedia* will be described as the *scientia scientiarum* only in the *Metaphysica vera*, as the outline of the system of the sciences that Geulincx puts at the beginning of the treatise: see Geulincx 1891–1893, vol. II, *Metaphysica vera*, p. 139. Originally as *Logica fundamentis suis restituta* (Geulincx 1662), and *Metaphysica vera et ad mentem peripateticam* (Geulincx 1691a).

22 "Mundus quidam est ipsum Belgium, si a munditia, quod volunt, mundus dicitur. [...] Sed si nitor [...] nonne indignum hoc foret, si domi quidem limpidi, foris nitidi volutaremur autem in schola?" Geulincx 1891–1893, vol. I, *Oratio de removendis parergis*, pp. 151–152. He was under criticism for the terse structure of his logic: "quis non aegre ferat, exstirpatis succulentis illis et floridis, relinqui tantum arida quaedam et stricta, quae saepe tetricum illud A, B, C mathematicorum affectant? Hae et similes rhapsodorum querelae [...] cum se ipsae satis explodant, non est quod a me pluribus refellantur," Geulincx 1891–1893, vol. I, *Logica restituta*, p. 172.

being adopted by another.[23] Actually, the expurgation of logic carried out in the *Logica restituta* is aimed against the influence of Aristotelian logic in metaphysics, where logical concepts have been mistaken for existing things, as Geulincx points out in his *Metaphysica falsa sive ad mentem peripateticam* (1691, posthumous edition).[24] However, this criticism also concerns the positions of some Cartesians, as it has as its direct consequence the disregard of epistemological discussions in logic. Logic, according to Geulincx, still has an architectonic i.e., preliminary, function:[25] it first functions as an instrumental discipline. This is carefully explained in the manifestos of Geulincx's philosophy: the introductory oration to the *Questiones quodlibeticae* (1652, 1665) and the dedicatory letter of the first edition of the *Ethica* (1665). Logic, indeed, only concerns demonstrations; thus it serves all the other disciplines, first of all mathematics, which adopts demonstrative proceedings.[26]

---

23 "Anomalia, congeries immensa sordium, Augiae stabulum dixeris, et Herculeus profecto labor est illud expurgare. Ita profunde subsidit, et per cuniculos in purissima quaeque surrepit illa colluvio. Anomaliam in scientiis voco cum obiectum uni, et tractandi modus alteri cuidam disciplinae accommodatus est: ut si res logicas physice, aut physicas logice contemplemur. Plerumque, qui in hac sentina volutantur, obiectum ex quavis disciplina nacti, modum tractandi mutuantur ex grammatica et metaphysica," Geulincx 1891–1893, vol. I, *Oratio de removendis parergis*, pp. 156–157. The very problem of method itself is to be considered, in order to avoid the confusion of disciplines, apart from logic, or by a *scientia scientiarum*, see the *Appendix* to the *Logica restituta*, Geulincx 1891–1893, vol. I, p. 454.

24 "[...] pronitas humane mentis ad affingendum modos suarum cogitationum rebus cogitatis," Geulincx 1891–1893, vol. II, *Metaphysica ad mentem peripateticam*, p. 200. The Preface of the *Logica restituta* recalls the same point: by the study of logic we cannot deduce anything in metaphysics. For instance, the use of *loci* – as examined by Ramus – is rejected by Geulincx as an improper use of metaphysical concepts in logic for the sake of the solution to all kinds of problems, even those about which scholars do not understand anything: "laudatam illam et arctam semitam nunquam eos ad causas ad effecta, subiecta, similiaque, quae metaphysicae tantum considerationis sunt, ducturam fuisse. Ego igitur conatus sum, quod titulus libelli promittit, eiectis omnibus alienis, quae iam totam fere logicam occupaverant, eam sibi restituere [...], e logica exterminare amplissimos illos locos [...] quibus iuventus instructa erat ad syllogisandum de omni proposito problemate, etiam in materia spectante ad disciplinam aliquam cui nomen suum non dederant, et de qua nihil intelligebant," Geulincx 1891–1893, vol. I, *Logica restituta*, p. 172. The use of *loci* as the cause of the confusion of logic and metaphysics is also noticed in the main text of the *Logica restituta*, Geulincx, *Opera*, vol. I, p. 385.

25 See the dedicatory letter of the *Methodus inveniendi argumenta* (1663): "pergit etiam ad vos libellus hic, via quam alter ei libellus anno iam vertente praeiverat. Logici sunt ambo, palos et rudera convehunt, futuris aedibus fundamentum," Geulincx 1891–1893, vol. II, p. 3.

26 Geulincx 1891–1893, vol. I, *Oratio de removendis parergis*, p. 162. See also the *Logica restituta*, Geulincx 1891–1893, vol. I, p. 382.

In the introductory oration to the second edition of his *Questiones quodlibet-icae* Geulincx outlines an ordered plan of studies beginning with mathematics, which has a pedagogical role, since it shapes young minds for the practice of demonstration. Then comes logic, the science of *consequentiae*, which includes mathematics.[27] After these, there is metaphysics, which provides all the other disciplines with their foundation. Metaphysics concerns the properties of body and mind, but not yet those of God.[28] Physics and ethics come next: in fact, in the *Oratio* the focus is more on physics than on ethics. The main topic of the text, indeed, is the eradication – in a Baconian fashion (Aalderink 2009, pp. 46–55) – of the causes of error in physics. This science, for which a "maturior stomachus requiritur," according to the first edition of the text (Geulincx 1891–1893, vol. I, *Oratio prima*, p. 41, n. 13), is put at the centre of the system and it is preceded by a preliminary discipline: natural history. Geulincx introduces a modern – Baconian – way for the observation of nature, no longer based on the textual discussions plaguing the Aristotelian commentaries, but on the use of "telescopia, [...] anatomica theatra, alembici, fornaces, magnetes" and other means revealing the miracles of nature to us, which are combined with our demonstrative skills for the sake of the formulation of physical hypotheses (Geulincx 1891–1893, vol. I, *Oratio prima*, pp. 41–42). These, at least according to the first edition, have a strongly provisional status.[29] Finally, ethics is at the end of the system, but no words are spent on it. Moreover, a concern with rational theology is not yet adopted by Geulincx. However, the empirical and hypothetical method for physics will find a strong basis in Geulincx's mature metaphysics and natural theology, as I am going to show. His empirical physics can not only be seen as a Baconian trace in the Dutch and Flemish context, nor just as the rejection of the bookish Scholastic

---

27 "[...] sine qua apodixes mathematicae non satis feliciter procedunt [...]. Quae quidem scientia non ita ex tempore et sparsim (ut hodie fit) velut silva et rhapsodia tradatur, sed ordine ac tenore geometrico," Geulincx 1891–1893, vol. I, *Oratio prima*, p. 42. Differences with the Louvain edition (1653) are presented in notes by Land.
28 "Tribus his [...] metaphysicam subnectant, sed probe repurgatam, mentis et corporis essentiam ac proprietatis apodictice perhibentem," Geulincx 1891–1893, vol. I, *Oratio prima*, p. 42.
29 "In hanc ne iuret discipulus, teneatur quoad phaenomenis omnibus respondeat. Ubi in puncto deficit, reiiciatur, et alia tentetur verum," Geulincx 1891–1893, vol. I, *Oratio prima*, p. 42. In the second edition, however, the accent is more Cartesian as reason is conceived as the faculty which decide upon the role of authority and experience, and corrects what experience suggests us: "quomodo enim ratione potior et antiquior auctoritas, experientia, aut aliud huius generis quodcunque, si cur ita sit, dicenda est ratio? Certe quod ratione aliqua suaderi probarique debet, totum id sub ratione est. Primum igitur, o homo, ratio est," Geulincx 1891–1893, vol. I, *Oratio prima*, p. 58. See also pp. 50–51, 54–55.

philosophy, but as a method fully consistent with the metaphysics which he later developed and which was mainly concerned with theology.

A similar agenda is proposed in the dedicatory letter of the 1665 *Ethica*, which starts off with a metaphor in which Geulincx compares the *encyclopaedia* of philosophy to a house. Logic is the foundation, mathematics and metaphysics are the columns, physics is the floor, and the decoration of the house and ethics is the roof which makes the structure complete.[30] Physics has here lost the status of the main discipline which it had in De Raey, even if it maintains an essential role as the floor of the House of Philosophy. Moreover, logic, mathematics, and metaphysics form the instrumental and epistemological basis of physics and ethics. Even if, in this plan, metaphysics comes after logic and mathematics, it is in fact the first science which provides the other disciplines with their epistemological foundation. Logic, on the other hand, is the instrument for inferences. Metaphysics also establishes some principles which are used by logic itself, such as those of whole and part.[31] Logic and metaphysics, despite all the efforts to keep them apart (in order to avoid the error of a metaphysical use of logical notions criticised in the *Metaphysica falsa sive ad mentem peripateticam*), are still presented here as being deeply connected, since Geulincx argues that a use of metaphysical *modi considerandi* is inevitable in every science. Both logic and metaphysics are about mental entities, which makes it difficult for them to remain detached; moreover, metaphysics provides logic with its concepts, thus posing the problem of a circularity in Geulincx's system. In the next sections, I will outline Geulincx's solution to this issue, clarifying how Geulincx defines the proper relations among disciplines.

---

30 "Ea re libellos vobis in lucem edidi logicos duos. Quorum alter palos et caementa, solidando paviendoque fundo, alter intritam et ferrumen conferret, quibus haec inter se durata vincirentur et coalescerent. [...] Iacta sunt encyclopaediae fundamenta. Interea vero, dum haec fundamenta sibi esse sino, ut, an dehiscant alicubi vel desciscant, an autem perstent et ferendo sint, explorem [...] quaedam, quae inter exstruendum usui futura videbantur, parabam, aptabam, dolabam: columnas, tigna. [...] Contuli me ad opus amoenum magis: futuri aedificii coronidem fabricare ingressus sum. [...] Coronis ea, de virtute et primis eius proprietatibus commentatio est. [...] Igitur in sapientiae sano laquearium et tectum est ethica. Ut a logica fundamentum sit firmum et bene fistucatum, a mathematica et metaphysica columnae robustae, parietes bene materiati, a physica pavimentum et opera intestina cuncta concinne eleganterque elaborata. Tamen sine ethica nunquam sartum tectum est hoc templum. Imo sine ethica non templum est sed impluvium: non ad sacra, non ad polluctum valet," Geulincx 1891–1893, vol. III, *Ethica*, pp. 3–4.
31 See his *Methodus inveniendi argumenta*: "principia generalia spectant ad metaphysicam. Metaphysica enim sola praecedit logicam inter scientias, quamvis nec sine logica tradi possit," Geulincx 1891–1893, vol. II, *Methodus inveniendi argumenta*, p. 6. These notions are the basis of the so-called logical containment theory, treated in Nuchelmans 1988.

4.4 Geulincx's threefold metaphysics: Autology, somatology, and theology

The proper foundation of physics and ethics, the two disciplines served by logic and mathematics, is provided in the *Metaphysica vera*.[32] Here metaphysics is described as the whole *corpus* of sciences, or the *prima scientia* from which all the others flow: geometry or the *excursus figurarum*, arithmetic or *excursus numerorum*, logic or *excursus in consequentias*, ethics or *excursus in mores*. Another distinction of disciplines comes through *miscellanea*, or the inclusion of hypotheses which give rise to physics, scriptural theology, law, medicine, and all the other arts (Geulincx 1891–1893, vol. II, *Metaphysica vera*, pp. 139, 266). Actually, whereas hypotheses concern something we cannot know with certainty, and *miscellanea* characterise provisional knowledge, all the *excursus* of metaphysics share the epistemological status of metaphysics, since all these *excursus* are based on purely intellectual principles, and are *scientia* or evident knowledge. This epistemological difference between *miscellanea* and *excursus* neatly matches the big difference in Geulincx between hypothetical physics and rational ethics. The point can be illustrated by examining the way in which Geulincx builds up his metaphysics.

Some notions in the *Metaphysica* are taken from Descartes's philosophy. Geulincx's argument, for instance, starts from the *cogito*. Thus, his metaphysics or *prima scientia* begins with doubt in order to make our mind empty of all apparent knowledge[33] and to attain the fundamental notions of mind, body, and God, referring to the objects of the three branches of metaphysics: *autologia*, *somatologia*, and *theologia*. Geulincx, however, includes some logical remarks in the 'introspective' intuitions of the *cogito*. According to him, for instance, the first truth we attain is inferential.[34] The supposition that everything is false, indeed,

---

32 Metaphysical considerations or summaries of his later *Metaphysica vera* are provided in his treatises on physics and ethics. They are summarised in the 1665 *Ethica*, where *humilitas* as virtue is deduced from the metaphysical account of the self, *inspectio sui* (see *Ethica*, treatise I, chapter 2, section 2, § 2: Geulincx 1891–1893, vol. III, pp. 30–37).

33 In the *Disputationes metaphysicae* Geulincx clarifies how doubt proceeds: that is, through the supposition of the falsehood of all knowledge. It is, in fact, a suspension of judgment: see Geulincx 1891–1893, vol. II, *Disputationes metaphysicae*, pp. 476–480. These disputations were first published along with his *Annotata maiora in Principia philosophiae Renati des Cartes* (Geulincx 1691b), along with other disputations on logic and physics that he held between 1663 and 1669.

34 "Suppositio haec, qua sic omnia falsa esse supposuimus, facit etiam ad clarissime demonstrandum primam veritatem," Geulincx 1891–1893, vol. II, *Metaphysica vera*, p. 142.

leads to the logical conclusion that it is true that something can be true or false.[35] The beginning of metaphysics confirms the connections between the different parts of Geulincx's system: logic is not used to formalise all the reasoning in metaphysics; however, logical inclusions are present in its arguments and support the validity of the first truth that follows the initial doubt. Logic, in some manner, teaches the metaphysician how to proceed by training his mind in the procedures of philosophy. On this account, it has a pedagogical, preparatory role within Geulincx's system.[36]

The formalistic attitude of Geulincx is mainly reflected, however, in his critique of the reliability of Descartes's criterion for evidence which, as for Regius, provides only a 'psychological' certainty. It is always possible, indeed, to suspect the propositions which seem to be 'evident'.[37] Intuitive evidence can deceive us: in some instances we can recognise deception only *a posteriori*, as in the case of the addition of the same number to an odd and an even number, which seem 'evidently' to be still single and double after the addition.[38] Evidences are corrected by the intellect, as this is the faculty which judges both the questions it raises

---

35 "Praeclarius enim demonstrari non potest propositio aliqua quam per dilemma, in quo ex falsitate propositionis demonstrandae infertur per necessariam consequentiam veritas eiusdem. Propositioni enim, quae sic demonstrata est, falsum subesse non potest, adeoque necessario vera est. Nam sive vera sit, vera est, sive etiam falsa sit, vera est. V. g. si dicam: aliqua propositio vera est, dico adversario: pone hoc esse verum vel falsum, perinde est, nam nihilominus demonstrabo esse verum," Geulincx 1891–1893, vol. II, *Metaphysica vera*, p. 142.

36 Geulincx assumes Descartes's 'argument' of the *cogito* as an example for his considerations on hypothetical syllogism: see the *Methodus inveniendi argumenta*, Geulincx 1891–1893, vol. II, pp. 88–89, 109–110.

37 "Suspectare etiam possumus in genere illas propositiones, quae nobis ante hac certissimae videbantur. Idque per rationes, ut apparet, convincentes, quamdiu intellectum avertimus ab evidentia istarum propositionum, quas tamen certo et evidenter sciamus," Geulincx 1891–1893, vol. II, *Metaphysica vera*, p. 142.

38 "Et videtur ratio aliqua convincens militare pro ista persuasione. Cum enim certo aliquo modo se habeant duplum et simplum, quamdiu utrique idem accedit, videntur in eodem statu permanere. Cum enim ea, quae certo modo se habent, mutantur, et modum istum amittunt, signum est, fortius aliquid uni eorum quam alteri accessisse. Sic duo parietes albi eodem modo albi manebunt, quamdiu aeque multum albedinis accedit ad utrumque, et aequalia semper manebunt aequalia, quamdiu idem aut aeque multa iis addentur. Unde videmus, falsitatem istius principii, duplum et simplum, cum idem et aequalia accedunt utriquae semper duplum et simplum manebunt, agnosci tantum a posteriori, applicando mentem ad exempla dupli et simpli. [...] Quamdiu autem a priori principium illud intuemur, fortissimum esse videtur, et nihil occurrit, quod veritatem eius suspectam reddere posse videatur," Geulincx 1891–1893, vol. II, *Metaphysica vera*, p. 143. This argument is used also in the *Disputationes metaphysicae* as one of the sceptical arguments, see Geulincx 1891–1893, vol. II, p. 484.

itself and the 'evidences' of the senses.[39] Therefore, it is our highest faculty. Its reliability seems to be justified only through its primacy. This critique of evidence as the criterion of truth, apparently provided by Geulincx *en passant*, is in fact a cornerstone in the avoidance of a foundation of *scientia* that might make superfluous an appeal to God. This is what Descartes does in the *Meditatio quinta*, grounding on God's benevolence only the reliability of the memory of past evidence, whereas the reliability of actual evidence is intuitive and self-grounded (AT VII, pp. 69–70). Geulincx, however, reinstates God in the role of a warranter of present evidence as well. This solution, in fact, is a *unicum* in the history of Dutch Cartesianism, with the exception of Regius's positions, which are however developed on the basis of the rejection of pure intellectual evidence. Accordingly, God's veracity is a guarantee to conclude from 'psychological certainty to logical certainty: it is this premise that turns Geulincx's metaphysics into theology. As will be shown in the next sections, intellectual evidence is grounded in intellectual truths perceived in the very mind of God.

4.4.1 An Aristotelian axiom

The parts of Geulincx's metaphysics are *autologia*, *somatologia*, and *theologia*, or the consideration of the self, of the body, and of God. *Autologia* is the first and main part, from which the others directly follow. Actually, the propositions or truths of metaphysics are attained by an introspective method (like Descartes's) and are not deduced in a strict geometrical way. Even if logic plays some role in metaphysics by being included in its preliminaries, Geulincx still follows a method very different from that of Spinoza.[40]

---

39 "Secundo intellectus noster corrigit sensum in suis evidentiis. [...] Quid scimus autem an, sicut sensus corrigitur in sua evidentia a facultate aliqua altiori, sic etiam intellectus in suis evidentiis non possit corrigi a facultate ipso altiore? Certe enim intellectus non evidentius percipit, duo et tria esse quinque, quam aspectus percipit circulum igneum in casu posito. [...] Sed quis dicit, intellectum sensu digniorem esse, nisi ipse intellectus ? Sensus enim hoc non dicit, sed ostendit et repraesentat modo obiecta, num ipse dignior vel indignior sit intellectu, in medio relinquit," Geulincx 1891–1893, vol. II, *Metaphysica vera*, pp. 143–144. Geulincx notices this also in the introductory oration of his *Quaestiones quodlibeticae*, see *supra*, section 4.3.1.
40 Rousset writes that "car c'est bien ainsi que va procéder Geulincx: alors que tout ce que nous avons vu reste de l'ordre des instruments extrinsèques, la méthode intrinsèquement philosophique, propre à sa Philosophie, à sa Métaphysique avec toutes les conséquences qu'on peut en tirer pour la suite, principalement pour l'*Ethique*, se ramène, comme nous allons le voir, à la simple *Inspectio sui*, sans aucun autre moyen pour constituer la *catena demonstrationum*, qui ne ne trouvera même pas sa forme dans ses raisonnements de nature déductive, comme chez Spinoza," Rousset 1999, p. 43.

Introspection or *inspectio sui* makes us aware of some truth, as "me esse," "varios habeo cogitandi modos," and "ego sum res una atque simplex" (Geulincx 1891–1893, vol. II, *Metaphysica vera*, pp. 147–149). This is the basis of the whole further development of metaphysics, along with the well-known axiom according to which

> impossibile esse, ut is faciat, qui nescit quomodo fiat. (Geulincx 1891–1893, vol. II, *Metaphysica vera*, p. 150; see Verhoeven 1973, Nadler 1999)

This axiom leads to the consequence that we are not the authors of our sensations and movements since we are not conscious of the way we produce these. Their cause must therefore be found in God. This axiom is a relic of the Aristotelian hierarchy of forms, in which those that are separated from matter and provided with a rational principle (namely, the human and divine intellects) are ontologically superior to those informing brute matter. Actually, Geulincx seems to develop his system upon an Aristotelian principle. Indeed, among souls, those that attain true knowledge are superior to the others as they realise their proper end. They are active, whereas the others, not aware of what they are doing, do not realise their potentiality by *theoresis*. The hierarchical principle of rational activity is rendered by Geulincx into a hierarchy of substances according to which only those substances which know how they act have a causal role. The Scholastic model of action and passion, based on the concepts of matter and form, is thus followed by Geulincx in his consideration of the relative power of bodies and souls. However, because substantial forms as centres of activity are banished from the Cartesian physical world (Van Ruler 1995, pp. 133–166) all activity is to be located in God and, to a certain degree, in human souls insofar as they are conscious of their internal acts:

> sapientia est, quam nemo videtur habere nisi qui rem illum effecerit. Talis est conscientia nostra amoris, odii, affirmationis, negationis, caeterarumque in nobis actionum, eo quod ipsi eas exerceamus et efficiamus.[41]

In fact, the communication of movement among bodies cannot be regarded as activity, for they have no consciousness and are passive.[42]

---

[41] Geulincx 1891–1893, vol. II, *Metaphysica vera*, pp. 192–193. See also the *Logica restituta*, Geulincx 1891–1893, vol. I, pp. 403–404, 459.

[42] This point is also discussed in the *Metaphysica ad mentem peripateticam*, where Geulincx criticises the attribution of active faculties to bodies: Geulincx 1891–1893, vol. II, p. 224.

This is the premise of Geulincx's 'occasionalism', or the result of the application of Aristotelian ontology of act and potency to a world deprived of substantial forms.[43] Geulincx develops a philosophical system concerning human beings as subject to God's activity in sensorial experience and bodily activity. As we are not aware of the ways by which we receive sensations or make movements, we are not actively involved in them. They come from God, a "sciens aliquis et volens diversus a me" (Geulincx 1891–1893, vol. II, *Metaphysica vera*, p. 150). Indeed, because I am a simple entity, for my nature is deduced from the principle of the *cogito*, thoughts cannot come from myself but must come from bodies as they are the occasions for sensory experience.[44] Moreover, because bodies are not conscious of their operations, and there is no communication between two different kinds of substances,[45] thoughts, and movements must come from God (Geulincx 1891–1893, vol. II, *Metaphysica vera*, p. 188).

---

43 The part of metaphysics devoted to the properties of body, *Somatologia*, presupposes the arguments concerning occasionalism. In accordance with Geulincx's comprehensive view on the passivity of body, it cannot move itself: it is just extension and does not imply motion in its essence (Geulincx 1891–1893, vol. II, *Metaphysica vera*, p. 176). Moreover, as it does not think, it is not active: "clarissime deducitur haec assertio ex isto axiomate quod in Autologia [...] asserebamus: quod nescit quomodo fiat, id non facit; nescit corpus (utpote res bruta) quomodo fiat motus; non ergo scilicet illum in se, non ergo motum habebit a se" (Geulincx 1891–1893, vol. II, *Annotata ad Metaphysicam*, p. 280). Thus, they come from outside, or from a mind, namely, from God as He is conscious of his operations (Geulincx 1891–1893, vol. II, *Metaphysica vera*, p. 176). Another argument supporting occasionalism comes from the notion of body. As it is infinitely divisible, that is, as it contains infinite parts to be divided, an infinite force is required for its division: see the *Annotata ad Metaphysicam* and the *Disputationes physicae*, Geulincx 1891–1893, vol. II, pp. 290, 502. Geulincx's *Disputationes physicae* were originally published in Geulincx 1691b. See Lattre 1967.

44 "Iam autem eas excitare debet aut mediante me, aut se ipso, aut tertio aliquo. Non excitat eas autem mediante me ipso, quia cogitationes sunt diversae, et ego sum res simplex, a quo diversae cogitationes emanare non possunt. Non se ipso, quia est aeque simplex ac ego: est enim aeque volens et sciens, id est cogitans, ac proinde simplex. Simplex enim sui qui idem cogitavi de variis. Restat ergo tertium, cuius interventu hoc faciat, quodque variarum mutationum capax esse debet, ut per hoc varia cogitationum obiecta exsurgant, illudque est extensum, quod potest variari, seu corpus. Tertium enim praeter cogitans et extensum nec novi, nec est," Geulincx 1891–1893, vol. II, *Metaphysica vera*, pp. 150–151.

45 "Ego partium omnium expers res sum, ut supra dictum est. Et qui incursus fiat in id quod partes nullas habet? Molem aliquam habere debet, et consequenter partes, in quod incursus fiet. Unde ego non proprie versor inter corpora, nullum ibi locum, nullum spatium occupo. Quantillum occuparem, extensus essem, et totidem haberem partes secundum molem, quot habet tale spatium (omne autem spatium, quantumcunque exile fingatur, infinitas tamen secundum omnem dimensionem habet partes, ut infra patebit). Ego enim hac ratione inter corpora versor, quod in res agant. Agant inquam velut instrumenta, non velut causae," Geulincx 1891–1893, vol. II, *Metaphysica vera*, p. 153.

Geulincx's occasionalism, as I am going to explain, results in a physics based on experience and hypotheses. Because physical processes are dependent on divine will and action – namely, they are God's very actions – there are no means besides experience to attain them, because no principle can explain them besides divine will. The continued intervention of God in the world and His absolute power are not within the scope of the human faculties. Consequently it is theology which determines the highest degree of knowledge we can reach. Before turning to this issue, however, I will spend some time on the problems involved in Geulincx's ontology.

4.4.2 The body

The following parts of the *Metaphysica vera* integrate the conclusions of the *Autologia*, establishing Geulincx's concepts of self, body, and God, and his account of the *conditio humana*.[46] *Somatologia* concerns the properties of the body and lends support to physics; however, it is not designed to ground a necessary knowledge of nature, but only to elucidate some concepts useful for the formulation of physical hypotheses. Their actual foundation is developed in the third part, *Theologia*. From this, Geulincx develops an empirical and hypothetical physics consistent with his account of God's causal power: divine providence is taken as warrant of its certainty.

The ontology of the body involves two problems for the theory of philosophical knowledge, showing the limits of natural knowledge and leading to a theological resolution of the truths of physics. The first problem is raised by the dualism between the realms of being and of becoming in the physical world. The way Geulincx deals with this is meant to help us understand the problem of necessary knowledge in physics. Secondly, the *somatologia* involves the question of the individuation of bodies or, as Geulincx interprets it, of their 'abstraction' from continuous extension. Such abstraction from singular bodies and their features is the main cognitive process in natural philosophy.

The first problem is to be evaluated in the light of the ontology of physics developed in the *Metaphysica vera*, as well as in his *Disputationes physicae*, in whose introductory or metaphysical section Geulincx defines the properties of the body necessary for the formulation of physical hypotheses (Geulincx 1891–1893, vol. II,

---

46 "Veniamus ad secundam partem metaphysices, in qua multa dilucidabuntur, quae spectant ad primam. Cum enim corpus pertineat ad humanam condicionem, quam nostram condicionem sine corporis, eiusque affectionum notitia non satis feliciter consequi possumus," Geulincx 1891–1893, vol. II, *Metaphysica vera*, p. 155.

Disputationes physicae, pp. 489–51). These notions, in fact, concern essences and properties whose actual existence in the physical world is to be discovered by the senses: its acknowledgment thus has a hypothetical status. At the beginning of the *Somatologia*, the first section of his *Metaphysica vera*, Geulincx introduces the idea of body, or the purely intellectual notion of extension (Geulincx 1891–1893, vol. II, *Metaphysica vera*, p. 158). Geulincx deduces some properties from it, such as that it is infinite, that its dimensions are related to each other,[47] that it is a simple entity and that no void can be admitted in it (Geulincx 1891–1893, vol. II, *Metaphysica vera*, p. 163). On the basis of these considerations on body, in fact, Geulincx introduces a second form of dualism. Besides the distinction between soul and body, he outlines a difference between the immutable and the mutable body, or between body-as-such (*extensio simpliciter dicta* or *corpus)* and the mutable body in motion, *mundus* (Geulincx 1891–1893, vol. II, *Metaphysica vera*, p. 188). Body-as-such cannot be divided because no void can be admitted in it (Geulincx 1891–1893, vol. II, *Metaphysica vera*, p. 167). Singular bodies, on the contrary, are the result of a division which does not involve the existence of a void. At the same time, such a division is not a mere mode of considering body-as-such in the mind, namely, as a concept without any real reference. The singular bodies can be truly abstracted by the mind:

> itaque cum corpus non sit divisibile, particularia tamen omnia corpora divisibilia sunt, tantum difficultas est, quo pacto corpora particularia habeantur [...]. Particularia corpora habentur per determinationem mentis seu abstractionem et praecisionem, sicut et linea et superficies et puncta tali aliqua abstractione proveniunt [...] nam verissimum est illud, quod scholae ex Aristotele hauserunt [...]: abstrahentium non est mendacium, seu, qui abstrahunt, non fingunt, non mentiuntur. Res enim quam abstrahunt, revera est, etiamsi non sub illo abstractionis statu sit. (Geulincx 1891–1893, vol. II, *Metaphysica vera*, pp. 168–169)

The existence of particular bodies, however, abstracted from the continuum of extension, is the result of the division of the body in motion, or *mundus*. They are not parts of the body-as-such but of the world in motion, the realm of becoming. Indeed, Geulincx defines motion as the combination of closeness and distance of parts of *mundus*: "vicinitas atque distantia duarum earundem partium inter se," and rest as the permanence of distance (Geulincx 1891–1893, vol. II, *Metaphysica vera*, pp. 175, 179–180). Division is thus motion itself. From motion and rest all sorts of figures and singular bodies flow, because figure is produced by

---

47 "Fieri non potest, ut intra sit sine extra, non magis quam [...] supra sit sine infra, pater sine filio, et generatim relatum sine correlato: nam quod intra est, intra aliquod fit necesse est, utique intra illud quod extra est," Geulincx 1891–1893, vol. II, *Metaphysica vera*, p. 162.

the rest among parts of matter and by the movement of one part with respect to other parts.[48] From this point of view, figures and particular bodies seem to be the same thing.

In Geulincx's *Disputationes physicae*, however, motion is further defined as *forma corporis* or *forma mundi*.[49] Besides the 'particular' body in itself (or the 'becoming' extension), only motion is real as its mode; by motion, parts of extension are united or separated. Their figures are determined by the motion of their closest bodies. Therefore, motion is the *modus realis* of the 'particular body', whereas figure is its *modus rationis*, as it results from a mental activity over parts of the body in motion among each other.[50]

The problem involved in these statements is the relation between body-as-such and the existing world of bodies in motion. According to Geulincx's *Metaphysica vera*, single bodies are not parts, but modes of the body-as-such, as lines, surfaces, and points are modes of particular bodies:

> non sunt figmenta, non entia rationis, non chimaerae [...] sed corpora particularia sicut sunt aliquid ipsius corporis simpliciter dicti, sic et revera superficies, lineae atque puncta sunt aliquid extra nos in corpore particulariter sumpto. Cum vero non sint partes [...] restat, ut modos esse dicamus. (Geulincx 1891–1893, vol. II, *Metaphysica vera*, p. 173)

However, they are parts of the 'particular body' subjected to motion, because motion is between *parts* of matter. As motion is what transforms the simple *corpus* into *mundus*, or body in motion, it affects the body-as-such as a mode. However, motion as division of parts can characterise only *mundus*, as body-as-such has no parts.[51] The problem is, thus, how to find a relation between these two kinds of reality: one is put into succession and time (as time is the measure

---

48 "Includit ergo figura et quietem eorum inter se, quae iunctim figura circumscribi dicuntur, et motum separationemque eorundem a quibusdam aliis," Geulincx 1891–1893, vol. II, *Metaphysica vera*, p. 181.

49 See Geulincx 1891–1893, vol. II, *Disputationes physicae*, pp. 490, 496. In addition to this, Geulincx follows the Aristotelian tradition of the *quantitates interminatae* as the essence of matter by defining extension as *forma corporis* in his *Disputationes physicae*: see p. 506. On the *quantitates interminatae*, see Donati 1988.

50 "Figuram esse modum-rationis ipsius motus: nam nec figura sine motu, nec motus sine figura esse cogitarive potest. At motus est modus-realis corporis, qui et abesse potest a corpore, et necessario aliquando etiam abfuit, nam corpus ab initio moveri non poterat," Geulincx 1891–1893, vol. II, *Disputationes physicae*, p. 517.

51 "Corpus est divisibile. Hoc intelligendum est de corpore particulari, nam de corpore generaliter sumpto intelligi nequaquam potest. [...] Si enim corpus, ut sic, divideretur, non nisi interiecto vacuo divideretur, atqui vacuum est impossibile, ergo et corporis generatim sumpti divisio," Geulincx 1891–1893, vol. II, *Metaphysica vera*, p. 167.

of movement, following Aristotle's definition (Geulincx 1891–1893, vol. II, *Metaphysica vera*, pp. 176–177)). The other is an immutable, extra-temporal entity. Outlining a difference between the consideration of body *sub specie aeternitatis* (from the perspective of God) and as it is in time (or as seen by men) is a traditional solution that cannot resolve these problems. According to this differentiation, we could say that man can truly recognise actual bodies as they are shaped by motion in the temporal realm. God, on the other side, contemplates the world as it is outside time, or a monistic extension. Actually, it is the same body under different viewpoints. What is lacking in Geulincx's analysis, however, is a theory of knowledge integrating these two perspectives: Spinoza will in fact solve this problem by showing how man can attain the same knowledge of God. Geulincx does not deal with an actual *adequatio humanae mentis et Dei*, even if he admits that single minds belong to God's mind as its *modi*.[52] This difference of perspectives, or the dualism between being and becoming, leads to a consideration of physics as a science of the apparent structure of the world, reflected by the contingency of natural laws or the impossibility of deducing them *a priori*.

The other problem raised by Geulincx's ontology for physics concerns the abstraction of single bodies as *modi rationis* and *aliquid corporis* at once. It involves the problem of individuation of bodies: "quid ergo de lineis, punctis et superficiebus censemus? [...] Per abstractionem seu praecisionem ea deradimus ex corporibus particularibus, sicut ipsa corpora particularia praescindimus ex corpore simpliciter dicto" (Geulincx 1891–1893, vol. II, *Metaphysica vera*, p. 171). Single bodies are not just in our thought, because they result from a consideration of body, not from a mere *cogitatio*. The difference between *cogitatio* and *consideratio* is grounded in our self-consciousness:

> dico in considerationem, non autem in cogitationem. [...] Si vero petas, in quo cogitatio atque consideratio differant? Respondeo, id dici non debere. Sunt enim actus operationesque mentis nostrae, quorum nobis per conscientiam intimam cognitio ac sensus est. (Geulincx 1891–1893, vol. II, *Metaphysica vera*, p. 172)

The only foundation which can thus be provided for the distinction of particular bodies is an appeal to our inner consciousness and intuition.[53] This problem results from the eradication of substantial forms from nature: within a

---

52 See *infra*, section 4.5.

53 The foundation of abstraction on self-consciousness is provided also in his *Physica vera*, where Geulincx discusses the definition of dimensions: see Geulincx 1891–1893, vol. II, p. 377. The *Physica vera* was originally published by Cornelis Bontekoe (see Bontekoe/Geulincx 1688).

mechanical worldview the problem of individuals is hard to solve. Figures are bodies themselves, as a body is abstracted according to the motion it has with respect to others. However, Geulincx does not mention the problem of the identity of a body through its geometrical modifications. In fact, the problem of individuation remains unsolved.

Geulincx's view is thus focused on a sort of inconceivability of body-as-such, which, in turn, characterises the *conditio humana* as being dependent on the way in which God has provided a way for us to understand nature. This *conditio* consists in the actions and passions that the soul has in itself or together with its own body (such as movements and sensations), no matter what the actual cause of their communications is.[54] These are recognisable in the coincidence of the motions in soul and body,[55] whose union is based only on God's will. As man does not know anything without the intervention of God, as far as experience is concerned, he is passive: this passivity leads to scepticism in philosophical knowledge. His theology is actually designed to solve this problem.

4.4.3 The freedom of God

Theologia is the culmination of Geulincx's metaphysics. Its problems are mainly discussed from an ethical perspective, for the definition of divine properties throws light on our position in the world and on our duties. Indeed, such definitions are not deduced from the idea of God but from the consideration of the human condition, or *a posteriori*.[56] However, they are relevant for epistemology as well, as they concern the status of essences or the objects of necessary knowledge.

---

54 "Id enim est hominem esse, a corpore aliquo pati, et vicissim in corpus illud agere [...]. Constituunt quidem plurimi rationem hominis in unione mentis cum corpore, imo omnes fere in ea unione humanitatem versari arbitrantur. Sed meminerint, unionem non esse primam notionem, sed secundam [...]. Iam autem ad humanam naturam videtur perinde esse, sive mens stabiliter a corpore patiatur, atque in illud agat, sive hoc ad momentum tantum fiat," Geulincx 1891–1893, vol. II, *Metaphysica vera*, pp. 154–155.

55 "Est igitur hoc quam clarissimum, me solutione humanae condicionis meae interire non debere. Nam quid hoc faciat ad interitum meum, si horologium corporis quoad motus suos non amplius consentiat cum voluntate mea, aut perceptiones meae non amplius pendeant a motibus corporis?" Geulincx 1891–1893, vol. II, *Metaphysica vera*, p. 157.

56 "Non exsequemur hunc tractatum procedendo ex definitione, et ex idea Dei ad eius proprietates descendendo (sicut id praecedenti tract., qui de corpore est, praestitimus), sed potius a posteriori; eo quod iuvet hanc scientiam connectere cum ea quae traditur parte prima, et gradatim a cognitione nostram descendere ad cognitionem Dei," Geulincx 1891–1893, vol. II, *Metaphysica vera*, p. 186.

Indeed, Geulincx states, in a Platonic manner, that ideas and essences are the divine archetypal models.[57] He thus supports the view that *scientia* comes only from the ideas of pure understanding. Following a traditional position, he admits in *Logica restituta* that the knowledge of ideas, essences, and definitions is necessary.[58] These are defined in *Metaphysica peripatetica* as leading to necessary consequences, whereas the products of God's will – namely, the created world – cannot be objects of *scientia*.[59] Indeed, the world is the object of sensory experience. In *Annotata ad Metaphysicam*, moreover, we find a comparison between sensory ideas in physics and passions in ethics.[60] Geulincx emphasises the pivotal role of innate ideas in the development of philosophy. Metaphysical *scientiae* thus

---

57 "Apud platonicos et veros philosophos essentia passim vocatur idea; et qui quidem hoc nomine usi sunt, arctius se ipsos ad contemplationem verae essentiae restrinxerunt, quam qui nomine illo scholastico, egregiam licet admonitionem continente, usi fuerunt. Sic idea corporis consistit in extensione, idea mentis (spiritus, dicunt scholae) consistit in cogitatione, idea globi consistit in certa figura, etc. Porro ex ideis illis proprietates et demonstrationes deducunt, v. g. quod corpus infinite extensum, infinite secundum partes suas divisibile sit," Geulincx 1891–1893, vol. II, *Metaphysica ad mentem peripateticam*, p. 263.
58 Geulincx 1891–1893, vol. I, p. 193 (*Metaphysica vera*), pp. 236–237 (*Metaphysica ad mentem peripateticam*). On these points, see Aalderink 2009, pp. 157–204.
59 "Essentia est praedicatum necessarium et primum. Primum est cuius non datur ratio per aliud pertinens ad idem subiectum. Necessarium vero est quod affirmari quidem de subiecto suo potest, negari minime. Non sufficit igitur ad essentiam ut attributum primum fuerit. nam huc omnia contingentia pertinent, quorum nec ratio, nec demonstratio, nec scientia ulla est [...] sed solum Dei arbitrium his causa est. Veluti quod mundus sit, sol, terra, caeteraeque eius partes, nosque homines, nulla horum ratio, nulla demonstratio, nulla scientia. Totum hoc est, quia Deo sic placuit," Geulincx 1891–1893, vol. II, *Metaphysica ad mentem peripateticam*, p. 261.
60 "Tum mens aut libera penitus aut saltem liberior, dici non potest quam facile ad veritatem intendat, quam sublimiter philosophari incipiat, solis iam suis ideis et innatis notionibus addicta et auscultans tota. In rebus ethicis simile quid contingit circa passiones," Geulincx 1891–1893, vol. II, *Annotata ad Metaphysicam*, p. 277 (the *Annotata ad Metaphysicam* were published for the first time by Land). In the *Ethica* he makes the same point, see Geulincx 1891–1893, vol. III, *Ethica*, p. 105. Moreover, pure ideas are distinguished from mere *phantasmata* and *schemata*, or sensory species and *modi considerandi* in *Metaphysica ad mentem peripateticam*, where Geulincx criticises the use of Aristotelian metaphysical notions ("loco idearum et notionum nostrarum, schemata et phasmata nostrorum sensuum surrogemus," Geulincx 1891–1893, vol. II, *Metaphysica ad mentem peripateticam*, p. 210), and in his *Oratio de abarcendo contemptu* (1665), aimed at criticising the neglect of the first principles and concepts in philosophy. In this oration Geulincx speaks against a way of reasoning according to which the notions of God and mind are known through the senses, as they must be derived from pure intellect. In fact, he is considering innate ideas such as those of immaterial things, not of bodies, as they cannot come from the senses nor can they be created by the mind itself (Geulincx 1891–1893, vol. II, pp. 132–133). The *Oratio de abarcendo contemptu* was originally published in Geulincx, *Annotata maiora*.

concern essences or the concepts in God's mind, no matter whether or not they match created entities in the world. Geulincx admits, for instance, that the idea of motion does not depend on the existence of motion itself (Geulincx 1891–1893, vol. II, *Annotata ad Metaphysicam*, p. 270). According to him, however, God is free in the creation of the world according to these ideas, for He is not constrained by any rule. The principle of goodness does not force God as it comes after His will.[61] Therefore, the created world is subject to contingency. However, God also follows the principles of justice and mercy, as Revelation states. These principles have no clear status because they seem to come after divine will but at the same time they influence the act of creation.[62] In any case, truths concerning the created, physical world are contingent: things could be different from the way they are as their existence depends on divine will.

In the light of this, Geulincx can state that some truths depend on God's will, whereas others are necessary, like those of mathematics.[63] The point is clarified by a comparison with the case of man. Given the concept of a triangle, it is necessary that its internal angles are equal to two right ones. But given the existence of man, it is not necessary that, if wounded, he suffers some pain:

---

61 "Sed dices: bonitas Dei necessitatem hic fecit, atque erat necessum ut mundum crearet, et homines conderet, qui tam bonus erat, resp. esto haec necessitas (si qua sit), non impedit contingentiam, non officit libertati [...]. Bonitas enim seu inclinatio faciendi hoc quod praestat, idque semper agendi, quod optimum est, non eripit libertatem, quia voluntate posterior est haec bonitas adeoque cum libertate bene compatibilis, et nulli de libertate decedit, si melior fuerit, atque ad id quod expedire iudicaverit propensior," Geulincx 1891–1893, vol. II, *Metaphysica vera*, p. 194.

62 "Creavit homines, sed inde non sequitur quod eos non creaverit libere. Ita, cum humanum genus peccasset, Deus redemit illud: potuisset id utique non facere, sed obstabat eius misericordia. Angelos lapsos non redemit, quod utique tam bene potuisset facere quam homines redemit, sed obstabat iustitia eius. Omnia autem ille nihilominus fecit libere. Quid enim, quaeso, officit libertati, quod agens determinatus sit nonnihil magis in unam quam in aliam partem? Eo magis libere certe id aget," Geulincx 1891–1893, vol. II, *Annotata ad Metaphysicam*, p. 296.

63 "Deus igitur liber est in condendo homine, in creando mundo. Nihil ipsum ad haec adegit, nulla hic necessitas est voluntate eius prior. Duo et tria ut quinque sunt, circulus ut aream habeat, mons ut vallem, necesse est. Deus hoc etiam vult, sed haec necessitas voluntate eius quasi prior est, et ex natura et intellectu eius dimanat. Sed nihil simile apparet in motu, nil simile in devolutione, qua Deus nos per motum de aliis cogitationibus in alias tam ineffabiliter devolvit. [...] Motus enim est de genere contingentium, cum et possit non esse, et aliquando non fuerit, imo et necessum sit ipsum aliquando non fuisse, ubi habetur tota definitio contingentis et plus quam ad illam requiratur. [...] Ex quo vides notabile discrimen inter ea quae necessaria sunt (tale enim primum a natura seu intellectu divino dependet, etiamsi voluntas Dei accedat et assentiatur) et inter ea quae contingentia sunt (id est, primum a voluntate divina Dei dependent, ex natura seu intellectu praecedente necessitate)," Geulincx 1891–1893, vol. II, *Metaphysica vera*, pp. 193–194.

> naturale [...] id est, quod pendet ab intellectu divino, antecedenter ad eius voluntatem, seu in quo tantum intellectus regula eluceat et nullum voluntatis decretum. Sic naturale est, triangulum habere angulos suos aequales duobus rectis [...]. Perverse autem scholae et populus haec ad quam plurima diffundit, ut cum dicunt naturalia esse, ut corpore laeso doleamus, [...] in quo vehementissime errant. Nam nulla necessitate ex antecedentibus illis haec consequentia deducuntur; sed tamen ex instituto divino libero atque arbitrario. (Geulincx 1891–1893, vol. II, *Annotata ad Metaphysicam*, p. 294)

Geulincx is comparing two different ontological levels: that of necessary mathematical ideas, antecedent to creation, and that of the created, contingent existence of man. Because man is composed of two substances which cannot interact, his properties – such as having pain – are in any case contingent, as they are fully dependent on the miracles of occasionalism. They cannot be deduced from any essence. Whereas the essence of extension leads to some necessary properties, and thus only its existence is contingent, man as the union of two substances has no essence before God's will. Moreover, even the world as a whole has no essence by which we can deduce all its properties and modes. Indeed, the world is made by motion, but bodies cannot move anything as motion is not included in their essence, nor they can be active, according to Geulincx's axiom. Therefore, the world's properties and modes do not depend on anything but God's will. Even the quantity of motion is subject to His will, as well as its laws (Geulincx 1891–1893, vol. II, *Disputationes physicae*, p. 510). An essence for the whole world and thus a rational deduction of its laws would require an independent activity of bodies; this, anyway, will contradict occasionalism. It is admitted by Spinoza, who accepts the identification of the world with God and an absolute necessity in all its modifications, and by Leibniz, who introduces in matter substantial forms involving a necessary connection among themselves. Furthermore, De Raey, who denies that we can refer the concepts of act and potency to bodies, still admits that they can produce effects, avoiding any occasionalist conclusion.[64] Geulincx, on the contrary, emphasises the passivity of the world and its being subject to God's action, which cannot be the object of evident knowledge (*scientia*). As motion depends on God's will, there can be different worlds and thus different ideas of it, as stated in *Annotata ad Metaphysicam*:

---

64 See his *De mundi systemate et elementis*: "quanquam enim nihil in se actuosum sit praeter naturam intellectualem, cuius substantia actus est, quamquam haec sola primum in se et hinc etiam in alia potestatem habeat, non impulsa ab alio, verum ex se et sponte sua. Hinc tamen non sequitur, quod natura corporea, quae tali modo actuosa non est, nullum proprius effectum habeat," De Raey 1677, *De mundi systemate et elementis*, p. 591. Originally published as a series of disputations held in Leiden in 1661 (De Raey 1661).

> ideam [...] mundi, partiumque eius habere possemus, imo [...] ideam huius mundi (plures
> enim esse possunt mundi, auctore naturae aliter atque aliter itemque remissius vel inten-
> tius, diffusius vel contractius movente materiam, haec enim omnia pendent ab eius arbi-
> tratu). (Geulincx 1891–1893, vol. II, *Annotata ad Metaphysicam*, p. 288)

However, these ideas do not seem to imply necessary consequences for the quantity and the laws of motion. God can change them because He is not forced to adopt the idea of one world instead of another. These ideas, in fact, do not involve a general essence of the world as they concern something which is arbitrary.

In the light of this, we can acknowledge to what extent the problem of divine freedom involves that of the necessary existence of things, and thus that of philosophical knowledge as it concerns essences or contingent entities. Indeed, as Spinoza and Leibniz, Gottfried Wilhelm von Leibniz's positions on the essence of the world are different from those of Geulincx, they give different accounts of divine freedom also. Leibniz, Gottfried Wilhelm von Leibniz admits a *potentia Dei ordinata* based on a divine freedom granted by the co-existence in God's intellect of infinite essences or ideas of world.[65] Spinoza, on the contrary, recognises the existence of one world, the actual existing one; every truth, according to him, is like those of geometry and can be deduced from its essence. In both cases, however, the world is provided with its own activity and does not depend on God's intervention to develop its states, independently inscribed in its essence. Because in Geulincx's opinion the existence of motion, its laws and degrees are contingent, we can acknowledge them only through the senses. Therefore, his physics is contingent from an ontological and epistemological point of view, because it concerns contingent objects and is based on experience. It is a hypothetical science, relying on some concept or idea matching essences and necessary properties (its 'metaphysical' part), and on some others known by experience, those concerning matters of fact (Geulincx 1891–1893, vol. II, *Physica vera*, pp. 422–423). According to Geulincx the senses are the only means to acknowledge the existence of corporeal things, because by pure intellect we can acknowledge only the essence of a limited range of things:

> cum motus sit contingens, et pendeat ab arbitrio moventis, hactenus etiam incertus est,
> quod ab essentia et a priori, id est ex ratione proprie dicta procedentibus, lateat. Non mirum
> igitur, si quidam hac sola lucerna gressum dirigentes, in naturae meridie, in mundi foro,
> mundum, id est motum, non invenerint. [...] A posteriori vero procedentibus explorata satis

---

65 A freedom warranted by the infinite number of possible worlds in God's intellect, and ordered by the principle of the richest production of phenomena according to the simplest laws. See his *Principes de la nature et de la grâce fondés en raison* (1714), § 10, in Leibniz 1875–1890, vol. VI, pp. 598–606.

et certa est existentia motus. Cum et successionem in nostris cogitationibus, et perceptiones nostras, inter caeteras, tales aliquas observemus, quas conscii sumus a nobis solis non pendere. (Geulincx 1891–1893, vol. II, *Disputationes physicae*, pp. 511–512)

The ontological contingence of the world finds its counterpart in the epistemic contingence of the hypothesis.[66] As a result, Geulincx's positions on the freedom of God integrate a view of physics according to which it has a provisional status and whose method consists of deductions based on experiences of motions, as first pointed out in his 1652 *Oratio*.[67]

4.5 The foundation of experience and intellectual evidence

The epistemic problem these positions involve concerns the reliability of the senses. It is solved by Geulincx by an appeal to divine providence. This is well explained in the *Annotata ad Metaphysicam*, where he admits that without the senses we can truly conceive the world as it is in itself, but only insofar as it is a possible world.[68] In other words, we cannot be sure that it is the actual world. In fact, there are two worlds: a world in itself, whose idea cannot be known through the senses, and a sensible world. This second one bears the marks of divine providence or of God's 'wisdom and goodness', as it is through it that we can attain the first and acknowledge it as really existing.[69] This is, actually, the only way by

---

66 "Hypothesium prima condicio est ut sint contingentes. Si nempe essent necessariae, ex illis, cum metaphysicae theorematis necessariis pariter, nunquam sequerentur apparentiae, quae contingentes sunt," Geulincx 1891–1893, vol. II, *Physica vera*, p. 422.

67 "Sufficere autem illum ad omnes illas perceptiones, quas de mundo eiusque partibus habemus, absolvendum, tum intelligenti satis demonstratum est, tum qui non satis intellexerit habet quod physicam adeat. In qua continua deductione per varios motus, varia atque adeo omnia naturae phaenomena abunde explicantur," Geulincx 1891–1893, vol. II, *Metaphysica vera*, p. 189.

68 "Sensibus etiam destituti, ideam tamen mundi partiumque eius habere possemus, imo etiam ideam huius mundi [...]. Nam habentes ideam mundi seu corporis in motu, possemus ad varias eius species descendere tandemque etiam ad speciem huius mundi appellere. Quo casu tamen hunc mundum non ut existentem, sed ut possibilem. Nec enim idea aliud de re obiecta [...] indicat," Geulincx 1891–1893, vol. II, p. 288. "Haec idea repraesentaret nobis hunc mundum eiusque partes ut sunt in se," Geulincx 1891–1893, vol. II, *Annotata ad Metaphysicam*, p. 288.

69 "Deus itaque duos quodammodo mundos fecit, alterum in se (et non est aliud quam corpus diversissime, ordinatissimeque motum), huius ideam habemus in nobis independenter a corpore, [...], alterum mundum fecit Deus in nobis sensibusque nostris miris elegantissimisque spectris et phantasmatibus praeditum. Et hic venustior est longe et magis artificiosus, plus sapientiae et bonitatis in illo spirat quam in alio isto mundo. Huius vero mundi nullam haberemus

which Geulincx can ground the reliability of the unavoidable use of the senses in physical explanations.

The importance of the senses is stated in the *Autologia* as well, in which he states that the succession of thoughts relies on the succession of bodily motions. Indeed, God cannot produce ideas in us without bodies as instrumental causes.[70] This point reveals a Platonic influence, because the body is conceived as the *organon* of the soul. Geulincx's appeal is indeed to Augustine and Paul, and it contains a critique of the immateriality of angels affirmed in the fourth Lateran Council (1215):

> non puto Deum posse successionem cogitationum in mentibus efficere nisi illas alliget ad corpora. Unde etiam Augustinus, ut salvaret successionem cogitationum in mentibus angelicis, dixit eos habere tenuia corpuscula, forte aeria et similia. Et forte sic est, neque enim contrarium est in Scriptura revelatum. Imo saepe de apparitionibus angelorum sub specie corporea in illa legimus. Certum est (quicquid hac de re sit), ecclesiam non posse temere hoc reiicere; Augustinus enim post Paulum optimus doctor Ecclesiae fuit et omnia eius ex intimis verae philosophiae penetralibus hausta videntur, tam mirabiliter consentiunt nobiscum. Scio tamen pontificios id reicisse. (Geulincx 1891–1893, vol. II, *Annotata ad Metaphysicam*, p. 282; see Denzinger/Schönmetzer 1991, § 800)

The relevance of bodily motions for the production and succession of ideas matches Geulincx's emphasis on the passivity of the soul, which depends on God (as the cause of ideas) and on bodies (as their occasions). Bodies in motion and the senses seem to be the only instruments by which God can cause thoughts in the human mind.

These statements seem to contradict the possibility of attaining purely intellectual ideas, according to a Cartesian theory of knowledge. However, another

---

cognitionem, nisi sensibus et corpore instructi essemus, atque id est, quod hic dicitur, nos de hoc mundo partibusque eius nisi per sensum nihil rescire posse. Hoc, inquam, de posteriori mundo intelligendum. Priorem autem mundum Deum voluit esse occasionem posterioris: voluit enim priorem illum mundum motu suo imprimere nobis diversas illas apparentias, imagines, phaenomena, phasmata (et quibuscunque tandem id libet nominibus exprimere), in quibus essentia posterioris illius mundi consummatur," Geulincx 1891–1893, vol. II, *Annotata ad Metaphysicam*, pp. 288–289.

70 "Sed Deus potest successionem causare in nostris cogitationibus sine motu corporum: tempus ergo potest esse sine motu. Resp. merito praesumimus Deum id non posse, ipse enim unus idemque, eodemque modo se habet. Necessum ergo est, ut instrumento diversimode affecto utatur, si diversos in nobis cogitandi modos suscitare certum habeat [...], atqui nullum est aliud instrumentum quod diversimode se habere potest, quam corpus," Geulincx 1891–1893, vol. II, *Metaphysica vera*, p. 177; "ineffabili illa operatione, quae per corpus et motum ([...] inepta et bruta instrumenta) cogitationes in nobis diversissimas excitat," Geulincx 1891–1893, vol. II, *Metaphysica vera*, p. 188.

point justifies their acknowledgment. Some sort of Platonism can also be found, indeed, in Geulincx's assimilation of human minds to God, as souls are modes of the unique divine mind. This assimilation leads to an illuminationist theory of knowledge. In this way, the appeal to God also guarantees the reliability of intellectual evidence, which is not an autonomous criterion of truth, as noted at the beginning of the *Metaphysica*. This appeal is the only means to grant the reliability of our faculties: even of intellect, whose faithfulness could otherwise be granted only through its primacy over the other faculties, as it decides about the truth of sense data (see above, on the notion of evidence). In fact, all thoughts come only from God: men cannot cause any idea, for they are not conscious of the ways they could do so. Therefore, besides innate ideas in the human mind Geulincx writes about the contemplation of pure ideas and eternal truths in the divine intellect:

> Ideae omnes et veritates aeternae, ut e.g. duo et tria sunt quinque, etc., sunt in mente divina, non in nostra, cum itaque nos consideramus ideas istas, consideramus eas in Deo. (Geulincx 1891–1893, vol. II, *Annotata ad Metaphysicam*, p. 287)

In this case, Geulincx follows the traditional, Averroistic strategy to guarantee the universality of knowledge by recognising its objects in the Divine intellect. As a consequence, he considers single minds as belonging to God, just as particular bodies are *modi* of extension.[71] For there are no means to understand how we perceive innate ideas, as they do not come from bodies, so it is necessary to explain their perception through God himself. Actually, Geulincx's solution is similar to that of Leibniz, Gottfried Wilhelm von Leibniz, according to which it is experience that awakens our innate ideas, or bodily movements are just the first step in our knowledge of them, and this is consistent with the illuminationist theory of perception of ideas.

This position, moreover, reflects a Cartesian principle of economy regarding substances: just as Descartes reduces individual substances to one unique matter, Geulincx reduces souls to God. He does not consider, anyway, the theological consequences of this point, nor the principle of the individuation of souls, as Spinoza does. In fact, Geulincx does not consider our presence in the divine mind in a wider perspective, as Spinoza will do according to the *adequatio humanae mentis et Dei*.

---

71 "Sumus enim modi mentis, ut corpora particularia sunt modi corporis [...]; si auferas modum, remanet ipse Deus," Geulincx 1891–1893, vol. II, *Annotata ad Metaphysicam*, p. 273. See also pp. 237–240, 269, 293.

4.5.1 The hierarchy of knowledge

In light of these remarks, can we regard physics as *scientia*? According to Geulincx's axiom, indeed, if we are not aware of the ways by which we attain knowledge we do not know anything. It seems, therefore, that even through the pure ideas guaranteed by God we cannot achieve any knowledge, as we do not 'cause' them. This problem arises from Geulincx's statements on the impossibility of knowing something without 'clothing' it in mental categories, on our being limited by a knowledge which does not go beyond the senses and modes of consideration. These statements are to be found in his outline of four degrees of knowledge:

> sapientia [...] est profunda aliqua penentransque cognitio rei coniuncta cum summa animi delectatione. Hanc sapientiam nemo habet in summo gradu circa rem aliquam, nisi qui rem illam effecerit, et in efficiendo intime possiderit. [...] Nota, varias esse perceptiones quae sapientiam non pertineant, ut imprimis est perceptio sensuum, quae minime rem ipsam attingit, sed tantum illa nobis, in quantum homines sumus, quid commodi vel incommodi afferre possit, demonstrat. Secundo cognitio certa etiam, sed rem non penetrans, seu sine evidentia, seu sine claritate (ut cognitio qua videmus Deum nos homines fecisse; etiamsi enim certa sit, cum tamen modum ignoremus et utique ignorare cogamur, obscura est, inevidens, et rem non penetrans). Tertio etiam scientia seu cognitio cum evidentia, sed quae haeret in cortice et rem non penetrat [...] v.g scientia qua cognoscimus et scimus res, prout substant operationibus intellectui nostri, seu modi illis ac externis denominationibus, quas ab intellectu nostro eiusque operationibus mutuantur [...]. Tandem est scientia illa, quae rem nude et abstractam ab omnibus modi cogitationum nostrarum denominationibusque proponit. Haec vero sapientia, quam nemo videntur habere nisi qui rem illum effecerit. (Geulincx 1891–1893, vol. II, *Metaphysica vera*, pp. 192–193)[72]

The lowest kind of knowledge is sense perception, which is neither certain nor evident. Then there is a knowledge provided with certainty, but not with evidence; it characterises some truths of metaphysics, such as that we have been created by God. We know with certainty but cannot understand how we have been created. The third knowledge is *scientia*, or the evident knowledge coming, however, through *modi considerandi*. The fourth knowledge is *sapientia*, or the knowledge proper to God and everyone who causes what he knows.

In his *Annotata* Geulincx clarifies these points, writing that the first two kinds of knowledge are neither *doctrina* nor *sapientia*, which are, respectively,

---

72 The fourfold distinction of knowledge, like the consideration of the last two as true, is analogous to Spinoza's theory of science as developed in the *Tractatus brevis*, and then turned into a threefold distinction in the *Ethica*. See Mignini 1984.

only the third and the fourth ones (Geulincx 1891–1893, vol. II, *Annotata ad Metaphysicam*, p. 291). What the latter have in common is their concern with ideas, and thus the real essences of things.[73] In fact, they are *scientiae*. *Sapientia* is the immediate knowledge of ideas without any *modum considerandi*. *Doctrina*, on the contrary, still concerns ideas but as they are 'clothed' by *modi*. Anyway, the acquaintance with things through these modes seems to involve some ignorance:

> non debemus res considerare prout sunt sensibiles [...] neque ut sunt intellegibiles [...]. Sed ut sunt in se, non possumus eas considerare; unde videmus magnam nostram imperfectionem. Hoc unum igitur restat nobis faciendum [...], ut iudicio mentis, quotiescunque rem aliquam sub modo aliquo cogitationis nostrae apprehendimus. (Geulincx 1891–1893, vol. II, *Annotata ad Metaphysicam*, pp. 300–301)

Plainly, only God has *sapientia* as He does not know things by abstraction, i.e., by *modi considerandi*, but immediately.[74] Man is concerned with abstraction and consideration, which are the same process, as I have pointed out.[75] They involve the meta-concepts of whole and part, which are necessary for the individuation of bodies or their apprehension *simul et semel* (Geulincx 1891–1893, vol. II, *Metaphysica ad mentem peripateticam*, p. 227). This mental process characterises the senses but also the intellect and reveals the problems of individuation raised by the abolition of substantial forms as the basic individuals.[76]

---

73 "Doctrina vero et sapientia ad ideas referuntur. Proprie tamen sapientia huc tantum spectat, nam doctrina versatur adhuc in considerationibus nostris. Unde sapientiam recte definies: cognitionem per ideam, seu cognitionem qua aliquid cognoscitur in idea sua," Geulincx 1891–1893, vol. II, *Annotata ad Metaphysicam*, pp. 291–292.

74 "Cum id quod ad praecisionem, abstractionem limitationemque pertinet, a nobis removerimus, clarissime Deum ipsum in nobis agnoscimus," Geulincx 1891–1893, vol. II, *Metaphysica ad mentem peripateticam*, p. 239.

75 See *supra*, on the difference between *cogitatio* and *consideratio*.

76 "Cum enim ratio totius duo involvat ad intellectum spectantia, nempe simul-sumptionem plurium aliquorum, et exclusionem aliorum ab eorum numero quae simul sumpta sunt (quae simul-sumptio et exclusio sunt modi cogitandi ad intellectum nostrum pertinentes), sensus quidem, qui aliter afficitur a mensa v. g. quam ab aëre ac pavimento adiacente, et speciem illam obiecto adscribit, exemplo suo quodammodo praebet intellectui, ut simul-sumptionem illam asserum atque palorum ex quibus mensam constare dicimus, simulque exclusionem aëris et pavimenti circumstantium, adscribat ipsi mensae. Mensam enim putamus ut tale totum extra nos exsistere in rerum natura, quod minime sic est, cum res quidem atque res sint extra nos, sed simul sumptae abstractaeque ab aliis seu aliorum numero (sub qua ratione tantum totum esse possunt) in rerum natura non sunt, sed hoc habent a modo cogitandi nostro," Geulincx 1891–1893, vol. II, *Metaphysica ad mentem peripateticam*, p. 211.

It is in the light of this that we should consider what is the epistemological status of physics and ethics. Actually, the highest, most certain knowledge seems to be found in metaphysics as it concerns the simplest ideas or essences. In his introduction to the *Metaphysica falsa sive ad mentem peripateticam* Geulincx is more open to a possible attainment of a *sapientia* or a knowledge which does not come through 'clothes'. In fact, the *Metaphysica falsa* is devoted to the individuation and criticism of the *modi considerandi* upon which the Scholastics have built their system by considering them to be ideas of actual existing things. It is the case of substances, accidents, relations, subjects, predicates, wholes, parts, that can lead us to regard them as existing things, whereas they are only mental categories (Geulincx 1891–1893, vol. II, *Metaphysica ad mentem peripateticam*, pp. 199–200, 204). This mainly comes from the confusion of logic and metaphysics, criticised in the introductory *Oratio* to the *Quaestiones quodlibeticae*. True metaphysics, on the contrary, recognises *modi considerandi* as mere instruments of thought and considers only the first notions concerning body and mind. Thus it is a *sapientia*, whereas Aristotelian knowledge is a *doctrina*.[77] Man can attain *sapientia* if he considers in a careful way the simplest ideas of the intellect and distinguishes them from their 'clothes'. Besides the limits imposed by Geulincx's axiom (which, strictly speaking, reduces human *sapientia* to the self-consciousness of internal acts), he seems however to admit a perfect knowledge of some metaphysical idea. Whereas for De Raey the human mind cannot think except with *modi considerandi*, Geulincx, who provides a justification of *scientia* by claiming that the highest truths are contemplated in the divine intellect, addresses the possibility for man of reaching a knowledge cleansed from the use of any *modum considerandi*, like God himself. In the *Ethica* metaphysical, introspective truths obtained through an *inspectio sui* are considered to be even more certain than mathematical ones: "quae ex [...] mei ipsius inspectione didici [...] ita perspicue didici, ut ad eam quam apud me [...] certitudinem et evidentiam habent, mathematicorum apodixes aspirare non valeant" (Geulincx 1891–1893, vol. III, *Ethica*, p. 36). Ethics is a rational discipline as it is based on the discovery of the self,

---

77 "Vera sapientia considerat res ut sunt in se, abstractae a modis nostrarum cogitationum, quibus circa illas versari solemus. [...] Res quidem sensibus subiectas vera sapientia abstrahit a speciebus et imaginibus, quae per sensum iis affingi et adscribi solent, easque sic abstractas contemplatur in physica. Res vero quae sub sensum non subiiciuntur, abstrahit vera sapientia atque praecidit a modis cogitandi nostrae intelligentiae, a phasmatibus et speciebus intellectualibus, [...] et res sic [...] considerat vera sapientia in metaphysica [...]. Doctrina autem peripatetica (quae ideo non sapientia est) considerat res quatenus inficiuntur modis nostrarum cogitationum," Geulincx 1891–1893, vol. II, *Metaphysica ad mentem peripateticam*, p. 199.

resulting in the *depectio sui*.[78] Geulincx's interest in ethics, the proper end of the system, leads him to find a ground for a Cartesian moral philosophy in truths provided with the maximum degree of certitude. Ethics, in fact, is the highest science. Physics also has a metaphysical, absolutely certain part concerning the basic ideas explained in the *Somatologia*. However, as it relies also on experience it has only a moral certitude, given by God's benevolence or by a principle coming after the divine will: therefore, it is not *scientia*. Intellectual truths, on the other hand, are granted by their being present and acknowledged in the divine intellect, as evidence is not an autonomous criterion of truth. In fact, whereas sensory experience is grounded only in divine will – and thus it is a matter of faith, as no reasons can explain His decrees – the knowledge of ideas is also given by the divine intellect. They can be truly grasped as they precede the indeterminacy of voluntary principles: plainly, intellectual truths have the strongest foundation.

In conclusion, these points reveal the tension, hidden in Geulincx's system, between the Cartesian demand for a rational ethics and the Christian perspective on human limits. In light of this we can fully understand the distance from De Raey's positions, developed to avoid the improper mixing of philosophy and theology of Meijer and Spinoza. In fact, Geulincx's philosophy is equally distant from De Raey's as from Spinozistic ideas. Geulincx is not concerned with a 'logical' epistemology, because, for the sake of a Christian moral philosophy, he builds his system on the relation between God and man and on his complete dependence on God's inscrutable actions. This approach has its premise in Cartesianism, as Geulincx adopts occasionalism to solve the problem of the interaction of substances in a world deprived of active forms. Moreover, it respects the commitments of Christianity, as it leads to an ethics based on the main virtue of humility. Despite his rationalisation of morality and religion, therefore, Geulincx's positions contradict those of Spinoza on the strengthening of *conatus* as the keystone of ethics. More than the intuitive knowledge of essences it is the second level of knowledge – that concerning the inscrutability of God's actions – that gives us the greatest delight.[79] Geulincx's ethics finds its end in a *docta ignorantia* more than in an intellectualistic way to freedom. This is reflected by an approach according

---

78 The *obligationes*, the rules for conducting life according to this view of man and God, result from the establishment of the first *scientiae* reached through the *inspectio sui:* "procedens igitur iuxta illud propositum, tam evidenter ex mei ipsius inspectione deductum, tam probum, tam legitime fundatum," Geulincx 1891–1893, vol. III, *Ethica*, p. 37.

79 "Nulla potest esse maior delectatio quam cum de Deo nostro aliquid incipimus intelligere [...]. Tunc enim ipsum Deum intuemur aliquo modo, et per aenigma vel in speculo, ut loquitur Apostolus. Unde concludere possumus, quanta futura sit illa delectatio, cum post hanc vitam Deum visuri sumus ut est," Geulincx 1891–1893, vol. II, *Annotata ad Metaphysicam*, p. 292.

to which metaphysics as the highest science implies the knowledge of God's role as the unique warranty of the truth of human knowledge. This perspective is embodied by a theological metaphysics and by a foundation of philosophical knowledge stating that man is not independent in the recognition of the world's features but relies on divine actions and ideas.

4.6 Physics de-metaphysicised

Providing a foundation for philosophy as a reflection and justification of its assumptions and method is the first step in Geulincx's construction of a philosophical system. Even if he admitted the crucial role of experience in physics before his adherence to Cartesianism, his development of a philosophical ethics through rational theology led him, during the Leiden years, to systematise his views on the method for physics. In accordance with his tenet of the arbitrariness of physical principles, Geulincx can claim that in physics explanatory principles are to be formulated only through a rational reinterpretation of experience. Accordingly, besides showing how a rational ethics can be provided with a foundation and besides showing what are the relations of such a foundation to the method used in physics, the case of Geulincx reveals two issues underlying early modern Dutch philosophy. First, Geulincx puts at stake the problem of the reliability of evidence, either empirical or intellectual. Given the misleading 'evidence' of some inference and the recourse to sensory experience in natural philosophy, evidence is questioned as characterised by a 'psychological' certainty. Secondly, it testifies to a 'de-metaphysicalisation' of physics: if metaphysics provides only the basic ontology to physics and explanatory models are formulated by a rational interpretation of experience, physical models cannot be drawn from metaphysical truths. In fact, such de-metaphysicalisation had already been explored – together with the problem of evidence – by Regius: however, in the case of Geulincx it results from a comprehensive foundational theory, whereas for Regius it was a consequence of the rejection of metaphysics and rational theology as such. This process will be noticeable also in De Raey's foundationalism – the subject of the next chapter – and by the next generations of Dutch natural philosophers, represented, above all, by Burchard de Volder and Willem J. 's Gravesande.

5 Foundationalism confronting radical Cartesianism around 1670

5.1 The 'misuse' and 'corruption' of Cartesianism

Thus far, we have seen that in the 1650s Cartesianism had gained acceptance in Leiden. However, it was only from 1662, i.e. with Geulincx's *Oratio de removendis parergis et nitore conciliando disciplinis*, that foundationalism officially entered at the university. Besides Geulincx, Cartesian foundationalism was developed, just before his move to Amsterdam, by De Raey, who presided over two disputations *De constitutione logicae* and *De constitutione physicae* in October 1668.[1] Once at the Amsterdam *Atheneum Illustre*, he gave his inaugural oration *De sapientia veterum*,[2] addressing the partition of philosophy as composed by logic or rational philosophy, physics, and (rational) ethics. These texts were not a direct answer to Geulincx; however, with his foundationalism, De Raey confronted misinterpretations of Cartesianism with which Geulincx's philosophy itself would have been associated at the beginning of the eighteenth century. What brought the Cartesian foundation of philosophy to the next step, indeed, was the use of Cartesianism outside the boundaries in which the Cartesian network of De Raey and the Cartesio-Cocceians worked, and the emergence of concurrent worldviews endangering the use of Cartesianism in the university.

De Raey's foundation, in fact, was aimed at discarding interpretations of Descartes's philosophy which he labelled 'corruption' or 'misuse' (*misbruyk*), which appeared from the 1660s.[3] Today, these forms of Cartesianism are labelled under the category 'radical Cartesianism': Tammy Nyden has considered Lambert van Velthuysen, the De la Court brothers and Spinoza as radical Cartesians as they used Cartesian concepts in political theories (Nyden 2007, chapter 2). Similarly, Wijnand Mijnhardt has used this category to describe the philosophical interpretation of the Scriptures of Spinoza, Adriaan Koerbagh and Lodewijk Meijer (Mijnhardt 2003). On the other hand, Tad Schmaltz has labelled Pierre-Sylvain

---

1 De Raey 1668a, De Raey 1668b. Also included in De Raey 1677 (pp. 707–721), repeated in Amsterdam in 1684, and finally printed in the Appendix of his *Cogitata de interpretatione* (De Raey 1692), pp. 596–618.

2 De Raey 1669. Also this text will be appended to De Raey 1677 and De Raey 1692.

3 In 1689 De Raey denounced the 'misuse' of philosophy in a pamphlet written with Ludwig Wolzogen and the Cocceio-Cartesian Gerbrandus van Leeuwen: De Raey/Wolzogen/Van Leeuwen 1689.

https://doi.org/10.1515/9783110569698-006

Régis and Robert Desgabets radical Cartesians as they elaborated some undeveloped aspect of Descartes's thought, such as that of the indefectibility of matter (Schmaltz 2002). Accordingly, 'radical Cartesianism' means today the use of Cartesianism in the political and theological fields, or a peculiar interpretation of Descartes's metaphysics. Yet, such a concept was used also in the seventeenth and eighteenth centuries: on the basis of the separation theses, Dutch Cartesians could reject the application of Cartesian philosophy to biblical interpretation carried out by Lodewijk Mejer, who studied in Leiden from 1654 to 1660, in his *Philosophia Sacrae Scripturae interpres* (1666). In 1668 Lambert van Velthuysen – a member of the Dutch Cartesian network along with De Raey, Andreae, Wittich, and Heidanus, whilst not himself a professor – argued in his *Dissertatio de usu rationis in rebus theologicis* that Meijer applied Descartes's criterion of clarity and distinction to matters where the principles of rationality do not pertain, such as the articles of faith. In the same year, this defensive strategy was adopted by Ludwig Wolzogen (professor of ecclesiastical history in Utrecht and Amsterdam, and part of the Cartesian network)[4] in his *De Scripturarum interprete* (1668). In 1669, Heidanus went further with this strategy, as he published an *Advijs* to the theological faculty of Leiden, rejecting the idea that Meijer's *Interpres* was drawn from Cartesian principles, as this text was written by a rogue. Similarly, the Cartesian reactions to Spinoza – whose *Cogitata metaphysica* appeared in 1663 and his *Tractatus theologico-politicus* in 1670 – were not aiming at showing his improper uses of Descartes's philosophy: rather, at using Cartesian arguments against Spinoza's determinism and at showing that the thought of Spinoza was truly independent from Descartes's, as in Van Velthuysen's *Tractatus de cultu naturali* and *Tractatus de articulis fidei fundamentalibus* (1680) (Van Bunge 2001, pp. 97–100, 111–113; see also Krop 1999). As noted by Wiep van Bunge, while Meijer was labelled a radical Cartesian even by Dutch Cartesians, Spinoza was not associated with Descartes's thought in Cartesian circles (Van Bunge 2001, p. 121). In fact, these polemics brought about a shared definition of the misuse of Cartesianism both as the misapplication of Descartes's method and as the rejection of his metaphysics. A first account of such a definition has been provided by Henri Krop, who has reconstructed the critiques of the Franeker Cartesian Ruardus Andala (1665–1727) to Willem Deurhoff, Pontian van Hattem, Frederik

---

4 Both Van Velthuysen and Wolzogen were, broadly speaking, part of the Dutch Cartesian network; more specifically, they were members of the informal *College des Savans* in Utrecht, active in the mid-1650s and led by Van Velthuysen. It included the theologian Frans Burman, Johannes de Bruyn (professor of natural philosophy, and defending Descartes's metaphysics in his *Defensio doctrinae cartesianae de dubitatione et dubitandi modo*, (De Bruyn 1670)), the professor of history Johannes Graevius. The circle had connections with Heidanus and Wittich. See Hartog 1876.

van Leenhof, and Arnold Geulincx himself as embracing some form of Spinozism. By applying the geometrical method to all the sciences and denying that experience is a source of knowledge and that particulars truly exist, they deprived words of their usual meaning, thus endangering the practical uses of language. They could be seen as pseudo-Cartesians, pretending to adhere to Descartes's thought but actually adopting "a paradoxical metaphysics caused by a neglect of experience connected with a concept of substance that leads to naturalism and to a rationalism with respect to religion and the Bible" (Krop 1996, pp. 63–65). Before Andala, such rejection of Descartes's metaphysics and the use of philosophy in theology was noted by De Raey, who in his *Cogitata de interpretatione*, published in Amsterdam in 1692, offers a retrospective view of Cartesianism and connects two main 'extremes' in philosophy: the rejection of Descartes's metaphysics, carried out by 'bad men' (*mali*) partly inspired by Hobbes's philosophy, and the application of Cartesian principles to practical disciplines, the endeavour of the 'good men' (*boni*), both leading to the impossibility of using a meaningful vocabulary:

> [...] ut hic iterum vitanda duo extrema sint, in quorum unum vel alterum deflectunt non pauci Cartesii sectatores. Siquidem mali, suo proprio vel Hobbesii errore seducti, prima philosophiae quam Cartesius tradidit, fundamenta evertunt, destruuntque communem inter homines sermonem. Boni philosophiae istius principiis sive fundamentis propriis aliena superstruunt, atque intellectum humani sermonis abstractum, quem admittunt, ut et illum nudum et simplicem, qui est proprius philosophiae, in communem vitam, in alias artes, et disciplinas, ipsamque theologiam intrudunt, quantum audent et possunt. (De Raey 1692, *Cogitata*, p. 215)

The identity of our *mali* and *boni* can be unveiled through the texts published as an appendix of the *Cogitata de interpretatione*. In a letter to Wittich written in 1680, De Raey provides an outline of the development of his positions on philosophy and practical knowledge. According to this letter, his thought was developed in five stages: (1) the acknowledgment of the difference between the Cartesian and Scholastic philosophy at the time of the *Dissertatio de cognitione vulgari et philosophica*, (2) the acknowledgment of the uselessness of Aristotelian and Ramist logic in the early 1650s, (3) the study of the iatrochemistry of Franciscus Sylvius in the late 1650s (mixing natural philosophy with medicine), (4) the controversy over the philosophy of Meijer and Spinoza in the 1660s (who is also referred to in the *Cogitata*),[5] and (5) the deepening of his epistemology during

---

5 In the *Praefatio* and *Notae* to the main text of his *Cogitata* De Raey mentions two occasions on which Cartesianism had been misused: one approximately fifty years before 1692 – thus, in the early polemics over Cartesianism, such as those involving Regius – the other, a bit more than twenty years before, around 1670: De Raey 1692, *Cogitata*, p. I (*Praefatio*, unnumbered)

his final years at Amsterdam (De Raey 1692, *Ad Wittichium epistola*, pp. 657–660). Moreover, De Raey mentions the improper mix of Cartesian and Aristotelian methodologies in the *Logica vetus et nova* of Clauberg, initially taught by De Raey himself as a replacement of Scholastic logic, and the application of the philosophical standard to the higher arts by Andreae (De Raey 1692, *Ad Wittichium epistola*, pp. 655, 658–659), who claimed that the Scriptures may be interpreted by means of philosophy in his *Assertio* (Andreae 1653–1654, p. 57). In a second letter, addressed to an anonymous theologian involved in the polemics at the University of Franeker[6] and dated 1687, De Raey blames the use of philosophy in theology by Wolzogen and Hermann Alexander Röell (professor of philosophy and theology at Franeker from 1686) who claimed that the truth of those biblical statements concerning the sun and the earth are to be interpreted by philosophy.[7] Yet, for De Raey the misuse of Cartesianism is more dramatically embodied by the rejection of Descartes's metaphysics. In the same letter De Raey uses as his main polemical target Henricus Regius, considered to be the first misuser of Cartesian philosophy and forerunner of Spinoza,[8] and admits his admiration for Gysbertus Voetius and Jacob Revius as they foresaw the radical consequences of Cartesian philosophy, what Revius, criticising Andreae, called 'Cartesiomania'.[9]

De Raey's map of Dutch Cartesianism, covering 40 years of intellectual history, takes into account both the misapplication and the misinterpretation or corruption of Descartes's philosophy, considered as two kindred errors and leading to the failure of linguistic communication among men. Against these consequences, De Raey first developed a Cartesian metaphysics vindicating the very basis of philosophy in late 1660s. Accordingly, the works which appeared in this and in the preceding decade are essential in understanding why foundationalism, in the hands of De Raey, entered into a new phase, in which metaphysics

---

and p. 338. Moreover, he explicitly refers to theology and law, and mentions the same years: De Raey 1692, *Cogitata*, *Praefatio*, p. IV (unnumbered). In 1685 De Raey attacked those aiming to apply geometry to every discipline in a disputation dedicated to the critic of Spinoza, Willem van Blijenbergh: see De Raey 1685, and Van Miert 2009, pp. 271–272.

6 Verbeek identifies him as Melchior Leidekker: Verbeek 1995, n. 146.

7 De Raey 1692, *Epistola ad theologum*, pp. 664–665. On Röell and the polemics over Cartesianism in Franeker in 1680s, see Bordoli 2009.

8 "Dixi et inculcavi ab initio muneris academici, [...] facilius cum Voetio quam cum Regio redibimus in gratiam. Quam verus in eo vates fuerim, experientia coepit longo tempore docere. Regius in corrumpenda philosophia antecessor fuit, Spinozae etc. a quorum ille erroribus infandis alienus non erat," De Raey 1692, *Epistola ad theologum*, p. 666.

9 For this reason, De Raey was labelled as 'voetianus' and attacked by some young scholars in Amsterdam: see De Raey 1692, *Epistola ad theologum*, pp. 663, 666–667. De Raey refers to Revius 1654, Revius 1655.

and logic became more and more detached from physics and assumed a meta-philosophical role, that is to say, assumed a reflective function over scientific practices. As indicated by De Raey, the seeds of the corruption of Cartesianism were already planted by Regius. Following his account, one can ascertain the emergence of different approaches to the problems opened up by Descartes, and so of different worldviews concurring with Cartesianism itself. The foremost cases had been those of Meijer and Spinoza, whose links with Dutch Cartesianism have been thoroughly explored in recent literature (Bordoli 1997, Douglas 2015): in fact, for De Raey their approaches were the most extreme consequences of attitudes he found at the core of Cartesianism – viz. in Regius, Andreae, Wolzogen, and later in Röell – and in the dissemination, in the Netherlands, of the ideas of Hobbes, whose materialism he links to Regius's. Indeed, it was precisely in 1668 that Hobbes's *De corpore* and *Leviathan* were published in Amsterdam.[10] Also, as De Raey refers to Hobbes's followers without mentioning them (De Raey 1692, *Cogitata*, p. 215, quoted *supra)*, he could also address Samuel Sorbière, who arranged for the publication of Hobbes's *De cive* in Amsterdam by Elzeviers in 1647 and by Blaeu in 1649 (as *Eléments philosophiques du citoyen)*,[11] and was also responsible for the publication in Amsterdam, in 1644, of Gassendi's *Disquisitio metaphysica*, where Gassendi maintains a materialist standpoint on the notion of soul, which has a corpuscular nature (Gassendi 1644, pp. 294–298; see Wilson 2008, pp. 122–124). Moreover, Sorbière was in contact with Regius himself: as shown by Vlad Alexandrescu, Sorbière likely influenced Regius on his positions on the nature of soul and on the decidability of metaphysical questions by rational means alone (Alexandrescu 2013). Moreover, De Raey may also have been reacting to Van Velthuysen, who published a defence of Hobbes's *De cive* (Van Velthuysen 1651a) and provided a combination of Descartes's and Hobbes's philosophy in the disputation *De finito et infinito* (Van Velthuysen 1651b)[12] as well as 'political' Hobbesians such as the De la Court brothers and Abraham van Berkel (the translator of Leviathan into Dutch), as he would note, in his *Cogitata*, the 'replacement' of meaning in words such as good or bad, such that what is

---

10 Hobbes's *De corpore* was first published in London in 1655: a second Latin edition appeared only in the *Opera philosophica* published by Blaeu in Amsterdam in 1668, including also the first Latin translation of *Leviathan* (republished in 1670). *Leviathan* was translated into Dutch by Abraham van Berkel and published in Amsterdam in 1667 and again in 1672. For full bibliographic details, see Schoneveld 1983, pp. 29–46, Van Velthuysen 2013, pp. 13–15.

11 Hobbes 1647 (1st ed.: Hobbes 1642); Hobbes 1649. Further Latin editions of *De cive* were published in Amsterdam by Elzeviers in 1657 and 1669, and followed by a Dutch translation in 1675: Hobbes 1657, Hobbes 1669, Hobbes 1675 (see again Schoneveld 1983 and Van Velthuysen 2013).

12 On the reception of Hobbes in the Low Countries, see Petry 1984, pp. 150–170, Sécretan 1987, pp. 27–46.

good by nature is replaced with a conventional good, as it is defined in Hobbes's *Leviathan* and *De cive*.[13]

The dissemination of Hobbes's thought may explain, in part, the development of foundationalism by De Raey, as he would counter the erosion of Descartes's metaphysics – and the application of philosophy to politics. The other, main intrusion of the philosophical standard in a practical field had been the medical thought of Franciscus Sylvius, which would have a long-standing influence on the Leiden scientific environment, and who assumed a chair in medicine in Leiden in 1658. His ideas have various points in common with Regius's. In his inaugural oration *De hominis cognitione*, he maintains the sensory origin of every idea, on which the mind works and forms 'second' notions such as those of genre and species, as well as notions common to other senses (as those of numbers, figures, and so on), and exercises ratiocination (Sylvius 1658, pp. 11–12). Only soul is perceived by means of pure intellect, insofar as it cannot be ascertained by any sense, and is for this reason regarded as spiritual. In fact, Sylvius maintains that soul consists of a soft corpuscle when we are newly born.[14] God, on the other hand, is conceived by Sylvius as a first principle, since mind cannot grasp an infinite series of things. On this basis, he distinguishes between *scientia*, which is the knowledge of sensible objects, and *intelligentia*, concerning insensible entities.[15] This theory of knowledge and metaphysics, actually, hardly fits with Descartes's. Moreover, although distinguishing between theoretical and practical disciplines, which are respectively aimed at truth and at utility (Sylvius 1658, pp. 18–19), Sylvius does not set a difference between the method of natural philosophy and medicine, as they are both to be based on reason and experience, namely on the solid and

---

13 "[...] *bonum, malum, honestum, turpe, more, atque voluntate hominum*, substituas, ut certe faciunt his temporibus plurimi, pro eo quod *est natura bonum, malum*," De Raey 1692, *Cogitata*, pp. 210–211; see Hobbes 1651, *De homine*, p. 5, Hobbes 1647, *Praefatio ad lectorem*.

14 "Nec mox ab ortu ingenio valeamus, iudicioque. Nam haec quoniam in caeteris infantibus observamus imbecilla, eadem similiter in nobis infirma fuisse opinamur. Nec quid primum, quidve secundum, et sic deinceps mox a nativitate nobis obvenerit cognoscendum, quave methodo in singulorum cognitione progressa fuerit in tenero corpusculo anima nostra satis constet," Sylvius 1658, p. 10.

15 "Atqui hoc ipsum est, quod animam vocamus, et mentem, utique partem hominis primariam. Quae cum in externos non incurrat sensus, insensibilis convincitur, soloque adeo intellectu perceptibilis, quam et spiritualem dicimus. Sed nec haeret homo quinpotius et suae, et caeterorum hominum existentiae initium cum vita observans, finemque in morte expectans, nec rerum finitarum in infinitum deductionem possibilem animadvertens, sistitur tandem in primo aliquo, adeoque aeterno rerum omnium itidem insensibili, sed et infinito principio et authore Deo, uno, vero, bono, a quo creaturae omnes et esse suum habent, et bene esse. Quemadmodum autem insensibilium obiectorum cognitio, scientia dicitur, sic insensibilium et spiritualium notitia vocatur intelligentia," Sylvius 1658, p. 14.

well connected *scientia* of natural things and the accurate history of the human body.[16] So, even if Sylvius retains some notions common to Descartes, such as the theory of blood circulation as the foundation of physiology (Sylvius 1658, p. 22; for a deeper discussion, see Schmaltz 2016a, pp. 262–267, Schmaltz 2016b), he rejects the adherence to any 'sect' in natural philosophy.[17]

In short, the map offered by De Raey aims to show how the 'corruption' of Cartesianism, which commenced with Regius, soon developed into spurious forms of knowledge, which turned to be were detrimental not only for the introduction of Cartesianism into the university, but also to civil peace,[18] and, above all, to the very possibility of the communication between men, as they deprive words of their usual meanings through the introduction of a 'category mistake'. In the following decades, indeed, De Raey would develop a comprehensive reflection on language aimed both at criticising the misusers and corruptors of Cartesianism, as well as at clarifying its conceptual basis in the light of his separation thesis. The first consequence of the radical Cartesianism disclosed by De Raey is the adoption of a materialist ontology entailed by the rejection of Descartes's metaphysics by Regius and Hobbes. Besides being philosophically untenable, materialism cannot account for our linguistic practices and makes everyday speech senseless, as one has to use a terminology often signifying sensory data and mere concepts rather than real modifications of bodily substance. However, this is not only the result of the rejection of Descartes's metaphysics but also the consequence of the application of a philosophical standard to practical disciplines, such that to comply with such a standard one should avoid referring to sensory qualities or beings of reason as logical categories. Both the misinterpretation and misapplication of Cartesianism, therefore, result in a corruption of speech, to

---

16 "Omnia veterum, recentiorumque, de rebus naturalibus et medicis scripta, ut et omnia a se ipsis perquisita, excogitata et observata ad veritatis trutinam rationem et experientiam revocantes et pensantes, solidam, concatenatamque construant naturalium rerum scientiam. Accuratam adornent humani corporis fabricae historiam. Describant exacte sanitatis et aegritudinem naturam et causas. Subiugant denique cito, tuto et iucunde ipsis medendi artem numeris omnibus absolutam," Sylvius 1658, p. 27.

17 "Quamvis rursus in naturalium rerum cognitione eruenda laborarint et olim et nunc quam plurimi subtilissimi philosophi, nem tamen ipsorum, quod sciam, principia sibi probata probavit unquam aliis ita, ut illorum cogeret assensum, atque illos demonstrationis suae evidentia in suam pelliceret sententiam. Quam ob rem etiam novae indies exurgunt et enascuntur physicorum sectae, plerisque alienas opiniones solicite magis et solide destruentibus et infirmantibus, quam proprias adstruentibus, confirmantibusve," Sylvius 1658, p. 24.

18 "Error pugnans cum veritate quam defendimus, infinitae dissensionis atque confusionis causa esse, atque pacem publicam turbare debeat," De Raey 1692, *Cogitata, Praefatio*, p. 11, note.

remedy which De Raey would aim his analysis of language in the 1670s-1680s.[19] Such analysis (on which I will focus later in this book), constitutes a late development of Cartesianism and a clear example of a descriptive and reflective approach to the conceptual apparatus of *scientiae* and *artes* at the end of the seventeenth century. Before its full-blown emergence, however, such analysis was preceded by a process of narrowing the purposes of foundationalism and the scope of *scientia*: from the 'Scholastic' foundational apparatus of Clauberg and the architectonic structure of philosophy of Geulincx, to a more essential metaphysical foundation aimed only at ensuring the scientific status of physics and at reflecting on actual 'scientific' practices: either in natural philosophy or in medicine. The first step of such 'simplification' in foundationalism is provided in De Raey's texts of the late 1660s, first of all, his *De constitutione logicae*. This text testifies to a further change in conceiving the relations between logic and metaphysics, which are unified as they both have a foundational function, and whose problems are dealt with by De Raey in the light of the pre-Cartesian logical and metaphysical tradition.

5.2 De Raey's foundation of scientific knowledge: Logic as metaphysics

As De Raey declares in the *De constitutione logicae*, the fundamental difference between practical and philosophical disciplines depends on the difficulty of applying Descartes's method to medicine, law, and theology, whose objects are complex and impossible to grasp with clarity and distinction, being thus beyond the capacities of mental faculties. For this reason, practical knowledge ascertains the observable connections among phenomena and is based on experience, opinion and authority, as well as on imagination and witnesses (De Raey 1692, *De constitutione logicae*, pp. 600, 605–606), whereas natural philosophy concerns intelligible causes[20] and deals with certain and evident knowledge,

---

19 "Adeo rarum et difficile est sobrie et modeste philosophari, intra certos se terminos continere, scientiarum fines vocabulorumque definitas significationes loco non movere, atque ulterius non provehere, neque etiam magis in arctum cogere quam id recta ratio atque usus in humana vita permittit. [...] Sicut his quoque temporibus fere inutilis et plena periculi suo insigni abusu facta est magni usus philosophia, quam ab autore cartesianam appellant, cuius fines dum conantur sine fine extendere, novis additamentis fundamenta bene posita evertunt atque nae intelligendo faciunt tandem, ut nihil intelligant," De Raey 1692, *Cogitata*, pp. 208–209.
20 "In artibus invenitur causarum cognitio et effectuum per causas [...] nititur observatione connexionis sive coniunctionis, quae ab una parte causa, ab alia effectus notionem atque nominationem parit. *Id quo tangente fit quo separato cessat effectus, rei causam nominamus,* in medicina

that is, *scientia*.[21] Moreover, the arts concern sensible bodies and have a practical purpose (De Raey 1692, *De constitutione physicae*, pp. 608–609). Therefore, in such fields one has to study phenomena *relate ad nos* instead of grasping their 'objective' nature,[22] in accordance with the separation thesis. Medicine, for instance, has to be based on a natural history cleared of the main errors of Scholastic physics, but still based on the use of the senses, while law and theology may still be based on Scholastic moral philosophy and metaphysics.[23] As a place for such reflection, De Raey's logic may thus be considered a meta-science, providing the other disciplines both with a justification of their status as well as a prescription of their methods and aims. Indeed, for De Raey logic consists, first of all, in the four Cartesian rules of method: these can be easily used in mathematics, whose objects are simple.[24] However, in addition to these four rules a *scientia logica* is needed in order to apply them to physics, a field obscured by

inquit Galenus. Haec solum notio causae ad communem sensum accomodata in omni arte supponitur. Ut necesse non sit artis exercitationem et propriam cognitionem quod attinet, distincte er clare intelligere, qua virtute agat causa et effectum producat, sicut in scientia physica id diximus necessarium esse. [...] Quamquam illa cognitio multum possit in vera scientia prodesse, quae cognitio intellectualis per causam est, in omnibus valde diversa ab ea cognitione quam a sensu habemus, in multis, ut videtur, etiam contraria. Ex quo perspicue sequitur artes omnes (quarum genitrix debet esse partim nostra propria, partim aliorum experientia) medicinam, agriculturam, fabrilem etc. separatas esse a philosophica scientia natura sua sive secundum naturam cognitionis humanae (quod notandum) ut non possint unquam ex natura sua pars quaedam physicae sive philosophicae naturalis scientiae esse," De Raey 1692, *De constitutione physicae*, pp. 616–617; see Galen, *De locis affectis*, I, 2 (Galen 1821–1833, vol. VIII, p. 32).

21 "Physica scientia dicitur, quatenus certa et evidens per naturae lumen notitia est, sive per causam et demonstrationem, sive alio quocunque modo comparata," De Raey 1692, *De constitutione physicae*, p. 609.

22 De Raey 1692, *De constitutione logicae*, p. 596. This thesis is also maintained in his *De cognitione vulgari*: see De Raey 1654, *Dissertatio de subsidiis, gradibus ac vitiis notitiae vulgaris*, p. 24.

23 De Raey 1692, *De constitutione logicae*, pp. 605–606. On De Raey's emendation of Aristotelian physics, in order to make it a natural history capable to lead medicine to consistent progresses, see Strazzoni 2012, pp. 262–264.

24 "Logica philosophiae propria, quam etiam philosophicam vocamus, imprimis quatuor potest regulis comprehendi. Quae quo breviores et pauciores, eo magis ad rectum, quem philosophia quaerit, rationis usum accomodatae sunt. Et praeter regulas has, quarum exercitatio artem parit, adhuc logicam scientiam requirimus. Regulas quod attinet, eas ispsismet authoris verbis tradimus, ex Dissertatione de methodo. [...] Quae totidem verae ac perfectissimae scientiae conditiones sunt. Deinde hae adeo breves atque tam paucae regulae imprimis insignem ac facilem usum habent in mathematicis disciplinis, quatenus circa res simplicissimas ac cognitu facillimas versantur, atque verae ac proprie dictae scientiae sunt, quarum finis in contemplatione veritatis consistit. Neque alia logica [...] in mathematicis scientiis opus est, et in ipsis quoque primus et maxime facilis harum regularum usus est," De Raey 1692, *De constitutione logicae*, pp. 598–599. On De Raey's theory of logic and language, see Strazzoni 2015, Del Prete forthcoming.

prejudices which need to be wiped out by logic itself.[25] Thus, logic is meant to be a science and a way to science at once ("scientia" and "modus sciendi") and the leading part of philosophy, "imperans ac praescribens obiectum suum, suo modo domina et architectonica": that is, a *philosophia prima*. Hence, it is to be defined as metaphysics itself, and is paired with Plato's dialectics, as it considers the first causes and principles and does not work but by immediate intuition, whereas even mathematics relies on suppositions and long chains of deductions (De Raey 1692, *De constitutione logicae*, pp. 601–603, 605). In fact, De Raey's logic i.e. metaphysics is the outcome of a process of rethinking the objects of these disciplines, which was prepared by the logical tradition prior to Descartes, and finally prompted by the Cartesian revolution in philosophy.

5.2.1 The intersections of logic and metaphysics in early modern philosophy

As already mentioned, the main logical theory in vogue during De Raey's studies in Utrecht and Leiden was provided in the *Institutiones logicae* (1626) of Burgersdijk, written by order of the States of Holland after the Synod of Dordt called for a reform of studies. Burgersdijk's main task was to provide a revision of Keckermann's *Systema logicae* (Keckermann 1600, Keckermann 1613a) and to make it more understandable by younger students (Van Rijen 1993). De Raey, indeed, comments upon Burgersdijk's *Institutiones logicae* through their *Synopsis* (Burgersdijk 1645),[26] in his *Specimen logicae interpretationis*, namely, a series of disputations he held in Amsterdam from 1669 to 1671, in which he provides his first reflections on language.[27] Moreover, in the *Specimen* he deals with the logic of Petrus Ramus, which

---

25 "Denique philosophiam quod attinet, et physicam imprimis quae praecipua philosophiae pars est, in ea hae regulae maiorem difficultatem habent, quam in mathematicis scientiis. Quia circa ea versatur physica ac tota philosophia, quae instar eorum quae mathematici tractant, simplicia et cognitu facilia non sunt, sed composita et difficilia. Quorum notitia idcirco, non videtur posse certitudinem et evidentiam habere, quae in mathematicis demonstrationibus est. [...] Haec difficultas ut superetur, quantum potest superari, praeter regulas logicas, logicam scientiam requirimus de principiis cognitionis humanae, quae prima scientia, in philosophia summe necessaria sit," De Raey 1692, *De constitutione logicae*, pp. 600–601. Later on he adds: "ut autem sit vera scientia, quantum potest, non sufficere regulas logicas, sed logicam diximus scientiam requiri. Atque eam ostendimus Platonis dialecticam esse, quae seposito sensu, sublata suppositione, et omissa fide, ad primas simplicissimasque veritates adscendit. Atque hae demum verae suppositiones in physica sunt, secundum quas facienda ratiocinatio est, ut vera scientia sit," p. 606.
26 Later commented in Heereboord 1650.
27 The full title sounds as *Specimen logicae interpretationis Amstelaedami 1669, 1670, 1671, octo comprehensum disputationibus, quae paulo post occasionem dederunt primis de interpretatione*

was a main subject of De Raey's pre-academic education.[28] Ramus, Keckermann and Burgersdijk had different views on the function and the relations of logic and metaphysics. In his *Dialecticae institutiones* (1543) Ramus treats logic as dialectic or *ars disserendi* (the art of discoursing) and reverses the traditional structure of logic by considering discovery (*inventio*) of the matters of reasoning (*loci*) as the first part of logic, for which he postpones the treatment of the formal organisation of judgments and scientific syllogisms (Ong 1958, pp. 182–183). Ramus's revisiting of logic goes along with a rejection of Aristotle's metaphysics. In *Scholae in liberales artes* (1569) he claims that Aristotle mixed logic and metaphysics, as in the fourteen books of *Metaphysica* Aristotle treated logical notions such as cause, opposition, comparison, genre and species, whilst he claimed, in various places in his logical and metaphysical books, that metaphysics is about first causes and beings and is not useful in learning and teaching. According to Ramus, Aristotle's mixing of metaphysics and logic was a result of the emulation of Plato, whose *dialectica*, dealing with notions common to every discipline, was considered by Aristotle and by modern Platonists as a metaphysics.[29] As a solution to Aristotle's misplacement, in his *Dialecticae institutiones* Ramus proposes a replacement of Aristotelian logic with his dialectic. Still, according to the first and second editions of this work (1543)[30] dialectic encompasses some sort of theology as it helps in finding the ends of arts and the Creator of all things in a 'third judgment', which pairs with Plato's dialectic.[31] Moreover, Ramus would not develop any metaphysics as an independent discipline: rather, his dialectic fulfils the role of a metaphysics as *sophia*, as it

---

disputationibus, anno 1673 et aliquot sequentibus, in De Raey 1692, pp. 535–596. For bibliographic details on the disputations on language mentioned by De Raey – on which he based his *Cogitata* – see Van Miert 2009, pp. 242–245, 377–380, 383–386, 389. Both the disputations on logic and on language were attacked in Amsterdam by the teacher of medicine Gerard Blasius, by means of some disputations held by the physician Van Lamzweerde (see Van Lamzweerde 1674, pp. 213–311).

28 See his letter to Wittich of 1680: "florem adolescentiae contriveram in studio logico, nec poenitet vel poenituit unquam ad haec tempora usque. Didiceram in scholis dialecticam Petri Rami, quod in hunc diem usque singulari soleo deputare foelicitati," De Raey 1692, *Ad Wittichium epistola*, p. 658. See Verbeek 2001.

29 Ramus 1569, *Praefatio*, Nn-Nn2. On Ramus's criticism to Aristotle's metaphysics, see Leinsle 1985, vol. I, pp. 21–30, Pozzo 2001, Frank 2012. On the relations of logic and metaphysics in the German context, see Pozzo 2004.

30 First edition as *Dialecticae partitiones* (Ramus 1543a), second as *Dialecticae institutiones* (Ramus 1543b).

31 Ramus 1543b, fol. 35r; see Ong 1958, pp. 189–190, Bruyère 1984, pp. 262–264, Goulding 2010, pp. 22–23.

concerns the rules of knowledge but also common essences and first causes.[32] It is clear therefore, that Ramus had been a forerunner of De Raey with regard to the objects of logic and metaphysics, although Ramus does not assign a foundational role to his dialectic as De Raey would do.

The unification of logic and metaphysics as figured out by Ramus underwent criticism by Keckermann, who, while considering in his *Compendium systematis metaphysici* (1609) logic and metaphysics as dealing with some common objects, such as substances and accidents as *entes primarii*, states in his *Systema logicae* that these are more properly dealt with in metaphysics, as logic considers only second intentions or concepts of concepts: i.e., instruments of knowledge rather than notions representing things (Keckermann 1613b, p. 60; Keckermann 1613a, p. 80; see Keckermann 1611, pp. 18–19). This conception of logic had been defended by Rodolphus Agricola and Julius Caesar Scaliger; moreover, in the *Problemata logica* of Rudolph Goclenius, himself deeply influenced by Ramus (as he defined logic as *ars disserendi* consisting of *inventio* and *dispositio*), but rejecting the idea that the notions dealt with by logic have real references in the world, as admitted by Ramus.[33] On the other hand, for Keckermann, metaphysics deals with *ens qua ens* and with its kinds (such as substance and accident), properties (truth, goodness, unity), and orders (as possibility and necessity). Yet, God is not dealt with by metaphysics, as He is above being itself (Keckermann 1611, pp. 17–23, 29–30, 66–69). Therefore, Keckermann can claim that Ramus improperly mixed logic and metaphysics, insofar as he dealt with the notions of truth, goodness, finiteness, and even God (as the universal cause) in his logic, inasmuch as these are common subjects and adjuncts of beings (Keckermann 1613b, pp. 27–28; see Hotson 2007, pp. 146–150).

Both Ramus's and Keckermann's ideas on logic are discussed by Burgersdijk, who sanctioned the existence of three schools in logic: the Aristotelian, which set the basis of all logic; the Ramist, which had a too narrow conception of logic, and Keckermann's, which combined Aristotelian logic and Ramist dialectic (Burgersdijk 1660, *Praefatio*, pp. III–X (unnumbered)). In proposing his own synthesis, Burgersdijk defines logic as the art by which the instruments for knowing things are developed. In fact, it can only imprecisely be labelled as *dialectica* or *ars disserendi*, since this is the task of the part of logic dealt with in Aristotle's *Topica*. Logic thus concerns *themata* or everything which can be grasped by mind, as well as words since these signify *themata* themselves (Burgersdijk 1660,

---

32 Ramus 1569, pp. 838, 864; Ramus 1543c, p. 18v. This approach was also adopted by Melanchthon: see Pozzo 2001, pp. 92–95.

33 Goclenius 1597, part I, problem 9, p. 60. De Raey briefly mentions Goclenius in his *Specimen*, as he maintained, like Ramus, that logic is an *ars disserendi*: De Raey 1692, *Specimen*, p. 541.

pp. 2, 10). This 'thematisation' of logic, as shown by Riccardo Pozzo, had begun with Agricola and Melanchthon. Still, Melanchthon maintained the real reference of Aristotle's categories to reality, as these enable the discerning of the *ordo rerum* and the different sciences: in this manner, he could replace metaphysics with logic, as logic is the means to treat things themselves. Building upon the 'thematisation' of logic, Keckermann made it a *scientia directiva*: not aimed at dealing with 'thematised' entities, but rather at preparing the mind to deal with any *thema* (Pozzo 2002, pp. 5–13). Eventually, Burgersdijk could divide logic into a *logica thematica* and *logica organica*, and maintain that the *themata* dealt with in logic are second notions (Burgersdijk 1660, pp. 5–6). For Burgersdijk, logic deals with first notions only accidentally and without scrutiny, contrary to metaphysics. Following Aristotle's tripartition of the theoretical sciences, for Burgersdijk metaphysics is the theoretical discipline concerning those things which cannot be dealt with in physics and mathematics, as: (1) immaterial and incorporeal substances: God, angels, demons, souls; (2) The general nature and species of accidents; (3) All the attributes of corporeal, incorporeal, infinite, finite substances and their accidents. Accordingly, metaphysics is about the notion of *ens* as the most common attribute of all that exists, and *ens* as it is immaterial, dealt with in general and special metaphysics respectively. As it deals with *ens qua ens*, metaphysics is the first discipline according to the *ordo naturae* but the last according to the *ordo cognitionis* (Burgersdijk 1640, pp. 3–4, 9).

De Raey's unification of logic and metaphysics and his interest in the ontology entailed by ordinary language are the result of his Cartesian interpretation of the objects and functions of such disciplines. First of all, De Raey reads Ramus's logic as an amelioration of Aristotle's and as an art devoted to the organisation of reasoning as this is expressed in language, and separated from 'true philosophy'.[34] For De Raey, Ramus's *loci* or *argumenta* are relations that mind figures between things themselves, that is, 'modi considerandi' or second notions used in everyday speech whose use in philosophy is allowed only through a prior analysis of the things to which they are applied (De Raey 1692, *Specimen*, pp. 538–540). Similarly, Burgersdijk's *themata* – as categories and every universal concepts – are all labelled as relations put upon things or as universal concepts which do not mean anything but themselves.[35] Building upon the Ramist definition of logic

---

34 De Raey 1692, *Specimen*, pp. 537, 540. In his letter to Wittich, De Raey distinguishes between vulgar logic, embodied by Ramus's, and Cartesian logic, offered in the four rules of the method and applied in the *Meditationes* and in the first part of *Principia*: De Raey 1692, *Ad Wittichium epistola*, p. 659.

35 De Raey 1692, *Specimen*, p. 543. De Raey assumes a moderate nominalist standpoint on universals, as he criticises the theory of universals expounded by Scaliger, who saw the foundation

of Goclenius, De Raey labels the whole logic of Burgersdijk as *ars disserendi*, as 'logica' means mental and uttered discourse.[36] The target of De Raey, rather than the particular uses of logical concepts, is the use of logic as a discipline which concerns only *themata*, being thus useless for philosophical knowledge as it is not aimed at the knowledge of things in themselves. His critique, however, is not merely a statement that old logic deals with second intentions; rather, he maintains that such notions are the results of a reckless use of experience and abstraction. Moreover, not only logical concepts are mere mental contents: it is also the case with all metaphysical notions as these result from the same kind of abstractive activity of mind (De Raey 1692, *Specimen*, pp. 545–546, 578–579). As he points out in another text which appeared in these years, his *De Aristotele et aristotelicis* (1669), the 'vulgar' logic pairs with metaphysics as they both concern the notions drawn from experience (*modi sentiendi*) and the mere ways to formulate and express concepts (*modi disserendi, predicandi* and *considerandi*), based on sense data (De Raey 1692, *De Aristotele et aristotelicis*, pp. 470–471, 484), i.e., on the *intellectum sibi permissum* described by Bacon and corrupting the whole philosophy: including metaphysics, physics, ethics, and politics, all based on logical categories (De Raey 1692, *Cogitata*, pp. 8–9, 15). Accordingly, De Raey considers all the metaphysical concepts – starting with *ens* – the result of such childish, linguistic generalisation and abstraction with no foundation *in re* (De Raey 1692, *Specimen*, p. 536 (quoting Bacon, *Novum organum*, aphorism 97), 566–567). In fact, De Raey's Cartesian metaphysics does not concern substance, duration, number considered in their abstract meaning, i.e., apart from any consideration of the actual entities these are to be applied to, but it takes into account things: namely, body, mind, and their actual modifications, of which universal concepts are predicated.[37] As De Raey assumes a Cartesian point of view on the sources of knowledge and on the ontology of mind and body, he can

---

of the predication of general concepts in the nature of things, rather than in our abstractive capacities: De Raey 1692, *Specimen*, p. 553; see Scaliger 1557, pp. 963–965. In his *De Aristotele et aristotelicis* (1669, in De Raey 1677 and De Raey 1692) De Raey distinguishes universals *ante multa*, roughly corresponding to Descartes's eternal truths, and *post multa*, or universal notions provided by abstraction from particulars: De Raey 1692, *De Aristotele et aristotelicis*, pp. 474–475.

36 "*Unde dicta est logica? A voce λόγος, quae tum rationem, tum orationem significat* [...] *est-que adeo ars rationis, non in se spectatae, sed ut oratione explicata est.* [...] *quare, ut pulchre* Goclenius Prob. log. parte I qu. VI, 'si id quod prius est, et fontem ipsum respicias, naturamque et essentiam logicae, rationalis ars est [...]'. Et qu. IV 'finem dialecticae recto usu rationis humanae, eoque universo ad bene disserendum definio'," De Raey 1692, *Specimen*, p. 541; see Burgersdijk 1649, p. 7, Goclenius 1597, part I, problem 4 and 6: p. 37 and 45.

37 See his *Pro vera metaphysica, quae de principiis humanae cognitionis tractat*, in De Raey 1677, pp. 424–425.

overlook the distinction of logic and metaphysics operated by Burgersdijk and Keckermann and emphasise the derivation of Scholastic metaphysics from logic. Also, having learned Ramus's dialectic in his pre-university education, and through the mediation of Goclenius, De Raey could interpret Ramus's dialectic as a logic without any metaphysical commitment. Still, De Raey came to consider, like Ramus, logic as dialectic or first philosophy, assuming Plato as his main fore-runner, as he conceived *dialectica* as *ars disserendi* but also as *ars intelligendi*, *rationis scientia*, and rational philosophy, which works by intellect alone.[38] In this way, he could ultimately reduce the 're-duplication' of metaphysics by Johannes Clauberg, also criticised for having improperly combined Cartesian and Aristotelian notions and methodologies in his logic, taught by De Raey himself in the early 1650s (De Raey 1692, *Ad Wittichium epistola*, pp. 658–659).[39]

5.3 The developments of De Raey's logic

After his *De constitutione logicae*, which is, according to its name, a program-matic text, De Raey develops his metaphysics in a more consistent manner in his *Pro vera metaphysica, quae de principiis humanae cognitionis tractat*, included in the second edition of his *Clavis* (1677). In this text he deals with the very con-tents and proceeding of logic or metaphysics. According to him, it consists of two parts. The first concerns the foundation of human knowledge through Descartes's metaphysics, that is, the demonstration of the reliability of evidence as the cri-terion of truth, and of the existence of God. The second is a careful analysis of our concepts, *summa rerum genera*, which was not fully carried out by Descartes and which is the main issue for De Raey. The two parts of logic are naturally con-nected, since the analysis of concepts is carried out from the point of view of Descartes's metaphysics, which is completed by De Raey through a full analysis of the contents of the mind, i.e., of the meanings of everyday and philosophi-cal terminology. According to De Raey, therefore, logic first has to study what is found in the intellect, such as the idea of God and other innate notions, to be considered as the first principles and causes. Hence, it serves to analyse all the concepts we deal with, and to justify their reliability.

---

38 Plato is recurrently mentioned as the source of De Raey's metaphysics in the *De sapientia veterum* (1666), *De Aristotele et aristotelicis*, *De constitutione logicae*, *Pro vera metaphysica*, *Cogitata* and in his letter to an anonymous theologian.

39 The same judgment could have likely been extended to the Port-Royal logic: according to Lamzweerde, Arnauld's logic was a main source of De Raey: see Lamzweerde 1674, p. 231.

The first part of De Raey's foundation begins with the Cartesian path of the *cogito*, proceeding by doubt, up until the demonstration of the existence of God as the guarantee for the truth of our knowledge. As it was for Clauberg, doubt enables a *repurgatio intellectus* of prejudices, anticipations, sensory notions and doubts, and allows us to do philosophy in an orderly way.[40] After doubt has prepared our mind, we can start to analyse its contents: first of all, the mind is aware of itself and of its being a thinking thing, whereas it is not immediately conscious of the existence of the body: in this way, the distinction of mind and body, and of mind and other incorporeal entities is demonstrated. In strict accordance with Descartes's metaphysics, thus, the mind discovers in itself the presence of ideas, and, from the idea of God, one can demonstrate His existence according to *a priori* and *a posteriori* proofs, to which the third proof, based on our very existence, which had to be provided by something different from ourselves, is added (De Raey 1677, *Pro vera metaphysica*, pp. 413–416). Eventually, the demonstration of the existence of God and the consequent acknowledgment of His attributes as the perfect being allow De Raey to ground the truth of clear and distinct knowledge on a Cartesian, metaphysical basis, since God is defined as *dator luminis* and the source of all knowledge,[41] or the cause of whatever clear and distinct ideas we can find in our perceptions (De Raey 1677, *Pro vera metaphysica*, pp. 421–422). Given this first, metaphysical basis of the truth of our knowledge, one needs, in accordance with the program set forth by De Raey in his *De constitutione logicae*, to analyse the very contents of our mind in order to distinguish obscure from clear notions, this being the main task of the *scientia logica*, i.e., of the *vera metaphysica*.[42]

---

40 "Praemittit primum praeparationem quandam humani intellectus, ut tam sublimis cognitionis capax sit. [...] Consistit ea praeparatio in dubitatione, quae quasi mortificatio quaedam veteris hominis est, aut si mavis corruptae rationis repurgatio. Cuius immensa, quoad veritatis contemplationem, est utilitas, et tanta necessitas, ut sine ea nulla esse possit vera philosophia. [...] Imprimis enim ab omnibus praeiudiciis nos liberat, atque adeo non solum praecipitantiam, verum etiam anticipationem in iudiciis nostris vitare [...] docet. [...] Deinde via facillimam sternit ad mentem a sensibus abducendam, quod tam ad rerum materialium ac sensibilium, quam ad mentis ac Dei cognitionem, et universim ad omnem rerum scientiam necessarium est: quatenus evidentia et certitudo in scientiis non tam a sensu quam ab intellectu pendet. [...] Efficit praeterea, ut possimus ordine philosophari," De Raey 1677, *Pro vera metaphysica*, pp. 412–413.
41 "Sic ordine philosophando, advertimus imprimis, nos existere, quatenus sumus naturae cogitantis, et simul etiam, et esse Deum, et nos ab illo pendere. Unde porro sequitur, ex eius attributorum considerationis, caeterarum rerum veritatem posse indagari, quatenus ille est ipsarum causa. Ut ita scientiam perfectissimam quae est effectuum per causas, acquiramus," De Raey 1677, *Pro vera metaphysica*, p. 417.
42 "Sic ergo satis non erit novisse, id omne quod clare et distincte percipitur, a quocunque demum percipiatur, verum esse: sed opera danda est, ut ea dignoscere possimus, quae revera clare percipiuntur, ab iis quae clare percipi tantum putantur. Quod non alia via fieri potest, quam

In his *Pro vera metaphysica*, De Raey performs a first analysis of the notions philosophers deal with. He distinguishes between the notions of *res*, which can exist outside the mind, and *veritates*, that is, propositions which cannot exist but in our mind, even if they express principles that are to be used in order to understand external reality itself.[43] De Raey rejects the Scholastic way of proceeding in metaphysics, adopted by Clauberg in his *Ontosophia*, as this consists of a consideration of the notion of being from its most abstract to particular, concrete notions.[44] The new metaphysics, as seen above, does not concern the ideas of substance, duration, and number apart from any consideration of the entities these are to be applied to, but it concerns things, namely, body, mind, and their modifications.[45] Also, the new metaphysics considers *veritates*, that is, the very principles, common notions, or axioms (as the Cartesian *praecognita*) to be used in philosophy. Having a propositional nature, these notions do not match any specific entity: yet, these *veritates* are to be adopted in order to understand reality itself, either mental or physical, and constitute the principles to be used in the other branches of philosophy (namely, in natural philosophy itself).[46] As a matter of

---

summatim enumerando simplices omnes notiones, ex quibus nostrae cogitationes componuntur, et quid in unaquaque sit clarum, quidve obscurum, sive in quo possimus falli, distinguendo. Quod antehac non fecerunt logici et metaphysici, ut facere debuissent," De Raey 1677, *Pro vera metaphysica*, p. 423.

43 "Quicquid cadit sub cogitationem nostram ad duo genera potest referri. Primum continet res, quae qualemcunque existentiam habent, alterum veritates, quae tantum in nostra cogitatione sunt," De Raey 1677, *Pro vera metaphysica*, p. 424.

44 "Ex iis quae tanquam res, consideramus, maxime generalia sunt, substantia, duratio, ordo, numerus, et si quae alia sint, quae ad omnia rerum genera se extendunt. Quae valde multa et operose tractat vulgaris metaphysica, quatenus pro obiecto assumit ens qua est, in latissima acceptione sua, qua idem est, quod in communi sermone res dicitur," De Raey 1677, *Pro vera metaphysica*, p. 424.

45 "Summa rerum genera, atque adeo particularia illa, quorum distinctas in nobis notiones habemus, duo tantum novimus: unum est intellectualium sive cogitativarum, ut sunt substantiae intelligentes sive cogitantes, una cum proprietatibus et accidentibus, quae referri ad eas debent. Alterum materialium sive extensarum, ut sunt substantiae corporeae, una cum suis proprietatibus et accidentibus. Sic enim intellectus et voluntas, omnesque modi percipiendi [...] et volendi [...] pertinent ad substantiam cogitantem, quae nomine mentis venit. Ad extensam vero, quae dicitur corpus [...] sive ipsamet extensio in longum, latum, et profundum, figura, motus, istius partium et talia," De Raey 1677, *Pro vera metaphysica*, pp. 424–425.

46 "Qui vero cogitat, factum non posse esse infectum, non de ulla re, sed de veritate cogitat, quae mente concipi quidem et ex recta rerum perceptione affirmari potest, non vero existere. Et quia pro diversa rerum inter se collatione, quam infinitis modis facimus, infinitae esse possunt affirmationes et negationes, infinitae veritates sunt, etiam de genere earum, quae vulgo communes notiones et axiomata dicuntur, quia adeo generales et obviae sunt. Quae generalia axiomata idcirco facile recenseri non possunt, sed nec etiam ignorari, quando in particularibus

fact, however, De Raey does not develop – neither in his *Pro vera metaphysica* nor in his *Cogitata* – an analysis of *veritates*. His physics, therefore, cannot be labelled as a 'metaphysical physics' like that of Descartes: in the *De constitutione physicae*, indeed, De Raey points out a difference between physics and metaphysics, as the former concerns secondary causes, while the latter focuses on God as the first cause.[47] For physics, to be a *scientia* does not mean that its truths are deduced from metaphysics, but only that it is characterised by *evidentia* and certainty as the criteria of truth, being thus acquired by a demonstration *ex causis* or by an immediate intuition of concepts. The epistemology devised by De Raey in this text is not different from that of his earlier *Clavis*, and is still based on the Aristotelian notion of *intelligentia* of axioms. Dianoetic knowledge is based on such axioms and consists of the deduction of effects from causes, which is the goal of physics as this deduction is the very explanation of phenomena.[48] Even in mentioning the laws of nature as what is more kindred to an end in the natural world – as they are the order according to which everything happens in nature – De Raey does not deduce or explain these in the light of the attributes of God, as he did not in his *Clavis*, where the justification of *praecognita* does not involve rational-theological considerations (see Strazzoni 2011). The detachment of physics and metaphysics by De Raey, therefore, could have been initially motivated by the necessity of teaching Descartes's physics without incurring polemics with theologians and

occurrit occasio, ut de iis cogitemus et praeiudiciis non excaecamur. Ut hac de causa etiam necessaria non possit censeri vulgaris metaphysica, ut multorum opinio est, peculiarem scientiam requiri, quae notiones communes exponat, atque omnium aliarum scientiarum principia demonstret," De Raey 1677, *Pro vera metaphysica*, pp. 427–428.

47 "Sic natura imprimis Deo opponitur, ut causa prima est, et naturale divino, ut est proximus atque immediatus effectus Dei. Atque ea ratione metaphysica et theologia de Deo ceu causa prima, physica de natura et causis secundis tractat. Et effectus quod attinet, qui pariter possunt ad Deum ac naturam referri, quatenus Deus universalis et prima, natura particularis et secunda causa est, eorum etiam alia consideratio debet in physica esse, quam in theologia et metaphysicam. Nam physicus, nisi velit scientiae suae fines transcendere, subsistit in natura," De Raey 1677, *De constitutione physicae*, p. 715.

48 "Porro physica scientia dicitur, quatenus certa et evidens per naturae lumen notitia est, sive per causam et demonstrationem, sive alio quocunque modo comparata. Ut non unus modus sciendi, atque intelligendi est. [...] Scientia quam sic in physica requirimus, imprimis potest unus ac simplex intuitus esse, sive rei unius et simplicis, sive veritatis, quae per se nota est. Ut quando intelligimus, quid materia, quid motus sit, quod ex nihilo nihil fiat in natura et c. Quae Aristoteli intelligentia dicitur, et est notitia principiorum, ut definitiones et axiomata sunt. [...] Plerumque vero scientia physica haud ita unis et simplex, sed compositus et multiplex intellectus est. Quatenus per notionum variam compositionem et connexionem unum ex alio deducimus et ratiocinando conclusimus. Atque ita imprimis possumus effectum per suam causam cognoscere," De Raey 1677, *De constitutione physicae*, pp. 716–717. See also p. 714.

Aristotelians. As his *Clavis* was published again in 1677, however, it is reasonable to assume that De Raey may have had not only extra-philosophical reasons for not mixing physics and metaphysics, i.e. that he truly believed in the possibility of developing a physics (like Regius) without recurring to the notion of God. On a broader level, this detachment led to the development of metaphysics as a reflection on science rather than as one of its parts.

5.3.1 A bifurcation in the academic curriculum

In sum, De Raey sets forth the basics of a new science that includes both Descartes's methodology and metaphysics, to the extent that this concerns the demonstration of the existence of God and of the reliability of clear and distinct perception. Moreover, such a science is aimed at pursuing a Cartesian analysis of all our notions, since these are formed according to a puerile, commonsensical worldview and make the understanding of reality intricate and obscure, especially in physics. This worldview, in fact, is characterised both by the errors coming from the use of the senses in philosophy, and by those characterising the functioning of intellect itself, that is, the consideration of *modi considerandi* as something existing outside the mind. These errors, ultimately, are reflected and increased by the use of language.[49] De Raey thus distinguishes between two kinds of logic and metaphysics: the 'vulgar' logic and metaphysics of the Aristotelians, which concern only the notions drawn from experience (*modi sentiendi*) and the mere ways to organise and express concepts (*modi disserendi* and

---

49 "Mirum non est, quod tanta [...] sequatur repugnantia in cogitationibus nostris, ut quod unice reale ac positivum et res subsistens est in rebus corporeis, pro nihilo habeatur, et contra illud nihil dicatur esse aliquid, longum, latum, profundum et c. [...] Atque hac de causa tota scholarum philosophia, saltem in physicis, una et perpetua sine fine disputatio et contradictio est, quam fovet vulgaris metaphysica, per eam simplicium notionum confusionem, multiplicationem et eversionem, quam primo ex sensu oriri, perspicuum fecimus. [...] Consequens est, ut paucis detegamus alteram simplicium notionum confusionem, multiplicatione et eversionem, quae ad intellectum referenda est. [...] Imprimis notari velim, intellectum haud ita primo et per se huius mali causam esse, per ipsasmet primitivas ideas suas, ut sensus per inanes et fallaces species, quibus res aliter quam sunt percipimus. Sed multiplex modus considerandi, cui imprimis per sensum assuescimus, huius mali causa in intellectu est: quatenus etiam per intellectum ad res ipsas referimus, quod non in rebus, sed tantum in nostra cogitatione est, uti id facere soliti fuimus per sensum. Quod maxime sit ex usus sermonis, quo tam considerandi et sentiendi modos, quam realia attributa praedicamus de subiectis suis. Quia id usus vitae exigit," De Raey 1677, *Pro vera metaphysica*, pp. 436–437.

modi considerandi),[50] whereas the new logic concerns concepts of things as they are, either mental or physical. The old logic grounds medicine, theology, and law, as these are based, according to the *Praefatio* to his *Cogitata*, on Aristotelian physics, metaphysics, and ethics respectively.[51] Thus far, De Raey's foundation of philosophy is a rethinking of the academic curriculum. According to him, one can gain clear and distinct knowledge (*scientia*) in metaphysics, mathematics, and physics. Metaphysics or logic – that is, first philosophy – works by intuition and has the highest degree of certainty, whereas mathematics works by chains of deductions and is subject to error to a greater extent, and physics concerns complex concepts, which need to be established and clarified by metaphysics itself. De Raey follows the traditional classification of *scientiae*: however, mathematics and physics may be as certain as metaphysics once they are provided with a foundation itself. To such disciplines – first philosophy and physics – De Raey added a further branch of philosophy, whilst not developing it. In his *Dissertatio de sapientia veterum*, the text of which had been read for his appointment as professor of philosophy at the *Athenaeum Illustre* of Amsterdam in 1669, and was later published in his 1677 *Clavis*, and in the *Cogitata*, De Raey mentions moral philosophy as one of the three parts of philosophy, along with physics and rational or first philosophy. Such rational ethics is described – with stoic overtones – as relying on the use of intellect alone, and teaching how to avoid the fear of death, the duties of man, and the depreciation of pleasure and pain.[52] This ethics is opposed to the common morals of men, as these pursue utility, and is supported by the *ius gentium* and *ius civilis*, based on the use of the senses,

---

50 See the previous note, and De Raey's *De Aristotele et aristotelicis*: "verum loco logicae istius posse ac debere aliam esse, quae [...] philosophiae propria, atque adeo prima pars eius est, et commune instrumentum. [...] Quod neque de metaphysica, neque de logica Aristotelis dici potest, quatenus in metaphysica generales et nimios abstractos modos considerandi, in logica modum disserendi ac disputandi tradit," De Raey 1692, *De Aristotele et aristotelicis*, pp. 470–471. See also his *De constitutione logicae*: De Raey 1692, pp. 597–598.

51 De Raey 1692, *Cogitata, Praefatio*, pp. XI–XII (unnumbered).

52 "Philosophia moralis ars bene beateque vivendi existens secundum verum intellectum, pariter docet nos in vita et moribus bonum a malo, virtutem a vitio secundum veritatem distinguere. Quatenus animus se ipsum cognoscens atque in se virtutis rationem intelligens, primum sui, hinc aliarum rerum potest pretium aestimare. Quo sit ut non amplius totus corpori inserviens, atque in eo sensibus et affectibus suis, etiam non sinat se seduci ab iis: sed rationem velit ac intellectus sequi. Quae prima, si non unica, virtus est, sub se omnes alias complectens. Ita fit, ut quisque apud se scientiam habens cum bona conscientia, inter alios viri boni officium praestare, apud se imprimis voluptatem et dolorem contemnere, hinc mortis metum effugere, et sic porro alia omnia quae cadere in hominem praeter voluntatem possunt," De Raey 1692, *De sapientia veterum*, pp. 378–379.

authority, and opinions instead of clear and distinct ideas.[53] De Raey does not, however, extend his consideration to such ethics.[54] He outlines a plan of philosophy consisting of logic or metaphysics, physics, and ethics, independent of the disciplines based on vulgar knowledge: these are the amended natural history, on which medicine must be based, ethics, which is the basis for politics, and 'vulgar' metaphysics, leading to theology. Yet, the practical end of rational ethics, insofar as it is prescriptive rather than descriptive or explanatory, seems to be at odds with such a dichotomy in the academic disciplines. If De Raey's projected plan of two parallel classes – practical or vulgar, and theoretical or philosophical – of academic disciplines was ultimately unfeasible, he was, in any case, one of the first philosophers who purportedly set a new function of logic and metaphysics. What led to the full development of metaphysics into a meta-science, however, would have been the emergence of experimentalism in Leiden, which led to the decline of Cartesianism in natural philosophy. The following parts of this book are devoted to such a final evolution of foundationalism.

---

53 "In iure gentium omnes homine, in iure civili populum, aut civitatem, una dici potest ratio movere, quae non tam veritas et verus intellectus, quo iuris originem et veram causam novimus, quam utilitas, opinio, et saepe vis et autoritas est ac potestas imperantis," De Raey 1692, *De sapientia veterum*, p. 381. Moreover, in his *Cogitata* he mentions ethics as part of philosophy, on which politics is based. In this case, he seems rather to refer to the 'vulgar' ethics: see De Raey 1692, *Cogitata*, pp. 8–9.

54 In his *Cogitata* he merely considers the case of the substitution of good according to nature with good according to men as the meaning of 'bonum', see *supra*, section 5.1.

6 Bridging *scientia* and experience: the last evolution of Cartesian foundationalism

6.1 Late Cartesianism in Leiden and Amsterdam

The assumption of a chair in philosophy at the *Atheneum illustre* of Amsterdam in 1669 by De Raey, and the death of Geulincx in the same year, did not bring to an end the teaching of Cartesian philosophy in Leiden, which was assumed in 1670 by Burchard de Volder (1643–1709). He can be considered as having introduced a more empirically oriented form of Cartesianism, paving the way for the upcoming Newtonianism of Willem Jacob 's Gravesande and Pieter van Musschenbroek.[1] In the same year, a former student of Regius in Utrecht, Theodoor Craanen (1633–1688), was appointed as professor of logic and metaphysics, but in 1673, discontent with his initial assignment, he assumed the chair of medicine, which he held until 1686, when he moved to Germany as physician to the Elector of Brandenburg. Craanen pursued Descartes's project of providing a mechanisation of the human body and explained the functioning of the body (the 'oeconomia animalis') by means of the notions of pores, particles, and fermentation.[2] Moreover, in 1669 Wolferd Senguerd – son of the Aristotelian professor Arnold – was allowed to give public lectures on logic, metaphysics, and practical philosophy on the condition that he would not expound novel theories.[3] Yet, in his *Philosophia naturalis* (1680, 1685) he would incorporate Cartesian notions in his overall eclectic philosophy, mainly taught by means of public experiments, also described in his *Inquisitiones experimentales* (1699). Their teaching thus brought about the appearance in the university of the figure of a natural philosopher more akin to that of a scientist emerging with the Scientific Revolution than to 'professional' philosophers concerned with metaphysical, logical, and moral problems. Yet, such *personae* were not yet fully separated: on the contrary, the progress made possible by the application of an experimental and mathematical approach to natural philosophy – as by Galileo, Huygens, Boyle, and many others – brought about new academic discussions over the method and conceptual apparatus of the natural sciences, which fostered the use of metaphysics and logic as a philosophy of science. Moreover, philosophical reflections on the progress in scientific fields came to be intertwined with the long-standing debates over the use

---

1 On him, see Klever 1988, Wiesenfeldt 2002, Lodge 2005, Nyden 2013, Nyden 2014. Biographical information is mainly provided in Le Clerc 1709, Gronovius 1709.

2 On Craanen, see Luyendijk-Elshout 1975.

3 On Senguerd, see Wiesenfeldt 2002, De Pater 1975.

https://doi.org/10.1515/9783110569698-007

of Cartesianism as the philosophy of the university. In the last quarter of the seventeenth century, further polemics took place in Dutch academies, namely the quarrel on the over-representation of Cartesianism in Leiden, involving Heidanus, Wittich and De Volder in 1674–1676; that on the use of Cartesianism in theology taking place in Franeker in 1686–1687, brought about by Hermann Röell, and, in 1689, the publication of the *Censura philosophiae cartesianae* of Pierre-Daniel Huet. All these factors are to be considered as shaping a final evolution of Cartesian foundationalism, on the eve of the emergence of Newtonian science. First, this evolution and the events shaping it are noticeable in the works of De Volder, who started his career as a student of Sylvius, and ended it as an admirer of Newton: in fact, whereas Craanen and Senguerd did not provide extensive reflection or a foundation for their scientific theories, De Volder devoted a large part of his works to assessing the metaphysical premises of the investigation of nature, which for him is a combination of Cartesian principles and an experimental-mathematical methodology. Secondly, the appearance of the last work of De Raey, his *Cogitata de interpretatione* (1692), although connected to his early polemics against the misuses of Cartesianism, is to be traced back to the concurrence of scientific innovations and longstanding debates over Cartesianism, as he provides a full catalogue of the concepts one may use in the different domains of knowledge, and assesses the conditions of the right use of sensory and abstract concepts in natural philosophy, namely, those he excluded in his earlier works.

6.2 Burchard de Volder's 'Cartesian empiricism'

Born in Amsterdam in 1643, De Volder studied philosophy from 1657 at the *Athenaeum illustre*. Subsequently, he matriculated at the University of Utrecht, graduating in 1660 as *magister artium*, and then in Leiden, where he graduated in medicine with the dissertation *De natura* (1664), dedicated to Sylvius. After some years spent in Amsterdam as 'physician to the poor', he came back to Leiden University as professor of philosophy in 1670. There, he was actively involved in the defence of Cartesian philosophy in 1674–1676. The quarrel concerned the refusal by the Aristotelian philosopher Gerard de Vries to assume a permanent chair in Leiden 1674, as he was discontent with the hostile environment he found at the university, as his lectures were often disrupted by Cartesian students. To this appointment, he preferred a position at Utrecht, which was less permeated by Cartesianism.[4]

---

4 Geulincx's pupil Cornelis Bontekoe, at that time a student in Leiden, was among those responsible for disturbances against the Aristotelian professors: see Van Ruler 2003a.

As a consequence of this event, the University Curators issued – two decades after the polemics between Heidanus and the Voetians (1656) – a list of 21 propositions not to be touched upon in academic teaching, including Cartesian metaphysics, the positions of Meijer and Wittich on the language of the Bible, and Cocceius's theology.[5] Moreover, in the resolution they remarked that Descartes's metaphysics should not be taught in the university, restating the prohibition of 1654.[6] To this resolution, Heidanus, Wittich, and De Volder reacted with the publication of the *Consideratien, over eenige saecken onlanghs voorgevallen in de Universiteyt binnen Leyden* (1676), commenting upon each of the propositions condemned by the Curators, and which they equated with the condemnation of Aristotelian propositions at Paris in 1210–1277. Their defence was aimed at supporting the Cartesio-Cocceian views in theology, and the consistency of Descartes's metaphysics with orthodoxy. This book, however, prompted the Curators to force Heidanus (who had his name printed on the book), to leave the university in the same year.[7] This was not the only intervention of De Volder in defence of Cartesian metaphysics, and of the role of Cartesianism as an academic philosophy achieved through years of polemics. Two decades later, indeed, he would provide an extensive rebuttal of Huet's *Censura philosophiae cartesianae*, in his *Exercitationes academicae quibus Renati Cartesii philosophia defenditur*, following the arguments of Huet and focusing on metaphysics (De Volder 1690–1693, De Volder 1695). Meanwhile, De Volder deeply involved himself in the study of different approaches to natural philosophy. In the early 1670s, indeed, he was in contact with Robert Boyle and other fellows of the Royal Society, whom he met during a journey to England in 1674. Impressed by

---

5 Among the condemned propositions, one can find "Scripturam loqui secundum erronea vulgi praeiudicia," "mundum [...] extensione infinitum esse ita ut impossibile sit dari plures mundos," "de omnibus rebus esse dubitandum, etiam de Dei existentia, et ita dubitandum ut habeantur pro falsis," Molhuysen 1913–1924, vol. III, pp. 317–318. See Fix 1999, p. 44, McGahagan 1976, pp. 344–346.

6 "Dat darenboven nogh publice nogh privatim in de voors. Academie sal werden gedoceert de Methaphysica [*sic*] van Renatus Descartes off van die geene, die desselfs opinien souden mogen hebben geamplecteert, nogh uyt deselve nogh uyt eenig gedeelte van dien eenige stellingen, theses ofte questien publice ofte privatim gedisputeert, geventillert ofte verhandelt, alls op pene dat de geen, die sigh hier tegens directelyk off indirectelyk, 't sy in 't publycq off onder de hand, sullen komen te vergrijpen, als wederhorigeende schadelyke leeden ende leeraers van de Universiteyt sonder eenige dissimulatie, verschoninge ofte conniventie, van hare ampten ende bedieningen sullen werden gedeporteert ende de leeden van deselve Universiteyt uyt deselve gerelegeert," Molhuysen 1913–1924, vol. III, p. 618.

7 Heidanus *et al.* 1676. The book had a Latin edition in 1678. See Le Clerc 1709, pp. 355, 368–373, Molhuysen 1913–1924, vol. III, pp. 291–294. Also the publication of the second edition of the *Clavis* by De Raey in 1677, actually, can be seen as a reaction to the resolution: this time, however, De Raey explicitly defends Cartesian metaphysics in the texts published along with his first work.

their experimental practices, De Volder asked for and obtained from the Curators of Leiden University the funding to establish an experimental cabinet (Le Clerc 1709, pp. 362–364, Molhuysen 1913–1924, vol. III, pp. 301–302). Opening in 1675, his *Theatrum physicum* was the first official experimental cabinet in any European university. In this cabinet, provided with instruments from the Musschenbroek workshop,[8] De Volder performed the experiments he found first in Boyle's *New Experiments Physico-Mechanicall, Touching the Spring of the Air and Its Effects* (1660), and then in Jacques Rohault's *Traité de physique* (1671). Mainly concerning pneumatics, the contents of his lectures are collected in his *Quaestiones academicae de aëris gravitate* (1681) and in some handwritten notes of an English student of his.[9] After having been appointed professor of mathematics in 1681, when he gave his *Oratio de coniugendis philosophicis et mathematicis disciplinis* (1682), De Volder became more interested in the application of mathematics to physics, as he carefully studied Newton's *Philosophiae naturalis principia mathematica* (1687) and made Christiaan Huygens acquainted with its contents.[10] At a late stage of his life, moreover, De Volder started a correspondence with Leibniz, Gottfried Wilhelm von Leibniz (between 1698 and 1706), although not accepting his views on metaphysics (Lodge 1998, Lodge 2001, Lodge 2013, Rey 2009a, Rey 2009b, Rey 2016), and with Newton, even though not embracing his physics (Hall 1982). Following the judgment of Jean Le Clerc, who reported De Volder as having become progressively discontented with Cartesian philosophy, some scholars have argued that he rejected Cartesian physics and metaphysics,[11] as his *Oratio de rationis viribus et usu in scientiis* (1698) is supposed to testify. This view has been corrected in recent years by other scholars arguing for a continuity in De Volder's thought, as he had always been open to the role of experience in physics (as a student of Sylvius); moreover, as late as 1695 he provided a defence of Cartesian metaphysics against Huet. Thus, one can recognise some original points in De Volder's Cartesianism, namely, the co-existence of Cartesian metaphysics (usually regarded as the basis of a rationalist or a speculative approach to physics) with the adoption of teaching

---

8 De Pater 1975, De Clercq 1997a, De Clercq 1997b, Wiesenfeldt 2002.

9 De Volder 1676–1678, De Volder 1681a, De Volder 1681c, Vinson 1676–1677. See Le Clerc 1709, p. 398, Dobre 2013b.

10 Le Clerc 1709, pp. 379–380. De Volder corresponded with Huygens: see Huygens 1888–1950, vol. IX, letters 2537, 2547; vol. X, letters 2798, 2799, 2800, 2802, 2803, 2861, 2862, 2701. Also, De Volder was the editor of the posthumous edition of Huygens's Κοσμοθεωρος (Huygens 1703), along with Bernhard Fullenius.

11 Le Clerc, sympathetic to Newtonian ideas, emphasised the empirical attitude of De Volder. According to him, De Volder was disenchanted by Newton concerning Descartes's vortex theory, and was annoyed at having to teach Descartes's *Meditationes* and Rohault's *Traité de physique*: see Le Clerc 1709, pp. 382, 398–399. This interpretation was followed in Ruestow 1973, pp. 89–112.

experimental practices, and the project to integrate experience and mathematics in physics. De Volder's case thus shows how Cartesian foundationalism met the methodological standards of the fellows of the Royal Society, who established what we can label as 'science' in the modern sense of the word and, more generally, the main scientific worldviews at stake from the 1660s to 1690s. In the next sections, I will show how philosophical quarrels and scientific developments in Leiden, as dealt with by De Volder, shaped such foundationalism.

6.2.1 From Descartes to De Volder: Iatrochemistry in Leiden

The first scientific approach De Volder embraced was the iatrochemistry of Sylvius, which is revelatory of the overall experimental approach he would adopt in his career. In his *De natura*, De Volder assumes the notion of *effervescentia* as the general principle capable of explaining the effects ascribed to the nature of the body by physicians.[12] For him, *effervescentia* is the cause of the circulation of the blood as this had been mechanically explained by Descartes and Cornelis van Hogelande; however, they did not distinguish it from the similar process of fermentation, as Sylvius did (De Volder 1664, pp. 5–6). Descartes, in fact, uses the notion of fermentation in his *De homine* and in his correspondence with the Dutch physician Vopiscus Fortunatus Plempius, in order to explain how digestion works: in the liver, particles of food are transformed into blood by a process he illustrates by an analogy with the fermentation of wet hay and wine, and consisting of the expansion and accretion of particles themselves (see Ben-Yami 2015, pp. 76–77; Easton/Gholamnejad 2016). Before Descartes himself, however, it was Regius who, with his *Physiologia* (written with the help of Descartes), published for the first time a Cartesian account of digestion and of fermentation. Regius uses Descartes's mechanical theory of blood circulation to explain *alitura*, the natural action of the body by which its heat and substance are continuously maintained.[13]

---

12 General *effecta* require a general explanatory principle: "naturam non in omnibus [...] mutationibus, sed in primaria, et maxime generali sitam esse. Cum enim generalia sint, quae a natura fieri dicuntur, effecta, et primario ad corporis nostri conservationem faciant, ipsam etiam causam talem esse necesse est," De Volder 1664, p. 4. The criteria in formulating hypotheses on particular phenomena are not addressed by De Volder. This kind of general principle is required to assess what (1) preserves and feeds our body, (2) causes its functions, (3) heals it, (4) excites fevers in order to recover it, (5) makes every medicaments active, 6) accustom itself to medicaments: De Volder 1664, p. 3.

13 "Alitura est actio naturalis, qua perpetuus caloris, substantiaeque corporeae defluxus ope sanguinis, praecipueque arteriosi, a corde in partes alendas impulsi, continuo restauratur," Regius 1641a, p. 17.

Alitura consists of nutrition and vivification, i.e. of the processes of assimilation by the parts of the body of the particles carried in blood circulation, and of the maintaining of spirits in the body by its renovation in the heart (Regius 1641a, pp. 29–30). *Alitura* is subserved by the natural appetites of hunger and thirst, which are dispositions of phantasy leading to eating and drinking, and by *coctio*, the process by which the insensible particles of foods are provided with a conformation adapted to the human body.[14] *Coctio* consists of three processes: the transformation of food into chyle, which takes place in the stomach, and then into chyme in the liver, and blood in the heart (Regius 1641a, p. 19). Descartes, actually, suggested that Regius should not mention three distinct processes, insofar as the three stages of *coctio* are not qualitatively different i.e. they all consist of the modification of particles.[15] Regius still provided a threefold distinction: in the stomach, food is dissolved by the heat of the heart and the action of humours coming from the arteries, and is transformed in chyle; in the liver, chyle is not attracted by any force but only by the pressure of the parts of the body, and is transformed into chyme by fermentation (whose mechanical process is however not explained by Regius),[16] while in the heart it is finally converted into blood (*sanguificatio*) by ebullition, where it is rarefied and causes the heartbeat.[17] So, fermentation was for both Descartes and Regius a mechanical, invisible process analogous to the visible fermentation of wine and hay. Moreover, Regius, under the suggestion of Descartes, does not mention effervescence as part of *coctio*, but refers only to the rarefaction of blood in the heart (see Bos 2002, p. 84). Similarly, Van Hogelande (whom Descartes opposed to Regius as a more faithful follower of his own principles, although he offered different views (Schmaltz 2017a, pp. 255–257), uses the notion of fermentation in his mechanical explanation of blood circulation, in his *Cogitationes, quibus [...] brevis historia oeconomiae corporis animalis, proponitur,*

---

14 "Coctio est adaptio particularum insensibilium ex quibus alimenta constant, ut ea conformationem humano corpori idoneam acquirant," Regius 1641a, p. 18; also in Regius 1640a, thesis 1. This process is commented on by Descartes in his correspondence with Regius: Descartes to Regius, 24 May 1640, AT III, pp. 66–70; also in Bos 2002, pp. 41–48.

15 Descartes to Regius, 24 May 1640, AT III, p. 67; also in Bos 2002, pp. 42–43.

16 "In hepate: cum chylus, primum per infinitos ventriculi et intestinorum poros in venas caeliacas, meseraicas, et lacteas, et ex his deinde in hepar, non aliqua vi attractrice, sed sola sua fluiditate et pressione vicinarum partium, ut diaphragmatis, musculorum abdominis, aliarumque, adiuvante sanguinis in corde ebullitione, delatus, sanguinique reliquo eo confluenti mistus, ibi fermentatur, et, ut chymicorum more loquar, digeritur, ac in chymum abit," Regius 1641a, p. 19. The notion of fermentation is added by Regius: see Bos 2002, p. 46.

17 "In corde fit coctio, cum chymus sanguini a reliquo corpore ad cor redeunti permistus, et simul cum eo in hepate praeparatus, in verum et perfectum sanguinem, per ebullitionem pulsificam, commutatur," Regius 1641a, p. 20.

atque mechanice explicatur (1646): in this account, fermentation takes place in the stomach, and is responsible for the transformation of food into chyle – which is a form of refinement or extraction of its essence – and in the heart, where chyle is rarefied and becomes pure blood (Hogelande 1646, pp. 49–53).

De Volder, on the other hand, embraced the account of Sylvius. In his *Disputationum medicarum decas*, held in Leiden (1659–1663), Sylvius provides an account of blood circulation based on two observable processes: effervescence and fermentation, respectively the composition and the resolution of the parts of matter, which are explained as the action of chemical principles (acid and alkali). Effervescence has two kinds, being either intestinal, and consisting of the action of three humours (alkaline bile, pancreatic acid, and saliva) in the duodenum, and vital, i.e. the effervescence of the blood taking place in the heart, caused by the acidity of the blood and the alkaline bile interacting with the innate heat of the heart (Sylvius 1663; see Beukers 1999). Thus, Sylvius developed an explanation of blood circulation by relying on the experimental principle of chemistry rather than on Cartesian reductionism. As shown by Evan Ragland, for Sylvius

> the only way to come to know the quantitative mechanisms of the world was through the senses, and especially though the witnesses of sight, touch, and taste in anatomical and chymical experiments. The experience of working with the sensible changes in bodies – animate or inanimate – moved Sylvius to endorse the approach of the chymists." (Ragland 2016, p. 192)

For De Volder, similarly, the role of *efferevescentia* as the core process of blood circulation – and thus of all the effects to be explained regarding the human body – is to be assessed through experience.[18] Against the more reductionist or speculative account of Descartes and Regius, therefore, in De Volder's thought a high value is attached to experience as the means of the discovery of the causes of bodily functions.

6.2.2 Experimental teaching in Leiden: De Volder and Senguerd

In fact, De Volder's openness to the use of experience is fully revealed by his *Quaestiones de aëris gravitate*, offering consistent evidence of the contents of his experimental lectures in Leiden, and of his actual views on natural-philosophical models. Even if it is reported that he carried out experiments on all the topics of

---

18 "Constans est omnium anatomicorum, et experientiae consentiens sentientiae," De Volder 1664, p. 7.

natural philosophy, and the equipment of the *Theatrum physicum* did not only consist of pneumatical devices (such as the well-known air pump built for him by Samuel van Musschenbroek),[19] De Volder mainly focused on hydraulic experiments. But such experiments had no heuristic function: they were primarily meant to criticise the Aristotelian prejudice for the *levitas* of the air by offering experimental evidence of the existence of air pressure or *gravitas*[20] – thus confirming a mechanical worldview, where air and water follow the same laws as they are two kinds of fluids.[21] Hence, such experiments had a didactic function, and served to increase the prestige of Leiden University. In his lectures De Volder repeats and comments on some experiments carried out by Torricelli with his barometer and by Otto von Guericke with his sphere, as well as those of Robert Boyle (De Volder 1681a, pp. 6, 8–9, 32–33). Actually, De Volder does not aim to provide new experiments but only those already known to the scientific community, as his purpose was only to teach (De Volder 1681a, pp. 9–10). These experiments are explained in the light of a mechanistic worldview, but this is never explicitly ascribed to a particular philosopher. In the first pages of his *Quaestiones*, indeed, De Volder exhibits his admiration for Galileo, Torricelli, Roberval, Pascal, Guericke, Boyle, and Huygens, adding that with their experiments they proved that air has a weight *mathematica claritate.*[22] De Volder's worldview is roughly Cartesian, as testified by his maintaining the circularity of motion, and his rejection of the void, filled with subtle matter, and the existence of any *suctio, attractio,* or *fuga* and *vacui metus* in bodies (De Volder 1681a, pp. 13–22, 37–50). However, he declares an openness to the theories of scientists with different approaches to the sources of philosophical-scientific knowledge and to the method of discovery:[23] for him, philosophy should be based on reason and on carefully performed experiments De Volder 1681a, *Lectori philosopho,* p. I (unnumbered).

---

19 See the *series lectionum* reported in Molhuysen 1913–1924, vol. IV, p. 45*. On Leiden equipments, see Molhuysen 1913–1924, vol. IV, pp. 104*–106*, De Clercq 1997a, p. 10.

20 De Volder 1681a, p. 7. 'Gravitas' and 'pressio' are used as synonyms by De Volder, who also uses the term 'pondus': De Volder 1681a, p. 50.

21 De Volder 1681a, p. 18. De Raey used Torricelli's barometer for the same purpose: see De Raey 1654, *Clavis,* pp. 193–198.

22 "Neque vero res dubia habita fuit, nisi postquam experimentis Galilaei, Torricellii, Robervallii, Pascalii, Guericke, Boylaei, Hugenii, aliorumque excellentium [...] virorum, gravitas ipsius aëris adeo manifeste demonstrata fuit, eiusque effectus adeo notabiles animadversi, ut qui eam nihilominus negare velit [...] tantae demonstrationum mathematicarum claritati tenebras offundat," De Volder 1681a, pp. 1–2.

23 "Plurimis amicus Plato, amicus Aristoteles, amicus Epicurus, amicus Democritus, amicus Paracelsus, amicus Helmontius, amicus Carthesius et paucissimis amica veritas," De Volder 1681a, *Lectori philosopho,* pp. I–II (unnumbered).

De Volder was not the only one using the *Theatrum physicum*: from the late 1670s, Senguerd also performed experiments for the sake of academic teaching. In fact, Senguerd was appointed by the Curators as he had the reputation of being a more traditional philosopher, and his appointment served to balance the teaching of De Volder: however, his positions were hardly Aristotelian. His *Philosophia naturalis* opens with a section on the principles of natural philosophy which extensively – but not exclusively – rely on Descartes. As the first principle of natural philosophy, Senguerd assumes a Cartesian notion of matter as extended substance, whose rarefaction is due to the presence of a subtle matter in the interstices of bodies (Senguerd 1681, pp. 15–17). Moreover, he provides a theory of motion compatible with Descartes's, as movement is intended as local motion whose cause is God: however, he admits that the principle of the conservation of the quantity of motion is only probable, as God has absolute power in changing it) (Senguerd 1681, p. 31). Also, he maintains the circularity of motion, which is intended not only as the successive replacement of bodies in the direction of the moving object, but also as the translation, in any direction, of the parts of matter surrounding the moving object, as happens to the water when a stone is thrown into it. On the basis of this principle, he argues that the union of two hemispheres does not depend on the pressure of air on them, but rather on the fact that once they cohere the circular movement of air through them is impeded, in the same way as smoke does not exit from a flue once one of its extremities is closed (Senguerd 1681, pp. 41–46). At the same time, however, Senguerd maintains the possibility of the existence of a vacuum, as this is consistent with the absolute power of God (Senguerd 1681, p. 109). In the second part of his treatise he overtly distances himself from Descartes, as he presents the astronomical models of Ptolemy, Copernicus, and Tycho Brahe, whose model is to be preferred as it is consistent with the Bible (Senguerd 1681, pp. 114–119). Given this model, the cosmology of Descartes, for which each vortex has at its centre a star made by subtle matter, collapses (Senguerd 1681, pp. 121–140).[24] Thus, Senguerd assumes an approach intermingling different sources, in an attempt at harmonising his scientific and theological concerns. In the second edition of his *Philosophia naturalis* (1685) and in his *Inquisitiones experimentales* (1690), moreover, Senguerd provides extensive descriptions of his experiments in pneumatics and of the new air pump, that Johan Joosten van Musschenbroek built in 1679. Such experiments served to illustrate the nature and features of air (such as elasticity, pressure,

---

24 In the second edition of his *Philosophia naturalis* (1685) he would add a description of Descartes's vortex theory, only, however, for the sake of criticising it. See Senguerd 1685, pp. 165–185.

and so on), given against the background of his natural-philosophical principles. That is, his experiments had a teaching function: so the first series of experiments serve to illustrate the properties of air (and vacuum) as these determine the life and death of animals, the movements of lungs (Senguerd 1699, pp. 13–58). In the main section of the work, *De aëre atmosphaerico*, Senguerd presents the features of air (gravity and elasticity) by means of a corpuscular explanation of its nature, and illustrates its effects by experiments (Senguerd 1699, pp. 83–158). As late as 1715, eventually, Senguerd provided a full-blown manual of the construction of pneumatic devices in his *Rationis atque experientiae connubium*. Thus, Senguerd did not provide his natural philosophy with a foundation, as only in the introduction to the *Inquisitiones* does he offer some reflections on the relation of experience and reason: his considerations do not go beyond a remark on the importance of the senses in the discovery of the features of the world, and of a sound use of reason, as maintained by Bacon in his *Novum organum* (1620), and as done by Harvey, Boyle, Willis, Guericke, Swammerdam, who used the senses as instruments, and reason as a guide in the discovery of truth (Senguerd 1690, *Manuductio*, pp. 6–10).

In sum, Senguerd showed a somewhat eclectic attitude, with no interest in the foundations (i.e. the justification of the validity of the premises) nor in reflections on his own scientific practices. At the same time, with his experimental practices he raised the problem of the role of experience in a Cartesian environment. It was De Volder who developed such a foundation: this was shaped through further developments of the Leiden scientific and experimental tradition in the late seventeenth century, namely, the debate over the role of experience in medicine set forth by the Cartesian physician Theodoor Craanen in 1685, and the appearance of Newton's *Principia*, as well as the general advancement of mathematical-experimental science at the end of the century.

6.2.3 From Cartesianism to Newtonianism

The progressive acquaintance of De Volder with experimental science is testified to by his *Oratio de coniugendis philosophicis et mathematicis disciplinis* (1682). Whereas his *Quaestiones de aëris gravitate* offer a highlight of his actual experimental practices, his 1682 *Oratio* is a programmatic text, expounding his views on methodology. This text is a critique to the Aristotelian distinction between mathematics and physics, since for De Volder, just as all *res* are connected, so all disciplines are connected (De Volder 1682, pp. 1–3). So physics and mathematics are about the same objects: extension, motion, size and shape, all capable of a mathematical consideration. Therefore, natural phenomena are explicable through mathematics, since they flow from motion, which can be mechanically defined,[25] as Galileo and Huygens did in the description of the acceleration of falling bodies and pendulum vibration.[26] In turn, such laws have an explanatory function as these are the real causes of phenomena. In particular, De Volder emphasises the importance of the laws of impact of bodies,[27] which were mathematically formulated by Huygens, Wallis, and Wren, who made public their discovery to the Royal Society in 1668. This discovery relied on mental experiments (as Huygens's experiment of the boat), actual experiments, and mathematical demonstrations, and discarded Descartes's formulation of the laws of impact in the *Principia philosophiae*.[28] So De Volder seems to be open to a formulation of physical laws on the basis of a generalisation and mathematical reinterpretation of experience, rather than on a metaphysical deduction of them *à la* Descartes, who did not provide mathematical formulations of them. Still, the overall cosmological model he assumed is that of Descartes, since De Volder adopts the Cartesian vortex theory as explanation for the overall structure of the universe. This can be seen in his later *Disputatio philosophica de mundi systemate* (1694), where the Copernican system is defended against the Ptolemaic and Tychonic

---

25 "Quae quidem omnia, ut generalem mathematicarum artium usum comprobant, ita proprie non pertinent ad eam quam primario mihi illustrandam proposueram rerum physicarum cum mathematicis affinitatem. Quid autem ego affinitatem dico? Cum revera una eademque sit scientia, et mathesis aut ipsa physica sit, aut certe physices pars maxime princeps. Considerat enim utraque corpus, eius figuram, magnitudinem, motum. [...] Nulla certe in physicis causa aut universalior, aut foecundior ipso motu, a quo nulla non exoriuntur phaenomena, omnes corporum fluunt varietates. [...] Proprietates vero motus, aut omnes aut praecipuas absque geometria cognosci posse [...] pernego," De Volder 1682, pp. 14–16.

26 "Nunquam magnus ille florentinus Galilaeus de Galilaeo admirabilem illam detexisset in motus acceleratione proportionem, nisi in geometricis demonstrationibus fuisset versatissimus. [...] Haec autem ea motus proprietas est, quae in rebus ad usum vitae pertinentibus spectatur plurimum. Nemo enim absque hac cognita motus indole, aquarum ex fontibus [...] erumpentium quantitatem, nemo proiectorum vim [...] definiet accurate. Hinc elegantissima pendulorum doctrina, et ex hisce accuratissima temporum observatio, sine qua in astronomicis [...] nihil exacti fiet unquam. Hinc accuratiora nuper inventa horologia, quae absque vibrationum in pendulis cognita proportione, absque cycloidis lineae contemplatione vere intelligentur neutiquam. Quod inventum ut illustri Hugenio debet orbis litteratus, ita illi debuisset nunquam, nisi caeteris cum scientiis [...] coniunxisse mathematicarum artium notitiam," De Volder 1682, pp. 16–17.

27 "Quae tamen illae leges sunt, quae corporum occursibus moderantes, omnium corporearum mutationum, atque adeo omnium physicorum effectuum verae sunt causae," De Volder 1682, p. 17.

28 See Murray/Harper/Wilson 2011. De Volder was the editor of Huygens's *Opuscula posthuma* (1703), containing his *De motu corporum ex percussione*, where Huygens explained the correct laws of impact.

models, and is overtly explained by means of Descartes's cosmology.[29] The Ptolemaic system is rejected as it contradicts Galileo's observation of the phases of Venus and his measurements of the variations of distance between planets.[30] Moreover, the Tychonic model does not justify why earth should be at the centre of the universe.[31] On the other hand, the Copernican system allows the explanation of the solar spots observed by Galileo, as well as of his other discoveries.[32] Its validity can be justified by means of arguments based on pure reason,[33] namely, through Descartes's vortex theory. According to it, since celestial matter is fluid and rotating, it moves celestial bodies like ships in a river: thus, it is implausible that any body would be at rest, like the earth or sun.[34] So it seems that De Volder was admitting the derivation of a cosmological model from purely intellectual ideas: namely, from the ideas of extension and its modes, and from some metaphysical principles. It is to be remarked that De Volder had also been the editor of Huygens's posthumous *Κοσμοθεωρος* (1698), which includes his critique of Descartes's cosmology. At the time of his *Disputatio de mundi systemate* (1694), although he could not have read Huygens's theory of the cosmos (as the *Κοσμοθεωρος* was finished only in 1695), he could however know Huygens's criticism of Descartes's cosmology contained in Huygens's *Discours de la cause de la pesanteur* (1690), which in turn relied on a manuscript *De gravitate* (1668). In his

---

29 "Haec sententia de terra mobili a Copernico invecta multos illustres sectatores habuit, inter quos maxime Cartesius omnium temporum philosophorum princeps, Cartesius, qui primus recta posuit philosophandi elementa, systema copernicanum illustravit et explicuit," De Volder 1694, p. 2.

30 "Ptolomaicum certe systema nullo modo admitti potest, quoniam non minus calculis, quam experientiae maxime adversatur. Primo calculis, quia Venus sexies a terra remotior est, uno, quam alio tempore, et Mars adeo suas varias distantias [...] ut testatur nob. Galilaeus in suo de Mundi systemate libro. [...] Secundo experientiae repugnat, nimirum phasibus Veneris, et Mercurii ex quibus constat planetas illos non semper citra solem," De Volder 1694, p. 3; see Galilei 1632.

31 "Praeter haec autem mundi systemata aliud a celeberrimo Tichone Braheo est effictum, inter quod et Copernici systema, non magna differentia est, omnibus enim phaenomenis tam tychonica quam copernicana hypothesis satisfacit [...]. Quamvis tychonica hypothesis ptolomaica multo probabilior esse videatur, non tamen omni plane defectu caret. Nam primo in eo laborat, quod falsa nitatur hypothesi, supponendo terram esse centrum universi," De Volder 1694, pp. 3–4.

32 "Praeterea illustris Galileus non terrena modo, sed et excelsa contemplanda natus observavit, solares maculas non perpendiculariter erecta, sed inclinatas ad planum eclipticae moveri," De Volder 1694, p. 7.

33 "Electo igitur copernicano systemate, restat, ut illud argumentis, non a praeiudiciis sensuum, sed a solo rationis lumine petitis stabiliamus, ac defendamus," De Volder 1694, p. 4.

34 De Volder 1694, pp. 5–6. Other arguments appeal to a principle of economy: for instance, according to De Volder it is more probable that only the earth moves, instead of the fixed stars: De Volder 1694, pp. 6–7.

Κοσμοθεωρος, Huygens rejects Descartes's vortex theory for two main reasons: first, the dimensions of the vortices set by Descartes would lead to their dispersion, and secondly, different vortices would hinder their reciprocal movements, and create irregular motion in the vortex (Huygens 1698, pp. 139–144). This criticism had already been set forth in the *Discours*, where he hypothesises that in the vortex, particles of subtle matter do not all follow the same direction, but have irregular motions as the vortex is spherical and not circular (Huygens 1690, pp. 160–162). De Volder does not accept such criticisms, as he adheres to Descartes's vortex theory well into the 1690s. The intellectual trajectory of De Volder – as to his ideas on method – was thus influenced by the rise of mathematical-experimental science at the end of the seventeenth century. However, his physical theories – at least in cosmology – were still unmistakeably Cartesian.

De Volder's integration of a mechanical view of the cosmos inspired by Descartes, with the use of experience and mathematics in establishing the laws of motion, seems to have been finally prompted by an internal movement of Cartesianism, i.e. by the use of the notion of *oeconomia animalis* by Craanen, namely, the last development of the applications of Descartes's physiology to medicine, which for De Volder has a highly speculative character. This is testified by his programmatic *Oratio de rationis viribus* (1698) where he takes into account the relation of medicine and anatomy, which for De Volder both have to be based on the use of experience. De Volder criticises those aiming to deduce the whole *corporis humani fabrica* from the first principles of physics, distinguishing themselves from the *empirici*. Since this attempt still deserves some respect, these philosophers nevertheless claimed to deduce the complex structure of the body from a few notions, as if one could deduce Archimedes's discoveries from Euclid's principles.[35] This is the case with Theodor Craanen's explanations as expounded in his *Oeconomia animalis* (1685) and his *Tractatus physico-medicus de homine* (1689). These build upon different sources, such as Descartes's *De homine*, which had been published for the first time by Florentius Schuyl

---

35 "Quapropter ad alterum propero cogitationum genus, quod in rebus est corporeis, in quarum, prout existunt, cognitione rationi soli ascribenda tantum fortasse peccatur, quantum in metaphysicis eidem abnegandis. [...] Quae ut nequaquam inficior, ita vereor non parum, ne qui ita ratiocinantur, nimium magnifice de nostra scientia sentiant [...]. Quoscunque enim physica recentiorum maxime industria hoc tempore fecerit progressus, tam parum ea provecta est hactenus, ut ex illius inventis ad corporis nostri effecta perpetuam argumentationem deducere qui tentant, multis partibus et absurdius et arrogantius facere nec iniuria videantur, ac faceret ille, qui perlectis omnibus Euclidis notionibus hoc solo instrumentum se putaret abunde, ad Archimedeas inde perficienda conclusiones," De Volder 1698, pp. 17–18. See also p. 30.

in 1662 (Descartes 1662);[36] on Regius's works, whose *Praxis medica* (1657) had been commented on by Craanen (Craanen 1686), and on Sylvius's iatrochemistry. Indeed, both in *Oeconomia animalis* (presented in the form of questions and answers) and in the more systematic *Tractatus physico-medicus*, the movement of the parts of the body is explained as a consequence of the movement given by subtle matter, but this explanation also includes some iatrochemical principles. The *Tractatus* opens with the statement of the substantial difference between soul and body – which allows a mechanical explanation of its functioning – and the comparison of the body to a clock, whose primary function is the *coctio* of food. This works by fermentation, which is generally defined as the separation and modification of particles of food and chyle. Food is first fermented in the stomach, and then sent to the intestines, where it becomes excrement, or in the heart, where it is transformed into blood by a further process of refinement. Fermentation takes place by means both of heat and of the actions of ferments present in the stomach, which are comparable to the action of *aqua regia* and *aqua fortis* on metals. These are nothing but acids and salts, mixed with alkali (Craanen 1689, pp. 1–5, 26–31, 37–38). If fermentation takes place in the stomach, in the heart its product, namely, chyle, is subjected to effervescence, which is just a faster fermentation, and leads to the generation of blood. The main factor in this generation is Descartes's first element or subtle matter, rather than acids, salts, and alkalis (Craanen 1689, pp. 136–141). The particles of blood are then fit to enter the pores of all the parts of the body. The state of health is determined by the fitting of the pores by the particles, whereas diseases are caused by their obstruction. This is the case, for instance, with inflammations, caused by the positioning of the wrong particles in the pores and tubules (Craanen 1689, pp. 273–275; see Luyendijk-Elshout 1975).

Such application of Cartesian concepts to medicine had been criticised, before De Volder, by the Scottish physician Archibald Pitcairne, who in 1692 became a professor of medicine in Leiden (a post he held only until the following year) and delivered his inaugural *Oratio, qua ostenditur, medicinam ab omni philosophorum secta esse liberam*. As the title declares, for Pitcairne medicine and philosophy have to be detached, for different reasons: first, they have different ends, as medicine is aimed at the preservation of health, and

---

36 Schuyl had assumed a chair in medicine in Leiden in 1664, before moving to the chair in botany in 1667. His philosophical manifesto was the *De veritate scientiarum et artium academicarum* (1667, published in 1672), in which he defends Descartes's dualism and his acknowledgment of truth criterion in clear and distinct perception. Yet, besides his edition of Descartes's *De homine*, in whose Preface he praises Descartes's rejection of animated principles from the explanation of living functions, he did not leave any treatise in medicine.

philosophy at the perfection of soul; secondly, medicine has to be based on well-acquired results, while philosophy proceeds by discussions and disagreement. Moreover, natural philosophy aims at discovering the first causes of phenomena, whereas medicine is based on the direct observation of the properties of medicaments (Pitcairne 1692, pp. 7–8). In sum, Pitcairne advocates the exclusion of any speculative principle from medicine. Some scholars have argued that Pitcairne was in fact following a Newtonian approach (Guerrini 1987); however, he does not apply any Newtonian notions to his medicine: in his *Elementa medicinae physico-mathematica* (1717), although proceeding by postulates and definitions, he does not employ mathematical proofs. Thus, as argued by Henri Krop, "the mechanical philosophy Pitcairne adopts is more like a general scheme, which leaves ample space for an empirical attitude" (Krop 2003b, p. 186). The same kind of argument by Pitcairne would then have been taken by De Volder, aiming at excluding any speculative principle from medicine. According to his 1698 *Oratio*, the shape of particles – which have to fit into pores – cannot be discovered by geometry or by experience and has no observable effects. Moreover, he is discontent with the use of iatrochemical principles such as fermentation itself, which is an effect to be further explained (De Volder 1698, pp. 22–24). Hence, De Volder advocates a method of discovery more attentive to the combination of experience and mechanism in explaining bodily functions. This method, actually, consists of the careful application of geometrical principles to observed phenomena, as Giovanni Alfonso Borelli and Lorenzo Bellini – whose *De urinis et pulsis* (1663) had been extensively read by Pitcairne, and who dedicated to him his medical *Opuscula* – practised in Italy.[37] Thus, De Volder embraces the iatromechanical approach, supported by a robust recourse to experience: for De Volder, this application is to be intended as a careful procedure of arguing for conclusions, and as the explanation of observed bodily functions through mechanical principles (De Volder 1698, pp. 25–26). In anatomy, explanations are to be provided through careful observation of the circulation or motion of fluids (as Borel or Harvey did), which can be mathematically described (De Volder 1698, pp. 28–30). In order to develop these mathematical explanations a new anatomy is thus required, based on vivisection and the observation of fluids in motion. The same method, however, is to be used in natural philosophy, as had been done by Huygens, but also by Leibniz, Gottfried Wilhelm von Leibniz and Newton, who discovered the laws of

---

37 See Borelli's *De motu animalium*, 1680, and Bellini's *Opuscula aliquot ad Archibaldum Pitcarnium* (1695).

motion in the same way.[38] In this way, one can collect those data allowing the formulation of explanatory hypotheses on bodily functions, from which phenomena can flow.[39] Such hypotheses must fulfil some conditions: not to contradict other assumed hypotheses, be open to correction by new experiments and reasoning, be consistent with experience, and allow explanations for newly observed bodily operations.[40]

In sum, De Volder's *Oratio de rationis viribus* thus shows some similarity with his 1682 *Oratio de coniungendis philosophici est mathematicis disciplinis*, testifying to the continuity in De Volder's thought, as he emphasises the role of mathematics in physics and appeals to a group of scientists concerned with mechanism, even if not sharing the same theoretical model. If any evolution is to be found in De Volder's philosophy, this is to be seen in the replacement of iatrochemistry by an approach more open to mathematical-experimental science, and to iatromechanics. So it seems that at least in medicine the method of scientific discovery

---

38 "Ostenderunt magna huius seculi nostri lumina Hugenius, Newtonus, Leibnitzius, ne simplicium quidem corporum motus, viresque investigandas unquam, non dicam absque notitia matheseos, sed addam absque recondita harum artium scientia. Qua qui instructus non est, in physicis hospes ut sit, necessum est. Tanta igitur cum inter has disciplinas sit affinitas, eo meliori iure inquiremus, num eadem methodo tractari queant," De Volder 1698, p. 26.

39 "Ea ergo experimenta anatomica et summo quidem cum iudicio facienda sunt, ex quibus patefiant corporis nostri non mortui membra, sed vivi actiones, qui fiant, qua partium operatione, quo fluidorum motu perficiantur. Quibus si ultimo accedat historia corporis affecti, quae morbos, quibus obnoxii sumus, eorumque singula symptomata, variasque periodos singulatim describat, in numerato habebimus, ut cum geometricis loquar, data, ex quibus de causis porro ratiocinemur. Huic denique aedificio ut fastigium imponatur, non secus ac astronomi hypotheses effinxerunt, quibus iam cognitos astrorum explicarent motus, rudes in principio, quas dein novis ex observationibus sensim emendando tandem perficerent. Ita et nobis necesse erit hypothesin excogitare, quae structurae partium, motui liquorum, efficaciae spirituum sensili conveniens causas in se contineant mechanicas, ex quibus, quae fieri per experientiam novimus, sequantur," De Volder 1698, pp. 30–31.

40 "Non secus ac astronomi hypotheses effinxerunt, quibus iam cognitos astrorum explicarent motus [...] ita et nobis necesse erit hypothesin excogitare, quae structurae partium, motum liquorum [...] conveniens causas in se contineant mechanicas, ex quibus, quae fieri per experientiam novimus, sequantur. Quod cum in fabrica totius corporis nimiam habiturum sit difficultatem, praestabit seorsum in singulis eius operationibus tentare, modo caveamus, ne quid in una hypothesi assumatur, quod alteri repugnet. Nec expectandum erit, eam, quae ita primo nobis in mentem venit, rei satisfacturam. Sed, ut in omnibus fieri solet, ea novis experimentis et rationibus limanda et perpolienda erit. Inquirendum scilicet porro, num, quae ex ea sequuntur, experientiae congruant, et num eadem paucis hinc inde pro re nata additis, demtisve, omnibus id genus in corpore operationibus adaptari queat. Quod si minus succedat, immutanda erit, donec tandem invenerimus hypothesin, quae in omnibus cum iis, quae fiunt, consentiat," De Volder 1698, pp. 31–32.

prescribes not going beyond empirical evidence about the functioning of the body. In natural philosophy, on the other hand, one can hypothesise the existence of insensible features of matter in order to explain the constitution of the universe. This is the case, indeed, with his *De mundi systemate*. The co-existence of different methodologies in De Volder's thought, on the one hand, and the hypothetical character of physical explanations on the other results in a standpoint according to which only few explanatory principles are provided with a foundation on Cartesian metaphysics: this, in fact, concerns only the basic notions of natural philosophy.

6.3 The quest for principles: philosophy of science without a foundation

The foundational arguments set forth by De Volder are presented in his metaphysical writings, namely, his *Disputationes contra atheos* (1685), and *Exercitationes adversus Censuram* (1695), which includes the *Disputatio philosophica de certitudine clarae et distinctae perceptionis* held in 1689 (De Volder 1689). Such texts are defences of Descartes's metaphysics not aimed, *per se*, at providing natural philosophy with a foundation: in fact, the *Disputationes* are aimed at defending Descartes's metaphysics against accusations of atheism. Although this is not directly mentioned, it can be interpreted as a vindication of Cartesianism against Spinozism. The *Exercitationes*, on the other hand, rebukes Huet's *Censura philosophiae cartesianae*, which appeared in 1689. Before defending Cartesian metaphysics against the polemics arising in the late seventeenth century, however, De Volder was interested in finding criteria for establishing such principles in a way independent of the development of metaphysics as a foundational discipline. Already in his *De natura*, in fact, De Volder stressed the importance of a *quaestio de principiis* as the crucial means for the development of the sciences, instead of a mere analysis of deductions and conclusions for each particular theory. Clearly referring to Descartes's revolution in philosophy, De Volder underlines that as doubt started to be systematically applied to philosophy, allowing the discovery of more reliable physical principles, natural philosophy underwent considerable progress.[41] Such a *quaestio de principiis*, eventually, is elaborated in his *Cogitationes de rerum naturalium principiis* (De Volder 1674–1676, De Volder 1681b, De Volder 1681c), in which the criteria for the choice of the first concepts of physics are set out in light of Descartes's theory of clarity and distinction as marks of philosophical knowledge. Borrowed from mathematics as

---

41 De Volder 1664, p. 1. This differentiation between metaphysical doubt and the endless analysis of every argument will recur in De Volder's other works, see *infra*, section 6.4.

the paradigm of *scientia*, indeed, clarity and distinction are the first criteria in De Volder's *quaestio de principiis*, prescribing the use only of clear and distinct principles in physical explanations[42] as the very first causes of phenomena.[43] This condition, nevertheless, is not taken as a guarantee of the truth of scientific principles by De Volder. Clear and distinct perception, actually, can be compared to the mere grasping of the meaning of a sentence: accordingly, it is necessary but not sufficient to assess the truth of scientific principles. Ultimately, it implies that the conclusions drawn from these would be indubitable, but not that such principles have a real explanatory value.[44] Hence, the second criterion prescribes that a scientific principle must not be the effect of some other natural or corporeal cause,[45] while the third dictates that physical principles must not involve the notion of mind.[46] The fourth criterion concerns the explanatory scope of scientific principles. According to this, every kind of natural phenomenon has to be explicable through them; moreover, one has to demonstrate that the human mind cannot attain any other explanatory principle.[47] Finally, the fifth condition prescribes that these principles have to be

---

42 "Quarum prima sit ut clare distincteque percipiantur," De Volder 1681b, p. 12. As in his *Oratio de coniugendis philosophicis et mathematicis disciplinis*, the failed application of mathematics to physics is regarded as one of the main causes of the underdevelopment of the latter, due to the Aristotelian prejudice of the difference of physics and mathematics (De Volder 1681b, pp. 10–11).

43 "Ut enim phaenomenum quodpiam explanem, nonne requiritur, ut eius causas ostendam? Quae aut primae erunt, et a nulla alia corporea causa dependentes, aut erunt aliarum causarum effecta. Priori in casu quid est manifestius, quam me ipsa demonstrasse principia? Sin vero aliarum causarum effecta sint, quis non videt [...] unquam huius phenomeni claram distinctamque [...] notionem acquiri posse, nisi huius causae iterum cognoscam causas, idque donec ad primas causas, sive ad ipsa rerum principia devenerim," De Volder 1681b, p. 2.

44 "Hactenus non requiro, ut demonstrentur, non ut certo vera esse ostendantur, sed illud tantum exigo, ut percipiantur, ut quae et qualia sint cognoscatur. Quae sane duo non parum differunt. Aliud quippe est percipere huius illiusve effati sensum, aliud eius veritatem cognoscere, de qua nisi prius illud percipiatur, constare nemini potest," De Volder 1681b, p. 12; see also pp. 13–14.

45 "Altera conditio est, ut prima principia non sint alterius causa naturalis, sive corporea effecta," De Volder 1681b, p. 14.

46 "Tertia sit, ut principiis hisce nulla ascribatur proprietas cogitationis aut mentis. Agitur enim hic non de natura earum rerum, quae cogitant sentiuntve, se de natura phaenomenum a corpore dependentium," De Volder 1681b, pp. 14–15.

47 "Quarta conditio sit, ut ex iis, quae pro principiis sumuntur, omnia mundi huius phaenomena queant deduci. [...] Neque tamen exigi putem, ut omnia revera deducantur, sed ut ostendantur deduci posse. [...] Sed hoc tamen fieri potest, ut iis utar principiis, ex quibus ostendam sequi certo et evidenter, primo omne qualecunque genus phaenomenum. Deinde in unoquoque genere infinitam phaenomenum varietatem, atque adeo maiorem, quam quae unquam hominum sensibus possit lustrari. Postremo: ut demonstrem humanum intellectum ultra ea principia nihil capere aut percipere, atque adeo me hisce principiis efficere, quidquid ab humano intellectu praestari potest," De Volder 1681b, pp. 17–18.

true, that is, that they are the real causes of experienced phenomena. De Volder does not regard this last criterion as necessary. In accordance with the traditional distinction between physical (or moral) and metaphysical degrees of certitude, De Volder states that the first four conditions are adequate to choose explanatory principles for physics, as these can serve to deduce every kind of phenomenon, and have indubitable consequences. According to De Volder himself, this would surprise those looking for the same degree of certainty in physics as in metaphysics – like De Raey: whereas metaphysics concerns mere concepts, in physics one has to rely on the senses and provide true hypotheses.[48] Nevertheless, this hypothetical status of physics is not only justified by our reliance on experience, or on the 'apparent' existence of the world: even in the case that the world exists, it is still doubtful whether it obeys the clear and distinct principles of mechanism. Phenomena, indeed, can have more than one possible cause: hence, one cannot ascertain their actual cause by reason alone, since different kinds of explanatory principles are conceivable. On the other hand, experience is not a means to the discovery of the first scientific principles or causes, since phenomena are the very *explanandum*. Two identical phenomena, for instance, can only lead to the same hypothesis as to their cause, even if they actually have different causes, but no differences can be inferred from their observation.[49]

----

48 "Quinta denique conditio est, ut principia certo demonstrentur esse vera. De qua tamen, an requiratur necessario, admodum dubitem. Me enim quod attinet, facilem concedam unicuique, ut assumat principia, quaecunque visa fuerint sine ulla ratione, ulla demonstratione, modo ea propribus conditionibus non repugnent. Quod forte mirum videbitur iis, qui putant omnia certo demonstranda esse, nullibi utendum hypothesi. Quod ut in metaphysicis, ubi omnia per ipsam rerum naturam determinata definitaque sunt, verissimum est, ita in physicis, ubi omnia ad sensus referuntur, paris sit evidentia, non immerito forte quid ambigat," De Volder 1681b, pp. 18–19.

49 "Humana [...] industria si inter varios quibus tunc mundum fieri potuisse supponimus modos, discrimen facere et verum eligere modum posset, profecto id benefici vel rationi deberet, vel experientiae. [...] Rationi autem hoc in negotio nullae reliquuntur partes, quippe quae suam explevit potestatem, si doceat tam hanc quam illam causam mundo efficiendo parem esse, neque vero plus potest. Nam quid evidentius si ex iisdem datis, animadvertamus problema variis modis posse dissolvi, rationem eiusque ad summum sese extendere, ut varios hosce modos enumeret, verum neutiquam ut demonstret hoc, non illo modo solutum esse mundum. Quod si faceret illud evidenter sequeretur, eo modo productum esse, nec alio posse, quod est contra hypothesin. Neque etiam huic difficultati enodandae auxilio est experientia, utpote quae tam ex hoc quam ex illo modo una sequitur eademque. [...] Nonne evidentissimum est, illum ad summum nihil aliud posse, quam ut ostendat, aut hoc aut illo modo ortam esse glaciem, non vero quidque etiam moliatur, cum idem sit phaenomenum atque adeo eadem ratione, ex quibus ratiocinetur suppeditet data, ut determinet hoc modo hanc, illo vero alteram lagenam concretam esse. [...] Eodem modo si homini huius artificii vel peritissimo duo proponantur horologia diversis ex rotulis confecta, eadem tamen externa facie [...] atque adeo eadem exhibeant phenomena, in vanum profecto ab ipso expectabimus, ut certo concludat hoc modo unum, illo vero alterum horologium confectum esse, nisi forte, quo nihil est absurdius, existimemus ex iisdem datis diversas conclusiones posse elici," De Volder 1681b, pp. 20–22.

Therefore, the fifth criterion is not required in the formulation of explanatory principles since reason cannot ascertain that some principles are true according to metaphysical certainty, that is, by showing that others are contradictory. Accordingly, clarity and distinction – that is, evidence – are not sufficient to establish the truth of scientific models.

De Volder, in sum, introduces a strong 'hypotheticism' in natural philosophy. As seen above, natural-philosophical explanations are provided by hypothesis based on the collection of sensory data and open to correction. Their hypothetical character, however, depends also on the limits of reason in deciding what is the actual model at work in the world, since different models are conceivable and no *a priori* reason can be provided for them. We have seen that for Regius and Geulincx the hypothetical character of physics depends on the use of experience in developing explanatory models. For Regius, as every kind of notion – including mathematical ones – is taken from experience and subjected to the power of God, any knowledge turns out to be subjected to a 'psychological kind of certainty. For Geulincx, physics is contingent as it relies on ideas matching essences and necessary properties (its 'metaphysical' part), and on some others known by experience, those concerning matters of fact. For Descartes, the hypothetical character of his physics (as he declares with respect to cosmology) has two reasons: from a theological point of view, the way in which he explains the construction of the world is contingent with respect to the Bible, sanctioning that God created the world and man in a state of perfection i.e. in their full-grown form.[50] Moreover, the ways in which the original continuum of matter broke up to form the universe of vortices which form solar systems cannot be deduced from metaphysics, but only hypothesised or acquainted by experience, given the limited powers of our mind.[51] In the

---

50 "Quinimo etiam, ad res naturales melius explicandas, earum causas altius hic repetam, quam ipsas unquam extitisse existimem. Non enim dubium est, quin mundus ab initio fuerit creatus cum omni sua perfectione: ita ut in eo et Sol et Terra et Luna, et stellae extiterint, ac etiam in Terra non tantum fuerint semina plantarum, sed ipsae plantae, nec Adam et Eva nati sint infantes, sed facti sint homines adulti. Hoc fides christiana nos docet, hocque etiam ratio naturalis plane persuadet. Attendendo enim ad immensam Dei potentiam, non possumus existimare illum unquam quidquam fecisse, quod non omnibus suis numeris fuerit absolutum. Sed nihilominus, ut ad plantarum vel hominum naturas intelligendas, longe melius est considerare, quo pacto paulatim ex seminibus nasci possint, quam quo pacto a Deo in prima mundi origine creati sint," AT VIII/1, pp. 99–100.

51 "At quam magnae sint istae partes materiae, quam celeriter moveantur, et quales circulos describant, non possumus sola ratione determinare, quia potuerunt ista innumeris modis diversis a Deo temperari, et quemnam prae caeteris elegerit, sola experientia docere debet. Iamque idcirco nobis liberum est, quidlibet de illis assumere, modo omnia, quae ex ipso consequentur, cum experientia consentiant," AT VIII/1, pp. 100–101.

case of De Volder, the most basic principles of physics are hypothetical because one cannot demonstrate that they are actually at work in physical realms. Indeed, the major part of De Volder's *Cogitationes* is devoted to a refutation of some philosophical hypotheses on the constitution of the world, and to the assessment of Cartesian principles as the best, although not demonstratively true, explanatory means. These hypotheses, actually, are traced back to four main philosophical schools classified by Francis Bacon in his posthumous *De principiis atque originibus secundum fabulas Cupidinis et Coeli: sive Parmenidis et Telesii et praecipue Democriti Philosophia* (De Volder 1681b, pp. 24–25; see Bacon 1653, pp. 208–284). Through a history of philosophical sects De Volder addresses some contemporary alternatives in philosophy. Thus, the first sect, which represents those who used one explanatory principle, is represented by Parmenides, Melissus, Heraclitus, Anaximenes, and Thales of Miletus, and in modern times by Jean Baptiste van Helmont.[52] The second school is that of the corpuscular philosophy: starting with Democritus, it inspired Gassendi, but also Bacon, Descartes, and Boyle.[53] The third is that of those who adopted multiple – but still determined – principles: the Aristotelians and the alchemists. The fourth category is that of those adopting infinite explanatory principles, such as Anaxagoras, embodying another form of corpuscolarism (De Volder 1681b, pp. 27–28). All these schools are regarded as providing principles not complying with the aforementioned criteria: the only principles to be admitted, eventually, are those of mechanical philosophy, although deprived of some corpuscular notion such as vacuum or *gravitas* as an essential property of the body (De Volder 1681b, pp. 29–144, esp. 102–121). Such philosophy, however, is not expressly ascribed to any author. Still, notwithstanding his placing Descartes among corpuscularians, the mechanical principles De Volder expounds are Cartesian. The principles matching the first four criteria are the notions of matter and motion.[54] Such concepts fit the first criterion: physical body is one with mathematical body, and the notion of motion can be mathematically

---

52 De Volder 1681b, pp. 25–26. Van Helmont, in fact, assumed that two elements, i.e. water and air, are the first principles of nature: see Pagel 2002.

53 "Ad secundam sectam pertinet Democritus, [...] quae principia dudum reiecta nostro demum saeculo in lucem revocarunt Gassendus, Verulamius, Cartesius, Boylaeus, et quantum est ingeniosorum hominum, qui corpusculari, ut angli vocant, addicti sunt philosophiae," De Volder 1681b, p. 26.

54 De Volder 1681b, pp. 44–145. Also, De Volder mentions rest, "de quo non laboramus," or what preserves bodies in their current status. He seems to be cautious in considering this notion as being grounded *a parte rei*.

described in terms of variation of distance between bodies.[55] The second condition is also respected; since matter has in God its only cause, neither motion nor rest are caused by matter itself, but by God alone. The metaphysical demonstration of the detachment of mind and matter, moreover, fulfils the third criterion. Finally, every kind of phenomenon can be deduced from such principles, in accordance with the fourth criterion (De Volder 1681b, pp. 148–151). Indeed, phenomena are *communia* or *propria* to each sense. As *communia* the size, shape and motion of bodies have to be explained through mechanical principles, grasped by intellect alone or by common sense.[56] On the other hand, sense data can be proper to each sense, such as colours and sounds. However, the metaphysical demonstration of the distinction between soul and body proves that no qualities such as pain or delight, nor colours and sounds can be found in the body, but only matter, figure and motion.[57] To that extent every kind of phenomenon can be mechanically accounted for.[58] Such principles fulfil the four conditions stipulated by De

---

55 "Quid enim clarius, quod distinctius cognoscitur ipsa materia? [...] Per ipsam enim nihil aliud intelligimus quam id, de quo agunt mathematici. Quod corpus mathematicum a physico distinguendo, immane quanto noxae, obscuritatis confusionisque scientiis scholastici attulerunt philosophi. [...] Non absimili ratione obscurum esse nequit, quid sit motus. [...] Quis enim profecto vel stupidissimus mortalium est, qui ignorat, quid sit corpus alteri vicinum ea ex vicinia recedere, et distantiam ab illo corpore continenter immutare?" De Volder 1681b, pp. 145–147.

56 "Phaenomena autem quae observantur vel plurium sensuum sunt communia vel singulorum propria. Quae pluribus sensibus conveniunt, vel in motu, vel in magnitudine, vel in figura, situ similibusque consistunt. Quae vero propria sunt manifeste spectant colorem [...] similesque [...] qualitates. Praeter quae nulla in rerum natura aut dantur aut dari queant phenomena. [...] Qua in re id occurrit primum, quaecunque mutationes vel in motu vel in figura vel in magnitudine occurrunt sensibus, eas qualescunque demum sint hisce principiis deberi. [...] Ex quibus itaque sequitur nullam in corpore aut motus aut magnitudinis aut figurae varietatem dari, quin ea ex iis, quae diximus, principiis sequatur," De Volder 1681b, pp. 152–154.

57 "Inter obiecta, quae ad singulos senus spectant, et ea quae plurium communia sunt, licet vulgo confundi soleant, permagnam esse differentiam negabit, ut opinor, nemo, modo attenderit, quam clare intelligat, quid sit in rebus extra se positis, motus, figura, magnitudo, situs [...]. Et quam obscurum ipsi sit, quid sit iisdem in rebus, color, odor, sapor, et c. [...] Ex hac autem distinctione sensuum et qualitatum, quae in corporibus sensum excitantibus revera sunt, id licebit animadvertere, qualiscunque ea corporea dispositio sit, eam ab ipso sensu omnino diversam esse. [...] Verum ne hoc generali ratiocinio, quamquam id vel solum puto rem conficere, solummodo niti videar, accedamus ad speciales sensus, eorumque peculiaria phaenomena. Cui rei non parum conducet annotasse, ex motu locali [...] nihil posse produci, quam varietatem in figura, magnitudine, celeritate, determinationeve ipsius motus. Hinc enim sequetur, si ostendam sensiles qualitates suam originem debere motui, eas vel in figurae, vel magnitudinis, vel motus diversitate consistere, atque adeo nostris principiis deberi," De Volder 1681b, pp. 154–158.

58 "Ex quibus ita constitutis non arduum est elicere, si singula phaenomenum genera nostris principiis debeantur," De Volder 1681b, p. 166.

Volder. Moreover, according to him mind cannot conceive anything beyond these principles, since it can grasp nothing but extension, size, shape and motion in the physical realm. So according to De Volder it is not required for such principles to be true, according to the fifth condition. First of all, no one can truly doubt that extension or motion exist. Moreover, the demonstration of their truth would belong more to metaphysics than to physics.[59] Therefore, his *Cogitationes de rerum naturalium principiis* may be considered as expounding a philosophy of science, whilst not providing a metaphysical foundation; in fact, their very evidence does not require it. However, as De Volder will provide reflections on the principles of mechanism in his metaphysical works, he justifies to what extent we can label them as true.

6.4 The foundation of the principles of nature: A vindication of Descartes's metaphysics

The metaphysical foundation of natural philosophy is provided by De Volder only indirectly, i.e. by defending Descartes's metaphysics against Huet and, previously, Spinozism. As stated above, it is first addressed in his *Disputationes contra atheos*, designed to defend Cartesian philosophy against accusations of atheism, and likely addressing any possible linking of Cartesianism and Spinozism (De Volder 1685, pp. 5–6; see De Volder 1680–1681). In 1684, for instance, Noël Aubert De Versé published a book whose full title is *L'impie convaincu ou Dissertation contre Spinoza, dans laquelle on réfute les fondements de son athéisme. On y trouve non-seulement la réfutation des maximes impies de Spinoza, mais aussi celle des principales hypothèses du cartésianisme, que l'on fait voir être l'origine du spinozisme* (modern edition De Versé 2015). In this work, De Versé traces Spinoza's theory of the uniqueness of substance to Descartes's equation of matter and space, which occupies every place and is, in consequence, the sole existing substance, which is nature and matter itself. Accordingly, for Versé Cartesianism has atheist consequences (see Hubert 1994, pp. 9–21; Kors 2016, p. 200). De Volder's *Disputationes contra atheos*, in turn, expounds Cartesian demonstrations of the existence of God, although the name of Descartes does not occur in the text. He

---

59 "Deinque, in quinta conditione hisce principiis applicanda multus ut sim necesse non est, cum nemo diffiteatur, haec in rerum natura locum obtinere, qualiacunque etiam principia sequatur. Quis enim est, qui aut extensionem non admittat, aut qui motum neget? Unde nec puto quenquam fore, qui accuratam huius demonstrationem severe exigat, quam conficere hic supersedeo, tum quia res planissima est, tum quia eius demonstratio metaphysici potius quam physici est fori," De Volder 1681b, p. 168.

relies on the indubitable existence of thought, which is confirmed by appealing to Augustine's *Soliloquia*,[60] and rests on two assumptions: 1) what necessarily follows from the nature of something is an attribute of such a thing, and 2) what necessarily follows from the idea of something necessarily follows from the nature of such a thing. The second assumption is grounded in the very nature of ideas as these are objective beings. Since ideas are represented natures, one can find the same connection between the nature of things and their attributes, and between ideas and what ideas entail.[61] Hence, as the idea of existence is entailed by the idea of God, existence necessarily belongs to the nature of God.[62] The second proof consists in the application to ideas of the principle of causality.[63] Whereas such application can be allowed by the common persuasion that every idea has a cause provided with those features it represents, like those perceived through the senses, De Volder does not want to rely on a commonsensical foundation of

---

60 "Verum omnis difficultas in eo est, taliane principia inveniri possint. Qui enim omnia negantibus, aliquid extorqueri potest quod non negent? [...] Quantumcunque enim dubitationi indulgeat, non tamen hoc efficiet, ut se cogitare nesciat. Quod ipsum es quod Augustinum impulit, ut, in inquisitione, quam instituit de Deo et mente, cogitationem pro fundamento poneret. Soliloq. I. 2," De Volder 1685, pp. 9–10; see Augustine 1970, book II, 1.

61 "Quae itaque paucis ut ob oculo ponatur, assumo, ea omnia quae per necessariam, et certissimam consequentiam ex rei cuiuscunque natura deducuntur, ea esse rei istius attributa, rei isti certo competere, nec absque iis rem illam aut existere aut concipi posse. [...] Assumo deinde, ea omnia, quae in idea rei alicuius continentur, sive quae ex idea istius rei necessario sequuntur, necessario quoque sequi ex ipsa natura. Nam quid aliud sunt quam naturae ipsarum rerum repraesentationes? Ex quibus quidquid sequitur, sequitur ex natura rerum quas repraesentat. Quis enim non videt, inter naturam rei, quatenus ab idea offertur menti, et attributa, quae ex eadem necessario fluere mens percivit, eandem omnino connexionem esse, quae est inter rei naturam extra nos existentem, et attributa quae producit? Ita quidem ut sicut omnino nequeo habere ideam rei sine illo attributo quod ex idea sequitur, sic res illa omnino nequeat existere absque eodem illo attributo. Nullam enim aliam ob causam ex idea id sequitur, quam quod cum ipsa rei natura necessario copulatur. [...] Manifesto siquidem eadem analogia et connexio est inter ideam rei et ea quae in idea continentur, quae est inter naturam rei, et ea quae ex illa fluunt," De Volder 1685, pp. 23–25.

62 "Profecto si quis ea quae [...] de connexione inter ideas, et rerum quae repraesentent naturas diximus, attente applicet ideae divinitatis videbit evidenter, istius naturae existentiam necessarium esse attributum, ideoque de ea non posse non affirmari," De Volder 1685, p. 26.

63 "Post eam existentiae divini numinis demonstrationem, quae suam efficaciam debet ispi naturae Dei [...] proximum est, ut videamus quidnam sequatur ex eadem illa idea, eam si consideremus, non in sua natura, sed tanquam causae alicuius effectum. Cui rei non inutile erit praemittere quaedam generalia ipsam naturam causae et effecti," De Volder 1685, p. 32. Cause is defined as "qua posita effectum ponitur et qua sublata tollitur." The necessary connection of cause and effect, however, is subjected to the actual agency of the cause: see De Volder 1685, pp. 33–34.

his proof.[64] Hence, he first considers the axiom according to which everything is *a se* or *ab alio*.[65] Given its evidence, he focuses on the nature of ideas: since these are, *ut obiectum*, 'natures', they differ from each other according to their representative being. In the case that the things ideas represent exist, therefore, the differences among things would correspond to those among ideas, insofar as these represent natures. As a consequence, the very connection between the natures of cause and effect can be found between the ideas of cause and effect.[66] We can grasp therefore the connection of cause and effect by means of ideas. In addition, one needs to admit that the difference between causes matches that between effects.[67] Otherwise, the conclusions drawn from identical data would be different, or the same conclusion would be drawn from different data.[68] The

---

64 "Cur enim quaeso dicimus terram, coelum, sidera, idem autem est de quibuslibet rebus existere? Nonne quia ea videmus aut aliis quibusdam sensibus percipimus? Visio autem haec nobis ne quidem persuadere, multo minus nos certos reddere posset, de rei alicuius existentia, nisi mentem afficeret. Quid enim evidentius, quam si corpus afficeretur, mens non afficeretur, nunquam nos visuros, nec ex ea corporis affectione si eius conscii non fiamus, nos nobis unquam persuasuris dari aliquid extra nos. Sola igitur idea est, quae persuadet [...]. Qua autem, quaeso, ratione hoc potest idea, nisi persuasissimi essemus ea causa requirere extra nos existentem, et talem quidem, quae illius repraesentatis perfectionibus respondeat? Quae non eo adduco, quasi ex hac persuasione argumentum petere velim, sed solummodo ut ostendam eos qui, ubi de idea Dei agitur existentia, hanc de causis idearum veritatem in dubium trahunt, eandem illam, ubi de rerum sensilium existentia agitur, extra omnem dubitationis aleam ponere, sive sibimet ipsis pugnantia loqui," De Volder 1685, pp. 49–50.
65 "Manifestum plane attendenti axioma est, omne id quodcunque rei alicui adest, adesse vel ab ipsa rei natura, vel a causa externa. [...] Cui equidem effato consectarium est, omne id quod existit, aut existere a se, a sua natura, aut existentiam suam mutuari ab alio," De Volder 1685, pp. 50–51.
66 "Quod si idearum nostrarum naturam vel obiter contemplemur, facile liquebit inter eas respectu rerum quas repraesentant eandem omnino diversitatem esse, quae foret inter ipsas res quarum sunt ideas, si eae forte existerent [...] Ex quibus nec difficile erit advertere eundem hunc nexum, qui est inter naturam operationemque causae, et effectum quod producit, esse quoque eadem omnino necessitate, inter causae, eiusque operationis ideam, et ideam effecti," De Volder 1685, pp. 36–37.
67 "Quibus omnibus consentaneum est, quantum inter diversas causas varietatis est tantundem necessario diversitatis inter earum effecta fore, et viceversa, quantum est inter effecta variarum causarum discriminis, tantundem quoque inter ipsas causas reperiri differentiae," De Volder 1685, p. 37.
68 "Ut enim in ratiocinio, fieri nequit, ut ex iisdem plane datis diversis concludantur eadem, sic nec fieri poterit, ut ex communibus iisdemque in utraque causa proprietatibus, operationes diversae sequantur, neque ut ex diversis proprietatibus, operationes diversae sequantur, neque ut ex diversis proprietatibus sequantur eaedem. Hoc etenim si fieret, sequeretur diversam plane esse inter causam et effecta connexionem ab ea quae est inter ideam causae, eiusque effecti, quod absurdum esse [...] evicimus. Merito igitur licebit concludere, eandem omnino causarum, quae effectuum, et viceversa varietatem poni," De Volder 1685, pp. 39–40.

validity of the principle of causality, therefore, is grounded in the nature of ideas as objective beings and our proceeding in thoughts: then, this is applied to ideas as these are effects of something else. This is the case with the idea of God, which differs from other ideas as God differs from things. Since the idea of God does not exist *a se*, and it is the idea of an infinite thing, it requires an infinite cause.[69] Still, this proof is based on the assumptions that everything is *a se* or *ab alio*, and that ideas represent things that can exist, namely, that we can conceive things as they are, in the case they exist.[70] Eventually, De Volder stresses that in the case we are not convinced that ideas require a cause external to them that matches its contents, we could not assess the existence of anything. Actually, we have no means besides mere ideas to grasp reality.[71]

The problem of the role of ideas in grasping external reality is re-examined by De Volder in his *Disputatio de certitudine clarae et distinctae perceptionis* (1689). De Volder defines clarity and distinction not only in terms of the immediate awareness we have of ideas,[72] but also in terms of the compulsion to the assent. Plainly, this compulsion concerns only the perception of the nexus of several ideas, that is, propositions, as in the case of mathematical and metaphysical principles like "totus esse maius sua parte" and "factum infectum reddi non

---

[69] "Quae quidem omina non difficulter evincent Dei existentiam, cum iam constet, ideam, quam habeo divinitatis, tantopere differre ab ideis aliarum rerum, quantopere ipse Deus sic existere ab illis quoque rebus existentibus diversus foret. Habeo ego ideam Dei, illa certe causam requirit. Nihil enim aut dari aut concipi potest quod existentiae suae essentiaeve causam non habebit, sive a semet ipso, sive ab alio. A qua igitur causa illa est? Vel certe ab ea quae est infinita et omnino perfecta, vel a finita: illud si quis dixerit, eo ipso fatebitur rem infinitam, hoc est Deum dari," De Volder 1685, p. 40.

[70] "A quo argumento antequam discedam, non inutile forte erit ex iisdem fundamentis [...] demonstrare conclusionem. [...] Quod ut fiat, illud primum considerari velim, omne id quod sub perceptionem nostram cadit, si forte non existat, existere tamen ex sua natura posse," De Volder 1685, pp. 41–42.

[71] "Verum, inquiet forte quispiam, ex cogitationibus nostris, quae multa comprehendunt, quae in rerum natura non reperiuntur, non licet concludere rerum existentiam. Imo vero existentiam concludere nisi ex cogitationibus nullo modo licet," De Volder 1685, p. 30. See also p. 36: "nam si omnes meae, quas habeo, ideae tales sint, ut nullam extra me causam agnoscant, de nullius quoque rei a me diversae existentiae certus fieri potero."

[72] "Quamvis ea sit mentis nostrae natura, ut suas operationes prae caeteris rebus clare et intime cognoscat, cum omnis cogitatio conscientiam sui involvat. Attamen verba idonea satis ad aliis indicandum, quae et qualesnam sint, vix excogitare potest," De Volder 1689, p. 1.

potest."[73] Clarity and distinction as *norma veritatis* are grounded in our absence of freedom to assent to such propositions: in fact, we cannot doubt that an external force is not deceiving us, as we cannot frankly work this hypothesis out.[74] This foundation of evidence as the norm of truth – or of the highest *scientia*, to be found in metaphysics and mathematics – is strengthened through an appeal to the existence of God, which is assumed in De Volder's argumentation. First of all, since *scientia* is something real and it is a perfection, it requires a cause. As God is the cause of positive beings, He is the cause of such knowledge. Moreover, since we are forced to assent to evident propositions, God would be the cause of error if He compelled us to assent to false principles. Finally, since truth is an attribute of God, we conceive evident principles in the same manner as God: otherwise, God will reveal something of Himself not matching His nature (De Volder 1689, pp. 2–3, esp. theses 3 to 5). Actually, De Volder proves that what we evidently conceive can truly exist by an appeal to the existence of God: whose existence, however, is proved in his *Disputationes contra atheos* by means of the same assumption. Apparently aware of this circle, De Volder underlines that the reliability of evidence as *norma veritatis* does not truly need a demonstration, like "duo et tria facere quinque." In fact, the evidence is so compelling that even past evidence cannot be put in doubt.[75] The argument based on the existence of God, therefore, merely confirms what has already been indubitably perceived. As stated above, we simply cannot doubt some principles: therefore, in the case that something external to the mind exists, we cannot deny that it must obey such principles.

---

73 "Ut ipsas voces aliquo modo determinemus, claram et distinctam perceptionem habere dicimur, ubi aut unam eandemque ideam tam evidenter percipimus, ut nullam ignorantiam cum illa commixtam cognoscamus. Aut idearum nexum et relationem ad se invicem, absque ulla confusione cum aliis ideis, tanta cum claritate et evidentia, mentis acie intuemur, aut non possimus, quin assentiamur cum plena voluntatis nostrae lubentia. Prout ex gr. (quia non de unius ideae, sed de idearum evidentia, impraesens agemus) intelligimus, totum esse maius sua parte, factum infectum reddi non posse, et eiusmodi sexcenta. Unicuique enim attendenti fit manifestum, se a talium veritati, assensu iudicium suum abstinere non posse, ex quo sequitur cum nullo modo errare, quod hocce exercitio serio et modeste paucis defendere aggredimur," De Volder 1689, p. 1.
74 "Hanc itaque veritatem edocemur primo ipsa experientia. Ubi enim clara distincta alicuius rei perceptio adest, tam plene de veritate rei perceptae convincimur, ut certi simus, nos errare ne per ullam quidem potentiam posse. Quod si fieri posset, iam semper aliquis nobis remaneret scrupulus, an non falleremur. Quotquot vero sumus, experimur, nos de talibus veritatibus praesenti illa clara et distincta perceptione, quicquid etiam moliamur, dubitare non posse: ut cum cogito totum esse maius sua parte, et c.," De Volder 1689, p. 2.
75 "Quia quis aliquando, absente illa clara perceptione, dubius haerere posset, [...] ideo hoc argumentum, adducitur [...] illud omne, quod aut unquam evidenter percipimus, aut in posterum sic percepturi, sumus, certum et inconcussum est," De Volder 1689, p. 3.

A metaphysical foundation of scientific knowledge is finally developed by De Volder in his *Exercitationes adversus Censuram* (1695). Whereas Jean Le Clerc wrote in his *Bibliothèque choisie* that the *Exercitationes* – also printed in separate booklets (De Volder 1691–1693) – were published without De Volder's permission,[76] it is beyond any doubt that De Volder actually embraced their contents, being encouraged by his students to write a defence of Descartes's philosophy against Huet's *Censura* (De Volder 1695, vol. I, pp. 1–4). The *Censura* addresses, like Revius's *Consideratio theologica*, different steps of Descartes's metaphysics and physics: namely, 1) Descartes's use of doubt and the validity of the *cogito*; 2) his criterion of truth as clarity and distinction; 3) his theory of mind and knowledge; 4) the demonstrations of the existence of God; 5) his theory of matter and void; 6) his explanation of gravity; 7) the overall value of Descartes's philosophy. Actually, most of the book is devoted to the first two points, as these are the foundation of the rest of Descartes's philosophy. Huet's main argument, as that of Revius's *Statera*, is that Descartes failed in providing philosophy with the degree of certainty he aimed at. He erred in the entire construction of his philosophy, starting with the doubt, which for Huet turns out to be a radical rejection of all knowledge as false, rather than a suspension of judgment or an analysis of the contents of the mind (Huet 1689, pp. 9–12; see Schmaltz 2002, pp. 60–61), and *cogito*, which is an inference subjected to the rules of logic, which are however rejected on the basis of doubt. As *cogito* is not clear or distinct but in words, Huet takes Descartes's criterion of truth as built *ad hoc* (Huet 1689, pp. 14–38; see Lennon 2008, pp. 79–80, 137–148; Schmaltz 2002, pp. 224–225). Also, the theory of mind of Descartes is vitiated by his errors on *cogito* and clarity and distinction, as by these he traces some conclusions on the nature of mind, which is completely distinguished by body and does not have relations with it (Huet, *Censura*, chapter 3; see Schmaltz 2002, pp. 226–227). Even the demonstrations of the existence of God are vitiated by his use of doubt, as it is taken as something real, and it cannot be overcame by such demonstrations (Huet, *Censura*, pp. 134–135; see Lennon 2008, p. 116).

Following the series of chapters of Huet's *Censura*, De Volder's *Exercitationes adversus Censuram* serve, rather than to enter into a dispute about what Descartes had truly said, or a dispute about persons, to defend the general

---

76 Le Clerc 1709, p. 383. Le Clerc was however right, as De Volder states, in writing that De Volder was driven by his students to answer Huet and that he was not supporting Cartesian philosophy as a whole, but only some of its general principles: see Le Clerc 1709, pp. 381–382.

principles of his philosophy.[77] Thus, it starts with an *Exercitatio de dubitatione universali*, concerning doubt as the grounding step of philosophy.[78] As a truth *per se nota* is required, that is, not relying on any other knowledge, doubt proves to be the only means to acquire it.[79] As doubt does not provide the annihilation of all knowledge, but only a strict examination of mental contents (De Volder 1695, vol. I, pp. 12–15), it is to be first applied to axioms such as "totum maius esse sua parte": thereby, it makes us aware that we cannot refuse to assent to this kind of proposition.[80] This is also the case with respect to the proposition "cogito ergo sum," the certainty of which cannot be refuted through Descartes's arguments concerning the unreliability of sense perception, of the difference between

---

77 "Quae disputationis ratio, ut ad personarum, ubi de iis agitur, defensionem multum potest, ita ad veritatis perquisitionem nihil confert. Sive hoc enim senserit Cartesius, sive illud, quid attinet inquirere? An verum sit quod senserit, ubi de sensu constat, hoc demum scientiarum et veritatis interest. [...] Accedebat et illud, quod hac occasione sperabam me eam philosophandi rationem, quam licet nequaquam in omnibus, in generalibus tamen sequendam existimo, ab innumeris liberaturum cavillationibus, et praecauturum hac via, ne tyronum, quibus haec solummodo destinantur, animi per illas avocentur a rei veritate exacte inquirenda et agnoscenda," De Volder 1695, vol. I, pp. 2–3.

78 De Volder distinguishes two methods in the foundation of science: namely, the Cartesian way, based on doubt, and a careful analysis of every opinion. This second way turns out to be an endless analysis of preconceived opinions: a foundation *ex novo*, actually, is the only suitable way to ground the new science (De Volder 1695, vol. I, pp. 7–9).

79 "Adeoque primam et praecipuam eius, qui nunc demum incipit philosophari, hoc est, certam scientiam quaerere, curam esse debere de fundamentis, quibis secure postmodum philosophiae suae superstruat aedificium. [...] Prima certe scientiarum fundamenta, si quae sint, per se nota sint, necesse est, adeoque ad sui agnitionem nulla alia re indigent. Non priori, siquidem sint prima, non posteriori siquidem omnium illorum cognitio a fundamentis pendet. Aut igitur omnis de fundamentis scientiarum ponendis exuenda sollicitudo est, quod nemo dixerit, aut si quae restat, id certe agendum, ut dum in fundamentis elaboramus, ita progrediamur, ac si nihil nobis cognitum foret, nullasque in mente haberemus opiniones. Quarum dein, si ordine progredi velimus, nulla assumenda erit, nisi quam fundamentis positis consentire, et ex illis necessario fluere certo animadvertimus. [...] Atqui hoc unice est, quod sibi vult haec tantopere a multis exagitata universalis dubitatio," De Volder 1695, vol. I, pp. 11–12.

80 "Progrediatur igitur noster hac methodo, ut primo generatim accuratius dispicienda sibi proponat omnia, ipsa etiam axiomata. Haec enim dum fundamenta futura sunt omnis ulterioris ratiocinii, cavendum summopere, ne in illis ulla fallaciae superesse queat suspicio. Quod dum agit, dum ad ipsa axiomata attendit, dum sibi horum aliquod ob oculos ponit, totum ex gr. maius esse sua parte, experitur statim hoc inter ea esse, de quibus, quantumcunque etiam dubitationi indulgere studeat, dubitare vel minimum non est in ipsius potestate," De Volder 1695, vol. I, pp. 18–19. Mathematics is therefore taken by De Volder as the paradigm of science, as mathematicians consider all the axioms and deductions carried out, until they find they cannot doubt of them.

sleeping and waking, and the hypothesis of a deceiving genius.[81] Whatever opinion we have of the power of God, consciousness convinces that propositions like "cogito ergo sum" or "duo et tria facere quinque" are indubitably true.[82] As there are no other means to ground the truth of our knowledge but our consciousness, our lack of freedom in assenting to certain propositions becomes the mark of the truth of such propositions. Accordingly, De Volder provides a strong, psychological foundation for clarity and distinction as the mark of truth, i.e. it redefines them in terms of the impossibility of negating them. Plainly, this canon of truth must not be applied to Revelation or to disciplines based on authority or experience, such as medicine: it merely concerns metaphysics and mathematics, which are both based on pure reason (De Volder 1695, vol. I, pp. 30–31). However, since De Volder considers both physics and mathematics as having the geometrical body as the common object, he provides physics with a metaphysical foundation with regard to its basic notions: extension, size, shape, and motion. Since we have no means to doubt their being capable of representing bodily features, no more foundational arguments are required to prove their reliability. In this way, a foundation of scientific knowledge can be provided without the demonstration of the existence of God, which is however taken into account in the justification of the possibility that something external to the mind can exist, and that ideas of mathematics are true. This is developed in his *Exercitatio de clara et distincta perceptione criterio veri*, which, in part, restates the contents of his *Disputatio de clara et distincta perceptione*. Besides remarking that we deal only with ideas, and that we rely on their properties – such as evidence – to assess their truth (De Volder 1695, vol. I, pp. 47–49, 52–55), De Volder deepens the point by analysing the very notion of truth. This is defined as *veritas rei*, or the very being of things what they actually are (De Volder 1695, vol. I, p. 57). Secondly, it regards

---

81 "Atqui tu nescis, an non fallaris in evidentibus: verum est, sed etsi hoc generatim nesciam, an hoc fieri nequeat, scio tamen hoc in casu me non falli, et experior illam generalem rationem me non posse abducere, quin his effatis, cogito, ergo sum, totum est maius sua parte, et c. absque ulla haesitatione assentiar. [...] Sed concedat tamen nobis, obsecro, si talis quis inter homines reperiatur [...] ut ex ea, quam in nobis experimur mentis nostrae constitutione argumentemur, et ea pro certis habeamus quae nobis certa sunt, et de quibus, quidquid sit de fide sensuum, de discrimine inter somnum et vigiliam, de genio deceptore, et si quid porro est, quod ad scientiam nostram labefactandam potest adferri, conscius ego mihi sum, me ita certum esse, ut quidquid agam, quidquid moliar, ut dubitem, dubitare tamen non possim," De Volder 1695, vol. I, pp. 24–25.

82 "Cum cogito duo et tria facere quinque conscius mihi sum, huius evidentiae, quae in mente est, hunc esse effectum, ut plane hac de rei veritate certus fiam, ut sciam, quidquid sit de illa potentia Dei, utut illa me forte in evidentissimis aliquando possit decipere, illam tamen hoc in casu me decipere nequam," De Volder 1695, vol. I, pp. 26–27.

ideas that do not involve any affirmation or negation, that is, non-propositional ideas. These may be true or false insofar as they represent something different from themselves, namely, an *obiectum*. Ideas of this kind are always true – since they represent something – but can be different according to their clarity and distinction.[83] Non-propositional ideas, therefore, are false whenever they do not represent anything at all, or whenever they represent just an affection of the mind, whereas they are believed to represent something else, or *vice versa*. This is the case of obscure and confused ideas.[84] The ideas we grasp in a clear and distinct way, on the other hand, do not represent mere affections of the mind: like the nature of a triangle, which is not a mode of the mind but can exist as something different from an idea, as its notion does not include that of mind. Moreover, even if no triangle existed outside the mind, its idea would still be true, thanks to the existence of God. Insofar as God exists, He can create whatever we evidently conceive as possible.[85] Thus, the power of God is the ground of the truth of ideas as these represent natures with evidence. The demonstration of His existence, on the other hand, is proved in the next *Exercitatio de idea Dei* and *Exercitatio de*

---

83 "Conceptus autem quod spectat, in illis omnibus verum est, ita me affici, me hoc cogitare, me hos illosve conceptus inter se coniungere, disiungereve, haec enim omnia aeque vera ac certa sunt, ac certum est me cogitare [...]. Manifestum equidem est [...] cogitationes [...] plures reperiri alias [...] quae praeter illam mentis meae affectionem, aliquis mihi repraesentant, quod ab ipsa mentis meae affectione, ipsa cogitatione mea plane concipitur distinctum. [...] Hoc quod ita mihi repraesentatur, obiectum huius mei conceptus dico, hoc rem reive modum voco, ipsam vero cogitationem huius ipsius, quod repraesentatur sive rei, sive modi, ideam voco [...]. In hisce autem rerum ideis hanc manifeste invenio discrepantiam, quod quaedam harum maximam et perspicuitatem et evidentiam sibi habent coniunctam, quas claras distinctasque vocat Cartesius, quaedam vero [...] confusas vel obscuras dicit. [...] In hisce autem conceptibus, utpote qui nullam affirmationem aut negationem continent, cum sola repraesentatio locum habeat, et semper verum sit mihi hoc illudve repraesentari, huius enim conscius sum, non video, quid in illis falsum dici queat," De Volder 1695, vol. I, pp. 58–59.

84 "Si tamen has veras, illas falsas, ut in vulgari usu est, dicere velimus, non video quaenam falsae dici queant, nisi vel quae videntur aliquid menti repraesentare, cum nihil offerant, vel etiam quae affectionem mentis repraesentant tanquam quid ab illa affectione diversum, et viceversa. Qua ratione si ideas in veras falsasque distinxero, clarae nihilominus et distinctae perceptiones omnes certissime erunt verae," De Volder 1695, vol. I, p. 60. In his *Disputationes contra atheos* De Volder considers the case in which the contents of ideas would not represent a nature: plainly, it would not be the idea of such nature: De Volder 1685, pp. 36–37.

85 "Erit fortasse triangulum sola mentis meae affectio? Nihil minus. Quod enim mihi repraesentatur nihil habet cum cogitatione commune. Sed forte nullum dabitur extra me triangulum, nec hoc affirmo, non dico dari, dico concipi. Quid ergo? [...] Siquidem ego, qui iam novi Deum esse, et illum omnipotentem et omniscium esse, facile etiam novi omnia illa, quae hoc modo a me percipiuntur a Deo etiam percipi, illum omnia ea posse, quae possum concipere," De Volder 1695, vol. I, pp. 60–61.

Deo, where De Volder restates the proofs already given in his *Disputationes contra atheos* (De Volder 1695, vol. II, pp. 1–7, 30). These are based, in any case, on the assumption that ideas can match something different from themselves.[86] This, actually, is the ultimate basis for the reliability of ideas, whereas the demonstration of the existence of God merely assures us that finite things may exist.

6.5 Foundation and philosophy of science go separate ways: De Volder and De Raey

In the light of this analysis, one can draw some conclusions with respect to De Volder's philosophy. According to him, physics is a hypothetical discipline. First of all, because it concerns the sensible world, whose existence can be proved only through the application of the causality principle to adventitious ideas. Moreover, the laws governing phenomena can be assumed only hypothetically, as their formulation also relies on experience. Finally, even the basic notions of physics are hypothetical insofar as the world could have been created according to different principles. Still, we have no other means to grasp the constitution of the world besides the notion of extension and its modes. In this sense, nobody can doubt that extension and motion exist – as he maintains in his *Cogitationes* – only in a non-metaphysical sense. Whereas Le Clerc voiced his unconcern about metaphysical problems as such, De Volder is to be considered an interpreter of Descartes's metaphysics, clarifying some of its implications. The hypothetical certainty of mechanical explanatory principles, indeed, results from Cartesian metaphysics as this does not allow us to demonstrate that the world actually follows certain physical principles, but only that this is our only means to explain its phenomena. Moreover, one can find a limited recourse to the notion of God both in defining the first principles of nature and in ensuring the reliability of our physical explanations. De Volder, in fact, limited the scope of his foundational theory to metaphysics, leaving no place for a recourse to natural theology as the basis of science. In fact, he also showed no interest in logic, namely, in an exposition of the methodology of natural philosophy, roughly set forth only in his two inaugural speeches. He does devote some words to logic in his *Oratio de coniugendis philosophicis et mathematicis disciplinis*, but states that even if Scholastic logic concerns the working of the mind, it does not ask how we can recognise the truth.

---

86 Andala would criticise De Volder in his 1718 *Apologia pro vera et saniore philosophia*, as he maintained a 'mechanical' correspondence of thoughts and physical impressions, not offering any account of their interaction: Andala 1718, p. 6; see Van Ruler 2003c, p. 110.

This can be taught and put into practice by mathematics alone, which turns out to be the only *organon* of philosophy.[87] Also, his difficulties in dealing with the methodological consequences of his openness to the use of experience are confirmed by his unconcern with a justification of the reliability of sense perception, which is briefly grounded in the application of the causality principle to sense data, as seen above.[88] This problem, in fact, will be consistently addressed by Dutch Newtonians, whose shift in epistemic paradigm will lead to a more attentive foundation of experience as a source of scientific knowledge, as well as to a more comprehensive theory of method.

From a broader perspective, De Volder's case shows the 'dead end' into which Cartesian physics came at the end of the seventeenth century. The emergence of an experimental-mathematical science rarely concerned with metaphysical issues, and whose foremost success was the discovery of the correct laws of impact by Huygens, Wren, and Wallis,[89] probably made De Volder adopt a

---

87 "Verumenimvero obstrepente huic meo sermoni videre mihi videor, ingentem dialecticorum turbam, qui, quod artibus adscribo mathematicis, soli dialecticae vendicent, hance eam esse praedicent, quae vera doceat a falsis distinguere, quae ratiocinii laqueos enodet, quid ex quolibet sequatur explanet, omnes denique ostendat et veri ratiocinii modos, et falsi technas. [...] Noscet enim accurate quisquis hanc excoluerit, quid sit demonstratio, quae sint conditiones ad veram demonstrationem requisitae, quales oporteat esse praemissas, qualem conclusionem, et quae huius farinae in dialecticis traduntur plurima. Itane vero? Quid sit demonstratio, quid rite sequatur, quid minus, id ego accuratius cognoscam ex illorum hominum tricis, qui nulla saepenumero demonstrationes percepere unquam, quam ex frequenti demonstrandi et clare sequelas percipiendi exercitio? Illa sc. ars docebit me certo ratiocinari, firmas formare conclusiones, quae ipsa vix ullas demonstrationes habet, [...] omnibus suis ratiocinandi regulis res nequidem suas demonstrare valet. [...] Quae quidem ita accipi nolim, ac si despectui plane haberem ea, quae a dialecticis tractantur, sed quod existimem, ad iudicii nostri comparandam firmitatem, hanc cum artibus mathematicis ne conferendam quidem esse. Neque enim dubito, si ex una parte habeamus eum, qui dialecticorum instructus sit dogmatis, ex altera, qui horum omnium ignarissimus solas artes didicerit mathematicas, quin in pari etiam intelligendi facultate, hic illo ad perceptionem certae veritatis multis partibus futurus sit habilior," De Volder 1682, pp. 9–10.
88 See also De Volder 1695, vol. I, pp. 124–126: "verum cum ex lege coniunctionis mentis cum corpore id fiat, ut nonnullos ex motibus corporeis certae comitentur perceptiones, has perceptiones non dubitat, quin debeatur menti, has, sensus proprie [Cartesius] vocat. [...] [Cartesius] inquirere voluit, an ex ideis suis alterius rei praeterquam sui existentia deduci queat, cumque non facile esset omnes ideas simul considerare, eas sibi primum examinandas censuit, de quibus probabilis esset coniectura, eas alterius rei existentiam comprobaturas." Actually, De Volder also avoids referring to moral certainty with regard to sense experience: see Nyden 2013, p. 243.
89 "The papers by Wallis and Wren (1668) and Huygens (1669) settled on a widely shared and recognized mathematical treatment of the rules of collision that were claimed to have high empirical confirmation and predicted surprising empirical results; this post-Galilean analysis of motion became an autonomous practice relatively insulated from metaphysical and theological concerns," Schliesser 2011, p. 109. See also Jalobeanu 2011.

broader attitude towards experiments, and to carry on a 'de-metaphysicalisation' of physics. However, his adherence to Cartesian physics and metaphysics prevented him from going further than merely announcing a new experimental science and a didactical use of experiments, aimed at demonstrating those truths consistent with Descartes's principles. Moreover, De Volder's case shows the incapability of Cartesian metaphysics to support this new science, as it offered no arguments to provide experience with a foundation, nor even – according to De Volder – a demonstration of the truth of the principles of physics. Therefore, the encounter of English science with the Continental Cartesian tradition apparently led to a reflection on the principles of science independently of their metaphysical foundation. Not surprisingly, in the same years (from 1668 onwards) De Raey was developing his reflections on language, which take into account the semantic value of the use concepts drawn from experience, as well as of the *modi considerandi* or second notions – which also include mathematical concepts – in (natural) philosophy. This analysis is in the line of his criticisms on the erosion of Descartes's metaphysics by Hobbes, as this would deprive the terms meaning such concepts of any possible use. Moreover, it reinforces his separation thesis, as his analysis is aimed at finding the proper place for use of different kinds of concepts. The *Cogitata*, indeed, came to light after the polemics in Franeker of 1686–1687, for which he blames Röell in his letter of 1687 – appended to the *Cogitata* itself. However, this work mitigates De Raey's earlier concerns with a physics based on intellectual principles only (as expounded in the *Clavis*), and rejecting the use of *modi considerandi* (as maintained above all in his *Specimen*). At the same time, as a text providing a reflection on the language of the sciences separated from metaphysical argumentations, it also signalled the emergence of a new genre of philosophical reflections on given scientific practices.

6.6 Philosophy of language as philosophy of science: De Raey's *Cogitata de interpretatione*

6.6.1 A novelty in the philosophical reflections on language

Like his unified logic and metaphysics, also De Raey's treatment of language constitutes a novelty in the history of philosophy. Indeed, De Raey was one of the very first philosophers to treat language in a separate body of literature, although formally connected to his logic, i.e. metaphysics, which is the discipline devoted to the analysis of the contents of the mind. De Raey's analysis was preceded by a long-term process of transformation in the logical approach to language, which was ultimately determined by the erosion of the Aristotelian worldview in the early modern age and brought about new reflections of the relation of language and philosophy. The treatment of language in the Scholastic and Renaissance traditions, as that of Keckermann, Burgersdijk, Lorenzo Valla, and Ramus, focused on the semantic properties of words and sentences without considering what kinds of entities these refer to, and for the sake of providing the formal rules of organisation of syllogism with well-defined matter. In the seventeenth century, on the other hand, the emergence of alternative worldviews brought attention to the philosophical consequences of the use of words, as these mean entities discarded from the new worldview. Bacon, Hobbes, Clauberg, Antoine Arnauld, and Pierre Nicole considered language as a possible cause of error in philosophy, while Géraud de Cordemoy provided a Cartesian, mechanistic account of its formation. Eventually, De Raey faced the semantic, ontological and psycho-physiological aspects of the formation of language, for the sake of the re-definition of the linguistic meanings and references of ordinary speech. With respect to the traditional ways of considering language, De Raey proposes a new kind of analysis against the background of Descartes's metaphysics. In the disciplines of the *trivium* language is considered according to the correct formulation of phrases (in grammar), to its *ornatus* (in rhetoric), or to the formulation of arguments (in Ramist dialectic and Scholastic logic). Moreover, all these disciplines deal with the 'vulgar' meanings of words: that is, with the basic concepts of the Aristotelian worldview, such as of sensory qualities, essences, forms, and particular substances, as if these would exist outside mind (De Raey 1692, *Cogitata*, pp. 1–6, 18–19). In fact, Ramus dealt in his *Dialectica* only with *notatio* and *coniugatio*, which are two of the topical arguments or *loci*, that is, the places in which to find the middle terms for syllogisms: *notatio* is the very definition of a term, while *coniugatio* is the finding of its synonyms (Ramus 1543b, chapters 23, 24). Keckermann, while maintaining that logic has *ratio* (i.e. the mind and its contents) as its primary object, while *oratio* (i.e. their expression in language) is its secondary matter, follows a linguistic criterion in distinguishing between a simple and a complex content of the mind. Indeed, Keckermann identifies simple concepts – which he labels 'notiones', 'cogitationes', 'conceptus' or simple 'themata' – by their being conveyed by simple terms. Before categories and any other *thema*, therefore, he deals with the notion of *vox* or *terminus*, which is defined as an articulated sound provided with a *significatio*. *Significatio* is the arbitrary relation of a word and a concept, and makes possible that such a concept is recalled and presented to mind when a word is uttered (Keckermann 1613a, pp. 68, 70–76). Reversing the order of treatment of Keckermann's logic (on the assumption that concepts are learned before words) Burgersdijk deals with *interpretatio* or speech after having treated simple *themata* – i.e. non-propositional concepts – but before explaining which are the instruments of logic, that is, definitions and syllogisms, which are the ways to use concepts in reasoning. He focuses on the kinds of speech: as *dictio*, or the utterance of names and verbs alone, and *oratio*, which is their union in a sentence. *Dictio* is analysed according to its meaning, which is the concept that words make known or recall to mind.[90] *Oratio* or complex *thema* is considered according to its being an *enunciatio*, that is, a truth bearer, and according to its being simple or composed by more sentences, which form a copulative, hypothetical, disjunctive, adversative, and relative complex sentence – according to the kind of their conjunction. Moreover, sentences can be pure or modal, as they express the kind of relation between their parts, and universal, particular, indefinite, singular, and so on. Eventually, the considerations of Burgersdijk on language are aimed at providing syllogistic reasoning with a foundation, as from the kind of sentences different kinds of syllogisms are formed, according to the depending on the kind of sentences as equivalent, subaltern, opposite or convertible (Burgersdijk 1660, pp. 126–142). In all these cases, accordingly, the treatment of speech is finalised to the development of a theory of reasoning centred on syllogisms, and little attention is paid to what words actually signify, since the Aristotelian worldview entails the correspondence between particular things and concepts.[91] Even if the Italian humanist Lorenzo Valla enquired into the entities referred to by ordinary language and traced the references of all the terms to substances, qualities and actions, redefining the terminological apparatus of Scholastic philosophy (Nauta 2009, part I), it was only with the appearance of alternative worldviews that language emerged as a philosophical problem.

Indeed, a substantial change in the way of conceiving the ontology entailed by Aristotelian language was brought about by Bacon and Hobbes, whose critiques of the linguistic signification of Scholastic terminology, according to De Raey, entailed a materialist ontology.[92] While agreeing with Aristotle that "words

---

90 Burgersdijk 1660, pp. 111–126. According to Earline Ashworth, "by the late fifteenth and early sixteenth centuries the standard definition of 'significare' was 'to represent something or some things or in some way to the cognitive power'," Ashworth 1981, p. 310. For Burgersdijk "voces articulatae significant animi conceptus, primo scilicet, atque immediate. Nam res etiam significant, sed mediantibus conceptibus," Burgersdijk 1660, p. 111.

91 "Hic est concinnus ordo et rerum ipsarum in natura et intellectionis seu cogitationis humanae," Keckermann 1613a, p. 70. See Dawson 2011, p. 25.

92 Hobbes is portrayed by De Raey as a misuser of Bacon and of Descartes himself, as he discarded the Aristotelian worldview: "[...] neque alia ratio aliorum nominum generalium est, quibus utitur Aristoteles in definitione animae, quae concedimus, non primae verum secundae notionis sive intentionis nomina esse [...]. Hinc vero non sequitur, ut multi putant, et forte Verulamius putavit [...] quod inania haec nomina sint, sive voces insignificantes, uti supra audivimus Hobbesium loquentem, atque suo hoc insigni errore abutentem ista Verulamii, et imprimis Cartesii observatione," De Raey 1692, *Cogitata*, p. 306.

are the images of cogitations, and letters are the images of words" (Bacon 1605, p. 59; see also Bacon 1623, book VI, chapter 1), Bacon assumes that language can truly express the order of external things only if words signify the forms and essences underlying the qualities one acknowledges by experience. Therefore, in order to speak meaningfully one needs to go through an *interpretatio naturae* enabling the recognition of essences by a method of induction, or the core of a new logic by which he aims to replace Aristotle's *Organon*. In fact, for Bacon, Aristotle's categories are nothing but badly abstracted concepts leading the whole of philosophy into confusion: in his *Novum organum* he acknowledges two main problems related to the idola fori or the errors conveyed by language: the use of terms which do not signify anything, and those which signify something obscure. Ordinary language, indeed, does not match the real essences of things, but only our immediate understanding of them (Bacon 1620, book I, aphorism 59, 60 and 127; see Losonsky 2006, pp. 42–45). Hobbes would further Bacon's criticisms of Aristotelian language in his *Elements of Law, Natural and Politic* (1640), *De corpore* (1655), and *Leviathan* (1651). He maintains that all cognition comes from sensation and results in different kinds of concepts impressed as images in the brain.[93] In language, one can recall a concept by another which is arbitrarily attached to it as a mark or name. Accordingly, science is nothing but the knowledge of names and concepts rather than of named things, and by names and by named concepts we are reminded of things that impressed our senses.[94] As a consequence, Hobbes considers as insignificant the majority of the terms of Scholastic metaphysics, as these are general abstractions from material entities, and do not have representative content. For De Raey, this makes communication impossible both in philosophy and in everyday life.[95]

Clauberg also confronted the problem of language from a new (this time, Cartesian) perspective. Maintaining in his *Metaphysica de ente* (1664) that words are signs as they make something known by prompting a concept – broadly conceived as mental contents (Clauberg 1691, *Metaphysica de ente*,

---

93 De Raey would quote from Hobbes's *Obiectiones* in this regard: "si pro cogitatione motum corporis, pro idea in cogitatis nescio quae simulacra corporea, denique pro ipsa in nobis mente corpus supponas, ac si ut aliqui fingunt *cogitatio opus corporis cogitantis sit, et similis esse possit in homine et bestia cogitatio, non quicquam amplius quam corpoream rei similitudinem complectens,*" De Raey 1692, *Cogitata*, p. 211; see AT VII, p. 182. Hobbes's *Obiectiones* were his first text translated into Dutch, in Descartes 1657.

94 Thomas Hobbes, *Elements of Law, Natural and Politic* (1640, handwritten; published without Hobbes's permission in 1650 (Hobbes 1650)), part I, chapter 2, §§ 2–3; 5, §§ 1–3; 6, § 4.

95 "Diminutio primum nos a multis cogit vocabulis abstinere, ac si sint voces insignificantes vel barbarae, sine quibus commode loqui possemus et philosophari, ut putant," De Raey 1692, *Cogitata*, p. 209; see Hobbes 1655, part I, chapter 3, and Hobbes 1651, *De homine*, chapter 5, § 5.

pp. 336–337) – Clauberg deals with language first from the point of view of the clarification of such meanings in the third part of his *Logica*, by the philological disciplines of lexica, *grammatica*, and *rhetorica*, through which one can find the meanings of simple terms and act as *media interpretandi: lexica* helps in the definition through etymologies, and finds out the different meanings of terms according to the different disciplines of use; *rhetorica* serves to find out the figurative meanings in a text; and *grammatica* helps in avoiding the fallacies coming from different ways in spelling and declining terms, as these can give rise to ambiguities in meaning. Yet, in the fourth part of the *Logica*, devoted to the analysis of the truth of speech, Clauberg underlines the difference between a philological and a philosophical approach, as a philosopher – i.e. *logicus analyticus* – scrutinises the actual references of single words, as these are supposed to refer to some kind of entity, either mental (as a *modum considerandi* or being of reason) or material. It is the case, for instance, of passions of the soul, which mean both a modification of the body and the mind, yet are expressed by a simple voice, entailing a composite meaning (Clauberg 1691, *Logica*, pp. 846–850, 869–870).

A further position on the problem of language is testified to by the *Grammaire* (1660) and *Logique* (1662) of Port Royal – though these texts are not directly mentioned by De Raey (see *supra*, section 5.2.1). In their *Logique*, Arnauld and Nicole assume the traditional theory of meaning according to which signifying is to make something known (Arnauld/Nicole 1683, part 1, chapter 14). Moreover, they maintain a traditional standpoint on what is signified by words and ideas in the *Grammaire*, where the objects of thought are divided into individual substances and accidents (Arnauld/Lancelot 1660, chapter 2). However, they replace *themata* with ideas such as the objective contents of mind, constituting the 'spiritual' component of words (Arnauld/Nicole 1683, part I, chapter 1), and, like Bacon, aim to replace the Aristotelian categorisation of the world with a new conceptualisation, as traditional categories are substituted by the concepts of mind, body, measure, position, shape, motion and rest, which actually match features of reality (Arnauld/Nicole 1683, part I, chapter 3). Moreover, they bring attention, in the first section of the *Logique*, to the fact that words such as 'sensation' have a composite meaning often overlooked in ordinary speech, and require a more stable definition (Arnauld/Nicole 1683, part I, chapters 11, 12).

The case of De Raey can be interpreted as a further development in Cartesian reflections on language and in the assessments over its entailments for logic, on the one hand, and for philosophy as such on the other. Following Clauberg, De Raey maintains that besides the common, Aristotelian meaning of words (De Raey 1692, *Cogitata*, pp. 8–9), it is possible to find a philosophical or Cartesian one: for this reason, he develops a fourth, new consideration of language – namely, a philosophical consideration of language (De Raey 1692, *Cogitata*, p. 87) – and addresses first the basic concepts of semantics. According to De Raey, words have a *sensus* or *intellectus*, and a *significatio*. *Sensus* or *intellectus* is the mental content conventionally associated to the 'body' of the words, that is, to the ink or the sound which exists outside mind and is perceived by mind through sensory experience. *Significatio*, as for Burgersdijk, is the act of meaning or making something known.[96] Hence, mental contents such as intellectual ideas and sense data are the *sensus* of words, and through them names signify those things ideas represent: if mental contents are the senses of words, things are their reference, both made known (i.e. signified or meant) by words.[97] This tripartite scheme of signification turns out to be necessary for De Raey to allow the use of many terms which do not have a reference in bodies even if they are supposed to. De Raey analyses terms according to an eightfold categorisation guided by a Cartesian ontology, and distinguishing between words signifying mental contents resulting from a movement of the body and words signifying intellectual ideas independent from such movement. The kind of philosophical analysis of language employed by De Raey is a clear and distinct definition of what words signify, even when such signification involves 'anticipations' or obscure concepts of what things are, as in most of the Aristotelian vocabulary (De Raey 1692, *Cogitata*, pp. 11–14). This analysis is grounded, therefore, not only on the theory of knowledge expounded by De Raey in his *De constitutione logicae* and *Pro vera metaphysica* but also on a physiological account of sensations and passions of the mind, which gives reasons for the meanings of words, for what compels men to talk (that is, for passions as these manifest themselves in discourse), and for what words actually are, that is, entities composed by modifications of matter, by our sensation of such modification and by the mental content (either an intellectual idea or a passion itself) arbitrarily attached to them (De Raey 1692, *Cogitata*, pp. 38–39). Thus, the causes of the mental contents meant by words are bodily motions in brain and heart which 'occasion'[98] sensory experiences (like Aristotle's five *sensibilia propria*), internal

---

96 De Raey 1692, *Cogitata*, pp. 825–829. In his *De cognitione humana*, printed in the Appendix of the second edition of his *Clavis*, De Raey writes that to signify means "potentiae cognoscenti [...] facere praesens," De Raey 1677, *De cognitione humana*, p. 244.

97 "Nomen [...] interventu ideae [...] refertur ad ipsummet corpus [...] in extantibus," De Raey 1692, *Cogitata*, p. 313.

98 In his *De forma substantiali et anima hominis* (1665–1667, in De Raey 1677) De Raey uses such terms as 'sympathia', 'harmonia', 'consensum', 'conspiratio' (De Raey 1677, *De forma substantiali et anima hominis*, pp. 569–570), stating that bodily motions prompt mind to produce its modifications (p. 524). Elsewhere, he states that that passions do not result from the union but are the very union, that is, the correspondence of the modifications of soul and body guaranteed by God: De Raey, *Disputatio philosophica qua quaeritur quo pacto anima humana in corpore moveat et sentiat* (De Raey 1663), in De Raey 1692, pp. 669–676.

sensations, as well passions like wonder, love, hate, cupidity, joy, and sadness (De Raey 1692, *Cogitata*, pp. 39–54). A 'physiological' theory of speech had been outlined by the French Cartesian Géraud de Cordemoy in his *Discours physique de la parole* (1668), following his *Le discernement du corps et de l'âme* (1666) and where De Cordemoy aims at discovering a criterion to identify the individuals provided with a soul. In fact, he restates the problem raised by Descartes in his *Discours de la méthode* and, like Descartes, maintains that the creative aspects of language cannot be explained in terms of mechanical processes but only by considering the 'creative' activity of an immaterial soul.[99] According to De Cordemoy, *significatio* is a thought arbitrarily joined to a sound or a line of ink: to signify thus consists of giving signs of thoughts (De Cordemoy 1668, pp. 122–123, 138). In his treatise on language he provides an account of how sounds are provided by considering human and animal anatomies from a mechanical standpoint – showing, for instance, how the sounds to which we allocate letters are produced, analogously to the mechanisms of musical instruments (De Cordemoy 1668, pp. 70–81). Accordingly, there is no need to assign an immaterial soul to animals, since their sounds are explainable mechanically: on the other hand, the novelty and creativity of human speech cannot be explained without recourse to a soul, which makes possible the process of signification (De Cordemoy 1668, pp. 109–114, 185–188). However, De Cordemoy does not display a full account of human passions nor of the concepts and things signified by words: he was interested in furthering Descartes's rare considerations on language only to provide a demonstration of the distinction of body and soul, therefore, he reinstalls Descartes's theory of body as a machine, rather than working on a theory of the formation of concepts and passions, and of their references. De Raey, on the other hand, displays a theory of passions as the foundation of his categorisation of the meaning of words.

6.6.2 The realm of sensibility

The first order outlined by De Raey is that of interjections, defined as "notae passionum inter loquendum." Their utterance evokes the concept of a passion, which is always brought to mind as someone uses them (De Raey 1692, *Cogitata*, pp. 63–66). The following orders, on the other hand, are guided by a more philosophical perspective, as they include names and verbs considered only according to their meanings. The second order contains names and verbs signifying

---

99 De Cordemoy 1668, pp. 8–21; see AT VI, pp. 58–59. See Ablondi 2005, pp. 80–86, 106–112, Cottingham 2008b, Chomsky 1966, p. 9.

passions by means of thoughts or concepts, which in turn are actions as they do not depend on the body for their creation, even if they are about a passion. Whereas interjections signify confusedly a passion and a concept, the terms of the second class signify properly the ideas of passions, and through these the passions themselves (De Raey 1692, *Cogitata*, pp. 66–67). In accordance with his theory of passions, signified passions are 1) the *affectus*, such as wonder, fear, hope, joy; 2) the natural appetites, such as hunger and thirst; 3) the sensations caused by something internal to the body, such as pain or pleasure. The words signifying affectus (1) can signify even the sole act in the soul, without the passion which comes after that of the body. Indeed, to act and to have a passion is the same thing in the soul, as it is a modification that we can consider in different ways (De Raey 1692, *Cogitata*, p. 73). Moreover, they can signify, according to their proper meaning, that modification of the soul which comes after that of the body. The case is analogous for the natural appetites (2), which can signify something pertaining to the sole mind, such as the *voluntas bibendi*, a modification of the body, or, more properly, a modification of the mind coming after a bodily motion. Also among the names of sensations (3) one find similar improper significations: that is, by 'hot' we can mean just a bodily modification (De Raey 1692, *Cogitata*, pp. 71–75). The terms of the third order are addressed in the same way, as they mean passions coming from a cause external to our body, like coldness, warmth, or Aristotle's five *sensibilia propria* (De Raey 1692, *Cogitata*, pp. 75–78). As in the case of the previous category, these terms have a proper meaning, that is, passion in the soul, and an improper meaning, or the sole bodily modification. The uses of these improper significations are legitimate, even if in different ways: indeed, the passions named in the second order have often a unique and determined cause in the body, whereas those of the third can have more than one cause, located outside our body. Thus, they are even less useful for speaking about bodies (De Raey 1692, *Cogitata*, pp. 81–93). Yet, expressions such as 'ortus' and 'occasus soli' can still make known something true, as they testify to a relative movement in apparentibus which is considered in practical disciplines.[100] Finally, the fourth category includes words which signify a passion (more properly, a sensation) and through it and along with it, something really existing in the physical world. These are the names of quantities, numbers, figures, positions and places, movement and rest, time (De Raey 1692, *Cogitata*, pp. 93–101).

---

100 De Raey 1692, *Cogitata*, pp. 107–108 (quoting *De caelo*, II, 8, 290a 23–24: "*nihil interest, sive oculus, sive id quod cernitur moveatur*").

6.6.3 Intellectual ideas and *modi considerandi*

If the treatment of the words of the first four orders clarifies the use of everyday language, tracing it back to its actual meanings, it does not fulfil De Raey's justification of the use of words meaning modifications of the soul in order to refer to bodies. He achieves both these ends by taking into account a second series of orders of terms, which mean the modifications of the mind as these are produced by the mind alone. This analysis is made possible by a different consideration of the soul itself, that is, from the point of view of its being active and independent from the body. He can thus take into consideration passions and sensations independently from their bodily cause, and the modi considerandi or second notions used by Scholastic logic and metaphysics, as these do not result from a bodily motion but from a mental activity. According to this perspective, in the same way as the first order includes the marks of passion, the fifth order includes prepositions, adverbs, conjunctions as the marks of the ways in which we pass from one thought to another (De Raey 1692, *Cogitata*, pp. 109–115). In turn, the sixth order includes all the words of the second and the third orders, as these are meta-names of passions – that is, they mean not only the very thought of a passion of the soul, but also names of purely intellectual passions, without a bodily cause (De Raey 1692, *Cogitata*, pp. 120–122). Finally, the seventh order contains names and verbs signifying thoughts by which we erroneously refer to some bodily reality, and the eighth order includes the name of things truly existing outside the mind. As to the former, one can find 'esse' and its derived terms 'ens' and 'essentia', and 'posse', 'potentia' (De Raey 1692, *Cogitata*, p. 212), and all the further second notions, κατηγορούμενα used in Scholastic logic and metaphysics: 'unum', 'verum', 'bonum', 'necessarium', 'contingens', 'substantia', 'accidens', 'quantitas', 'qualitas', 'causa', 'effectum', 'totum', 'pars' (De Raey 1692, *Cogitata*, pp. 146–147), and less general terms signifying relations, which do not match anything outside the mind, like the concepts of divisibility, which is only an expectation that bodies can be divided (De Raey 1692, *Cogitata*, pp. 157, 196–197). Still, these notions, which are 'added' to other notions as their subjects,[101] make known some kind of reality along with those terms signifying *res extantes*, which

---

101 De Raey 1692, *Cogitata*, pp. 191, 203. De Raey clarifies their use through Boethius's definition of eternity as "interminabilis vitae tota simul et perfecta possessio": if 'vita' is a first notion, its possession and qualifications are just ways of considering it: see p. 193. Another comparison is with the shadow of a body (p. 192), also used by Keckermann to distinguish first and second notions: Keckermann 1613b, p. 61. In his *Anti-Spinoza, sive Examen Ethices Benedicti de Spinoza, et commentarius de Deo et eius attributis* (Wittich 1690) Wittich criticises Spinoza as relying on second notions in some propositions of his *Ethica*: see Douglas 2014.

are included in the eighth order: as the names of motion, figures and magnitudes, i.e., the geometrical properties of matter, considered according to the intellect as abstracted from a *concretum* or composite subject, and those terms referring to 'individual substances', such as names of men, animals, plants, what he calls the *supposita substantiva separata*, which means a modification within the continuum of matter (De Raey 1692, *Cogitata*, pp. 180–181, 188). Through his linguistic study De Raey thus confronts a crucial problem in Cartesian philosophy, already discussed by Clauberg and Geulincx: that is, the definition of individual objects within the continuum of extension, deprived of substantial forms. The collapse of this ontology led to the emergence of the problem of how to find a reference for those terms usually taken as names of substances. De Raey solves this problem by taking into account their inner mechanical structure and shape: the world is still composed of forms, which have lost the feature of substantial forms and are mechanical constitutions, constituting a *totum physicum* or *essentialis* (De Raey 1692, *Cogitata*, pp. 285–288). This form, however, is abstracted by mind from a continuum: the justification of mathematical and physical abstraction is provided insofar as the entities we refer to are present in actual bodies as parts in a whole. According to De Raey, abstraction is made possible because the mind, as for Aristotle, is the place of forms, τόπος εἰδῶν.[102] As to the names of *modi considerandi*, these are necessary in order to carry out any research in mathematics – which consists of mental operations as equations (De Raey 1692, *Cogitata*, p. 188) and in physics – where one uses terms as 'facultas', 'vis', 'actio', 'natura', which cannot be easily replaced by 'motus' and 'materia':[103] that is, their *sensus* – like those of names of sensations – cannot be substituted and restricted as to mean only a bodily modification (De Raey 1692, *Cogitata*, pp. 209–210). The semantic value of such notions is abused by the Aristotelians, who consider most of

---

102 De Raey 1692, *Cogitata*, pp. 188–189, 213, 216; see *De anima*, III, 4, 429a 27–28.

103 "Ac si nomen [...] facultas a facere, actio ab agere, nomen inane sit, quia haec singula ita praecise [...] non significant [...] quid rei ὄντως sit, in nobis vel extra nos. [...] Putamusque horum nominum significationem neque ab humano sermone, quo vel in communi vita, vel in disciplinis utimur ad huius vitae usum spectantibus removeri (ac si, ut loquitur Hobbesius, voces insignificantes sint) neque per substitutionem everti debere, ac si non amplius facultas, vis, actio, natura, vita, anima, verum motus, materia primi elementi, globuli coelestes, particulae striatae, dicere, aliisve debeamus novis nominibus uti, propter hoc unum, quod usitata illa non significent, non faciant notum in extantibus id quod in iis philosophus desiderat," De Raey 1692, *Cogitata*, pp. 212–213.

the entities meant by language as existing outside mind,[104] and is rejected by radical Cartesians.[105] De Raey, aiming to find a *via media*, attempts to 'save' the use of all such notions, both of sensory and intellectual origin. His analysis, even if conducted from a Cartesian standpoint, cannot be regarded as a rear-guard battle, at least, no more than that of De Volder, who was faithful to Cartesian metaphysics notwithstanding his interest in the new sciences. Rather, De Raey's analysis is symptomatic of the adaptation of traditional disciplines, namely logic and metaphysics, to new needs. This happened at two moments: in the 1660s, he unified logic and metaphysics for the sake of the defence of Cartesianism against its misuses, maintaining their Cartesian or foundational function. Secondly, with his analysis of language he further detached metaphysical and logical consider-ations from their foundational function, and assigned to them a reflective role. Moreover, by justifying the use of *modi considerandi* he set the ground for a more ontologically liberal approach to natural-philosophical investigations, as for the sake of acquiring the truth one can use concepts which mean nothing but themselves.

6.7 Dutch Cartesian philosophy at the turn of the century

De Volder and De Raey belonged to two different generations of Dutch Cartesians and had different views on the method to be used in natural philosophy. De Raey, consistent with his early views on the separation between practical and philo-sophical knowledge, still maintained in his *Cogitata* the necessity of using two different conceptualisations of the world in practice and in philosophy. De Volder, even though embracing Descartes's vortex theory and using experiments as a means to teach and to confirm the theory, maintained an openness to the use of experience in discovering the laws of motion and the mechanisms of the human body. However, the positions of these two representatives of Dutch Cartesianism came to intersect, as they both used metaphysics, at the end of the seven-teenth century, more as a means of reflection on the conceptual assumptions of

---

104 The kinds of linguistic errors described by De Raey are the mistaking of properties of names for properties of things, as substantive names are considered as names of substances (De Raey 1692, *Specimen*, pp. 561–582, and *Cogitata*, p. 314) or of 'real accidents' (De Raey 1692, *Specimen*, pp. 581–582). Moreover, it is the case of the abuse of the term 'actus', which properly applying only to voluntary actions of the soul can be used to describe every movement in nature, leading to the error of universal soul: De Raey 1692, *Cogitata*, pp. 136–142.
105 "Errore ab una parte in Aristotelis, ab alia opposita, in Cartesii sectatoribus notatus, qua-tenus illi multiplicant, hi minuunt entia sine necessitate," De Raey 1692, *Cogitata*, pp. 207–208.

natural philosophy rather than to establish the principles of motion. With their death – in 1702 for De Raey, and in 1709 for De Volder – Cartesianism (which had already come to a dead end in physics) ceased to be a major driving force in the Netherlands. At the University of Franeker it would be still defended by Ruardus Andala, who assumed a chair in philosophy in 1701 and who can be labelled the last Dutch Cartesian. In Andala's inaugural oration, he assumes positions akin to De Volder: Cartesianism is intended as a means against atheism and scepticism, and does not imply a rejection of experience in physics. Indeed, Andala praises the efforts of scientific societies in Italy, France, and England for progress in experimental physics (Andala 1701; see Caroti 2014). Yet, the main text to which Andala refers is still Descartes's *Principia*, of which he provided a defence – including a vindication of its first, metaphysical part – in 1709 and a *Paraphrasis* in his *Syntagma theologico-physico-metaphysicum* (1711). Moreover, as late as 1718 he warned students not to follow experimental science such as that of De Volder himself, Boerhaave and Newton (Andala 1718, p. 163). The defence of Cartesianism of Andala was in fact functional to his theological interests, as he conceived natural theology as the necessary basis for revealed theology and as a means of solutions in biblical interpretation, and against its 'misuses', such as those of Spinoza and Geulincx. It would therefore be tempting to analyse in detail how Descartes's metaphysics, in the hands of Andala, became not only the basis of natural philosophy but of revealed theology as well. However, for the sake of the comprehension of the evolution of philosophy as the foundation of the natural sciences, it is more valuable to see how foundationalism was assumed in Leiden after the demise of Cartesianism, and with the emergence of Newtonianism. This will be the subject of the next chapter.

7 The aftermath: The Cartesian heritage in 's Gravesande's foundation of Newtonian physics

7.1 Leiden University in the early eighteenth century

The scientists – namely, natural philosophers and physicians – active in the first decade of the eighteenth century in Leiden did not embrace a fixed standpoint on the principles and method of the investigation of nature. Besides Senguerd, this is the case with the most important figure dominating the Dutch scientific environment at the beginning of the eighteenth century, namely, Herman Boerhaave. In 1687 and 1688, under the presidency of Senguerd, Boerhaave gave disputations on the cohesion of bodies, which he explains as the effect of occult qualities, and on the nature of mind, of which he gives a Cartesian explanation.[1] In 1703, two years after having started to lecture in medicine, he gave an oration *De usu ratiocinii mechanici in medicina* in which he embraced a mechanist position on the functioning of the human body, mixed with – like Cartesian positions, as he compares the functioning of the body to a clock: Boerhaave 1703, p. 29). Yet, these positions changed over the coming years, as he became more discontent with a purely mechanical interpretation of physiology, as for him most effects in the body cannot be explained by means of mechanism, and even have the possibility of drawing universal laws from the observation of the powers of particular bodies (Boerhaave 1715). Accordingly, Boerhaave came to reject both the Cartesian and Newtonian standpoints on the investigation of nature: he instead adopted chemistry as the discipline capable of explaining the powers of bodies, once it is purged of the principles of the alchemists and based on observation alone (Boerhaave 1718). For this reason, in 1718 Andala (together with De Volder) accused him of Spinozism, as he had not provided any first principle for the study of nature, thus discarding Descartes's metaphysics and easing the acceptance of Spinozist ideas (Andala 1718, part IV; see Knoeff 2002, p. 47). Boerhaave had some philosophical interests, as his whole scientific enterprise has been interpreted as aimed at 'peace of mind' (Cunningham 1990). However, he did not provide any systematic treatment of the metaphysical premises or the scientific methodology of his investigations, his only consideration being in his inaugural orations. This can be explained as the consequence of his lack of interest in natural philosophy or physics, being concerned mainly with medicine, and of his scepticism on the very possibility

---

1 Senguerd/Boerhaave 1687, Senguerd/Boerhaave 1687–1688. On Boerhaave, see Lindeboom 1968, Knoeff 2002, Knoeff 2003.

https://doi.org/10.1515/9783110569698-008

of finding universal laws ruling phenomena. Therefore, he had no need nor was he able to justify or explain his approach to the study of nature in the manner of a systematic natural philosophy, as the Cartesians did. As soon as a new, full-blown approach to the study of nature appeared in the university, however, foundationalism and the use of philosophy as a reflection on science came again to the fore.

In the early years of Boerhaave's academic teaching at the faculty of medicine, Willem Jacob 's Gravesande (1688–1742) – from 1704 to 1707 a student at the faculty of law in Leiden – began to take interest in the physics of Newton. When he started to teach it in 1717, as holder of a chair of philosophy, he developed a systematic logical and metaphysical introduction to the new physics, in the same ways as his Cartesian predecessors had in the 1640s. The foundation of Newtonian science was required in order to validate its conceptual premises and methodology, and to enable its introduction into the university as certain and secure knowledge. As during the introduction of Cartesian philosophy, the establishment of Newtonian physics at the beginning of the eighteenth century was characterised by a justification of its epistemic assumptions, given its groundbreaking impact on the Cartesian and Aristotelian framework of academic culture. 's Gravesande provided Newtonian physics with an introduction and foundation in his *Introductio ad philosophiam, metaphysicam et logicam continens* (1736), a comprehensive metaphysical and logical treatise in which the use of experience in physics is justified as provided with moral, but indubitable evidence. The need of providing natural philosophy with a foundation was thus not specific to Cartesianism, but signals a change in the function of other branches of philosophy – i.e. logic, metaphysics and rational theology – as laying the premises of natural philosophy. Furthermore, given the specificities of Newton's physics, i.e. its detachment from metaphysical conceptualisation, with 's Gravesande one can find the complete – and self-declared – detachment of physics from metaphysics. His philosophy, accordingly, assumes the role of a philosophy of science, intended as a reflection on the limits and purposes of scientific knowledge, and a foundation for its reliability on the basis of actual scientific practices.

7.2 The introduction of Newtonianism in Leiden by 's Gravesande

Whereas most Cartesian natural philosophers had a medical background, which often reflected their philosophical standpoint, 's Gravesande[2] had a juridical

---

2 On 's Gravesande's life, see Allamand 1774, and Marchand 1758–1759. See also Gori 1972, pp. 64–159, Van Besouw 2016.

training, and in 1707 he started his practice as a lawyer in The Hague. However, 's Gravesande was first and foremost interested in mathematics, as he used his mathematical skills as a cryptographer in the last phase of the War of the Spanish Succession and helped the government to solve some economic questions. In fact, he started to be known in 1710, when he entered into a debate raised by John Arbuthnot in an article published in the *Philosophical Transactions* on the role of divine providence in maintaining the ratio of male and female newborns.[3] In 1711, moreover, he published his first scientific treatise, *Essai de perspective*, which gave him a reputation in Dutch and English mathematical and philosophical circles. Finally, in 1713 's Gravesande along with Justus van Effen and some other friends founded the *Journal littéraire*, hosting articles on literature and politics, but also on law, ethics, philosophy, mathematics and physics.[4] Given the proximity to the English intellectual context shown by 's Gravesande and Van Effen in their journal (Gori 1972, pp. 76–77), they both participated in a diplomatic mission by the Baron Wassenaer to London in 1715, where they became members of the Royal Society. Given the esteem for 's Gravesande in Dutch and English intellectual circles, and under the recommendation of the Baron of Wassenaer, 's Gravesande finally assumed the chair of mathematics and astronomy at Leiden University in May 1717, giving an inaugural speech, *De matheseos* ('s Gravesande 1717). As a teacher, he gave lectures in geometry and on Newton's physics. Eventually, his teachings resulted in the publication of his most important work, *Physices elementa mathematica, experimentis confirmata. Sive Introductio ad philosophiam newtonianam* (1st ed. 1720–1721), which had a huge distribution in continental Europe and in England also, as attested by their many editions, reprints, and translations, which definitely determined the acceptance and dissemination of Newton's theories in Europe.[5] This work also had an abridged version for students, namely, the *Philosophiae newtonianae institutiones* ('s Gravesande 1723;

---

3 Arbuthnot 1710. 's Gravesande defended the effectiveness of providence by means of particular laws in his correspondence with Bernoulli – who admitted a probabilistic explanation of such a phenomenon – and in his *Démonstration mathématique du soin que Dieu prend de diriger ce qui se passe dans ce monde, tirée du nombre des garçons et des filles qui naissent journellement*, which circulated in manuscript before being published in 's Gravesande 1774, vol. II, pp. 221–236.

4 See, for instance, 's Gravesande's *Examen des raisons de Mr. Bernard contre le mensonge officieux* ('s Gravesande 1721), and his *Remarques sur la construction des machines pneumatiques & sur les dimensions qu'il faut leur donner* ('s Gravesande 1714).

5 's Gravesande 1720–1721; see his *Physices elementa mathematica, experimentis confirmata. Sive Introductio ad philosophiam newtonianam. Editio secunda, auctior et emendatior* ('s Gravesande 1725), and *Physices elementa mathematica, experimentis confirmata. Sive Introductio ad philosophiam newtonianam. Editio tertia duplo auctior* ('s Gravesande 1742). On their other editions, see Gori 1972, pp. 311–312.

the book had various editions). After having published a new edition and a commentary on Newton's arithmetic ('s Gravesande 1727), 's Gravesande became professor *totius philosophiae* in 1734, delivering an inaugural speech *De vera et nunquam vituperata philosophia* (1734).[6] In 1736 he published his *Introductio ad philosophiam* ('s Gravesande 1736, 's Gravesande 1737). He kept his position at Leiden University until his death, in 1742, after having refused a chair at the Royal Academy of Berlin founded by Frederick the Great.

Being neither the first nor the only Dutch scientist to embrace Newton's physics, partially accepted by Bernard Nieuwentijt – a non-academic scientist – and then fully embraced by Pieter van Musschenbroek (1692–1761) in Leiden,[7] 's Gravesande is to be considered the most important teacher of the new physics in the Netherlands and Europe, as he adapted the contents of Newton's physics to the academic audience in his *Physices elementa mathematica*, aimed at teaching the new physics by means of experiments rather than by mathematical demonstrations. As the dissemination of his works testifies, 's Gravesande is to be ranged among the most important expounders of Newtonian physics in Europe: in fact, he made possible its acceptance in a European context both by adapting its barely comprehensible mathematical structure to a wide audience and by providing it with a justification and an introduction through logic and metaphysics. Eventually, 's Gravesande made Newton's physics fit the needs of academia.[8] The interconnectedness of 's Gravesande's philosophical introduction and defence

---

6 In 's Gravesande 1734a and 's Gravesande 1734b.

7 On the dissemination of Newtonianism in the Netherlands, see Ruestow 1973, pp. 113–139, Jorink/Maas 2012, Van Bunge 2013, Van Bunge 2017.

8 Scarce attention was devoted to 's Gravesande until the appearance of Gori 1972, who offered the first deep overview of 's Gravesande's foundation of science by rejecting Cassirer's views on 's Gravesande's supposed biological and sociological account of the certainty of physics (Gori 1972, p. 254, Cassirer 1951, p. 61). This work has been followed by the more recent studies of Cees de Pater, focusing on 's Gravesande's notion of moral evidence and interpretation of Newton's rules of philosophy (De Pater 1975, De Pater 1988, De Pater 1989, De Pater 1994, De Pater 1995). Paul Schuurman – who has highlighted the place of 's Gravesande's theories in the logic of ideas established by Descartes and Locke (Schuurman 2003b, Schuurman 2003d, Schuurman 2004, pp. 129–155) – Steffen Ducheyne and Jip van Besouw, who focused on 's Gravesande's methodology and on the epistemological and theological implications of his theories (Ducheyne 2014a, Ducheyne 2014b, Van Besouw 2017a, Van Besouw 2017b). These studies have highlighted the sources of 's Gravesande's arguments and the specificity of his approach with regard to Newton's. Accordingly, I will assume their conclusions while considering 's Gravesande's philosophy in the light of the interplay of his physics, metaphysics and his foundational arguments, arguing for the strength of the division between various possible foundations of philosophy and science – logical, metaphysical, and theological – insofar as even after the demise of Cartesianism 's Gravesande adopted these three possible ways.

of Newton's physics, in fact, can be explained by taking into consideration the demands of academic culture in the early eighteenth-century Dutch context. Since this was dominated by a Cartesian stance on philosophical knowledge – *scientia* – 's Gravesande attempted to provide the empirical knowledge of natural laws with a certainty equal to mathematical evidence. This attempt was designed to fulfil the need to provide physics with a mathematical or absolute certainty, while avoiding its development on the basis of metaphysical notions. Eventually, 's Gravesande pursued this objective by considering experience as a primary means in the accomplishment of God's providential plan, since it allowed men to lead a good life as the end which God has placed on His creation. 's Gravesande thus aimed at giving Newton's modern science the status of *scientia*. However, before dealing with 's Gravesande's justification of Newtonian physics an outline of his methodology and of the structure of his academic manuals on Newtonian physics is required, in order to illustrate the specificity of his defence of Newtonianism. In fact, 's Gravesande's foundation of Newton's method went along with its reworking.

7.2.1 The didactic of Newton's physics

The function of 's Gravesande's *Physices elementa mathematica* – whose contents consistently increased across the three main editions (1720–1721, 1725, 1742)[9] – is to teach the contents of Newton's *Principia* through the description of experiments rather than by mathematical demonstrations, which are confined to a few *scholia* since the second edition of 's Gravesande's book (1725): indeed, the experimental teaching of physics in Leiden by the precursors of 's Gravesande (De Volder and Senguerd) paved the way for the acceptance of Newtonianism by an academic audience more versed in experiments than in the complex mathematics of Newton's *Principia*. Actually, 's Gravesande was not the first in presenting a Newtonian physics based more on experiments than on mathematics. A similar attempt had already been made in 1702 by John Keill in his *Introductio ad veram physicam* – still provided, however, with a complex mathematical backbone

---

9 The first edition, which appeared in 1720 and 1721, includes four books devoted to the notion of body, to the movement of solid and fluid bodies, to the explanation of light and to celestial mechanics ('s Gravesande 1720–1721). In the second edition ('s Gravesande 1725), 's Gravesande added several *scholia* containing those mathematical demonstrations missing in the first edition. Finally, the third edition ('s Gravesande 1742) was enriched with the addition of two further books – enlarging those sections already included in the previous editions – and introduced by his *Oratio de evidentia* (1724).

(see Gori 1972, pp. 94–95) – and by Francis Hawksbee and Jean-Théophile Desaguliers, mentioned in the third edition of 's Gravesande's *Elementa* as attempting to teach Newton's physics 'without geometry' ('s Gravesande 1742, vol. I, *Praefatio tertiae editionis*, p. XVI). However, the structure of 's Gravesande's *Elementa* reveals not only their didactic and propagandistic purposes, but also some of the peculiarities of 's Gravesande's approach to Newton. While maintaining in his *Philosophiae newtonianae institutiones* (1723) and in his *Elementa* a rejection of the use of hypotheses and replacing the mathematical deduction of the laws of nature with the observation of phenomena, "ex phaenomenis, reiectis hypothesibus conclusiones deducuntur,"[10] 's Gravesande embraced a method of scientific discovery and exposition different from that of Newton, as observed by Steffen Ducheyne (see Ducheyne 2014b, pp. 104–105).

First of all, the structure of 's Gravesande's *Elementa* and his exposition of Newton's physics consistently differ from those of Newton's *Principia*. Whereas Newton provides an axiomatic consideration of the laws of motion and centripetal forces in the first and second book of his *Principia*, and systematically applies such laws to phenomena from book III onwards,[11] 's Gravesande begins his *Elementa* with a consideration of phenomena – those concerning attraction, fluidity and repulsion of bodies (book I, parts I, II) and then expounds the laws which explain such phenomena: namely, those of Newton, Galileo, and Huygens (book I, part III, chapters 17 to 20). Moreover, in his *Elementa* 's Gravesande does not mention Newton's fourth rule of philosophy, added in the third edition of Newton's *Principia*, admitting that the formulation of laws can be falsified by some counterexamples or replaced with others with a larger explanatory scope.[12] Indeed, 's Gravesande defines a law of nature as the rule by which God regulates the course of phenomena in every case,[13] and aims, as I am going to show, to provide their knowledge with a certainty equal to that of mathematical demonstrations.

In the third place, the very consideration of scientific method itself provided by 's Gravesande is different from Newton's. Both, actually, recalled the traditional

---

10 's Gravesande 1723, *Ad lectorem*, p. VII (unnumbered). See 's Gravesande 1720–1721, vol. I, *Praefatio*, p. X (unnumbered). See also the third edition of his *Elementa*: 's Gravesande 1742, pp. 24–25, and his *De matheseos*: 's Gravesande 1717, pp. 14, 15, 17–18, 21.

11 See Ducheyne 2014b, pp. 104–105, and Gori 1972, p. 101; Newton 1726.

12 See Ducheyne 2014b, pp. 100–101; "in philosophia experimentali propositiones ex phaenomenis per inductionem collectae, non obstantibus contrariis hypothesibus, pro veris aut accurate aut quamproxime haberi debet, donec alia occurrerint phaenomena, per quae aut accuratiores reddantur aut exceptionibus obnoxiae," Newton 1726, p. 389.

13 "Naturae lex ergo est, regula et norma, secundum quam Deus voluit certos motus semper, id est, in omnibus occasionibus, peragi," 's Gravesande 1742, vol. I, p. 2.

differentiation between analysis and synthesis. Newton, in his *Opticks*, defined analysis as the method of discovery by means of observation and mathematical generalisation, and synthesis as the application of the conclusion reached by analysis to the explanation of phenomena (Newton 1704, pp. 404–405). On the other hand, in the section devoted to the consideration of method of his *Introductio ad philosophiam*, that is, his *Logica*, 's Gravesande heavily relies on the methodological and logical rules of Descartes and Malebranche, providing a Cartesian account of analysis and synthesis, insofar as in analysis one proceeds from the complex to the simple, and in synthesis from the simple to the complex. Analysis, in fact, can concern *a priori* reasoning – having mathematical certainty – or *a posteriori* reasoning, having a more moral certainty. Analysis is the very method of discovery and explanation of phenomena. Synthesis, on the other hand, concerns the mere exposition to other people of the knowledge acquired by analysis. 's Gravesande, furthermore, sets forth five rules concerning *a priori* analytical reasoning, to which he adds a sixth rule concerning *a posteriori* reasoning, based on experience and providing moral evidence.[14] To these rules, he adds six further rules concerning the use of hypotheses in science – included for the first time in his *Introductio* – whose use is allowed in order to acquire certain conclusions whenever they are confirmed by experience (rule V) or can explain new phenomena (rule VI) ('s Gravesande 1736, pp. 292–300). Ducheyne, actually, has shown that whereas Newton admitted that a hypothesis is true when it expresses the sufficient and necessary cause of a phenomenon,[15] 's Gravesande merely requires that a hypothesis has to be confirmed by experience. Thus, 's Gravesande adopted a looser approach to the use of hypotheses, given the influence of Huygens's use

---

14 's Gravesande 1736, pp. 278–292, 314–327. See Gori 1972, pp. 145–148, Schuurman 2004, pp. 150–152.

15 "In order to avoid arbitrary speculation Newton required that the causes to be adduced in natural philosophy should be constrained by imposing the demand on them that they should be shown to be the necessary and sufficient causes of certain effects given the laws of motion, i.e. given a set of non-arbitrary principles which have been shown to be promising in the study of motion and which remain neutral with respect to the *modus operandi* of gravitation. Put differently, according to Newton not just any cause will do in natural philosophy: true causes in natural philosophy are those causes which have been shown to be necessary and sufficient given a set of prioritized theoretical principles, *in casu* the laws of motion. Furthermore, he demanded that independent measurements of causal parameters obtained from phenomena of the same kind should converge and that, given his focus on the systematic dependencies between causes and their effects, a theory should provide accurate measurements of its parameters from the phenomena they serve to explain," Ducheyne 2014b, p. 111. See Newton 1726, book I, propositions 1 and 2. Such differences in views, actually, are confirmed by the actual method 's Gravesande adopts to solve the *vis viva* controversy: see Ducheyne 2014b, pp. 111–112.

of them in discovery (see, for instance, 's Gravesande 1736, p. 298, § 985) and the fact that from the 1730s onwards Newtonian physics was well established in the Dutch context: the supporters of the new experimental philosophy no longer banned hypotheses from the process of science in order to distinguish their theories from the speculative, metaphysical physics of Descartes, according to which the constitution of the universe can be derived from a few innate principles (Gori 1972, pp. 48–63). Eventually, the acceptance of the use of hypotheses can be noted in the *Praefatio tertiae editionis* of 's Gravesande's *Elementa* ('s Gravesande 1742, vol. I, p. XV).

's Gravesande's approach, then, reveals some discrepancies with Newton's, that is, some original points that can be traced back not only to 's Gravesande's didactic aims but also to his peculiar standpoint on the method of physics. Such differences are to be appreciated in light of 's Gravesande's metaphysical and logical considerations, which aimed to provide science with a foundation. In fact, 's Gravesande's use of observation as the first source of knowledge in science, his retaining of the first three rules of philosophy expounded by Newton (quoted *infra*, section 7.4), and his use of hypotheses as a means to acquire probable knowledge found their justification in his views on the functioning of the human mind.

7.3 Mathematics and experience in the discovery of natural laws

First and foremost, the foundation of Newtonian physics is given by 's Gravesande in his 1736 *Introductio*. However, the problem of the reliability of the use of the human faculties and of the certainty of empirical knowledge is also addressed in his orations *De matheseos* (1717), *De evidentia* (1724), *De vera philosophia* (1734), as well as in the *Praefatio* to the first and third edition of his *Elementa*. The recurrence of the topic and the different editions these texts went through, testify that the problem of a foundation was at the top of 's Gravesande's philosophical agenda.

According to his *De matheseos*, physics is to be based on the observation of phenomena because natural laws rely only on the will of God: we can grasp them only through experience, without any recourse to hypotheses about the first constitutions of things.[16] Given the fact that motion is the basic phenomenon in

---

16 "Physica phaenomenorum naturalium causas tradit, id est, examinat quibus legibus Creator voluit universum adstringere, ut continuata motuum successione quaedam mutentur, et mutata maneant, alia semper ad primum statum redeant, et quo modo ope illarum legum phaenomena producantur, haec ars explicat. Hae leges a sola Creatoris voluntate pendentes, cum nulla divina

nature, mathematics is the basis for physics, as it enables the quantification of motion, whose study applies to all fields of natural philosophy and astronomy ('s Gravesande 1717, pp. 15–22). The problem of the relation of experience and mathematics in the discovery of natural laws, briefly addressed in his *De matheseos*, is the main topic of 's Gravesande's *Praefatio primae editionis* and of the first chapter of his *Elementa*, where he defines the end and the scope of physics. Since physics is a sort of mixed mathematics, as it concerns things that exist outside the mind (whereas pure mathematics concerns abstract ideas of figures), physics explains how everything happens according to the laws of nature without considering the genesis of the world but providing a descriptive, mathematical account of natural regularities.[17] First of all, echoing Locke's *Essay Concerning Human Understanding*,[18] 's Gravesande's denies the possibility of knowing material substances in themselves. Even if one can know some properties of matter, the knowledge of their subject is beyond our faculties, since the body may have properties that we do not know. Besides those properties flowing from the essence of matter as an extended and solid substance, one can admit that God provided matter with other properties not essential to the body. Eventually, 's Gravesande dismisses Descartes's view on the perfectly evident knowledge of material, extended substance, from which one can deduce all its properties as necessarily belonging to it.[19] Indeed, he admits the possibility of

---

revelatione nobis denegantur, ex ipsis phaenomenis sunt quasi exhauriendae. Hypotheses fingere, illasque pro fundamento systematis habere, hominum est in errorem lubenter decurrentium et verae physices ianuam sic claudentium," 's Gravesande 1717, pp. 13–14.

17 's Gravesande 1720–1721, vol. I, *Praefatio*, pp. I–III. See also p. VI. On 's Gravesande's criticisms of Descartes's genetic physics, see also his *De vera philosophia*: 's Gravesande 1734a, *De vera philosophia*, pp. 21–22, and Gori 1972, pp. 48–63.

18 See Locke 1690, book II, chapter 23. For a comparison of Locke, Newton, and 's Gravesande's ideas, see Schuurman 2003d.

19 "Substantiae quid sint inter nobis ignota referendum est. Quasdam ex. gr. materias proprietates novimus, sed in quo subiecto haereant has nos omnino latet. An corpori non multas alias tribuendas fini proprietates, de quibus nullam habemus ideam, quis asserere potest? Cui etiam enotuit an, praeter corporis proprietates, quae a materiae essentia profluunt, non dentur alias a Dei libera potestate pendentes, substantiamque extensam et solidam (haec enim a nobis corpus vocatur) quibusdam, sine quibus existere posset, proprietatibus ornari. De ignotis nihil affirmandum aut negandum est. Quantum ab hac regula aberrant illi, qui, quasi omnia quae ad corpus pertinent plenissime perspecta haberent, in physicis ratiocinantur, paucasque corporis proprietates notas ipsum corpus constituere asserere non dubitant! Quid obsecro sibi vult proprietates substantiae ipsam constituere substantiam? An quae separatim subsistere non possunt simul iuncta subsistent? An extensum, impenetrabile, mobile esse, et c. concipi possunt, sine subiecto cui has proprietates competant? Et an huius subiecti ullam habemus ideam? In dubio relinquendum quod certum non est, hoc ne ignorantiam fateri pudeat, neque timendum de

the existence of a void through the sole analysis of the ideas of extension and matter: one can imagine a non-solid extension, because the idea of solidity is gained by the senses, whereas that of extension is independent of touch. Hence, 's Gravesande can reject Descartes's identification of the notions of matter and extension or space. In any case, extension and solidity are two essential properties of the body, along with mobility and inertia.[20]

Moreover, the discovery of natural laws does not rely on the consideration of such properties. It is unknown, indeed, whether natural laws flow from the essence of matter, or if they can be deduced from properties that may depend on the will of God (being not essential to the body), or if such laws depend on other, unknown causes.[21] In the main text of his *Elementa*, finally, 's Gravesande will declare the immediate dependence of every natural law on the will of God, and the possibility that phenomena flow from mediate causes or from the direct action of God. To that extent one can grasp natural laws only by induction.[22]

---

ignoto nimium affirmari, dum subiectum omnino ignotum quibusdam incognitis proprietatibus forte praeditum esse asserimus. Qui vero cum hoc axiomate se nixos dicunt, quod de incognitis non sit ratiocinandum pro ratiocinii tamen fundamento habent, nil circa corpus ignoti dari, nisi forte fortuna errorem non vitabunt. Corporis proprietates a priori detegi nequeunt, corpus ipsum ideo est examinandum, huiusque proprietates exactissime perpendendae sunt, ut possimus determinare quid , in rerum phaenomenis, ex illis proprietatibus sequatur," 's Gravesande 1720–1721, vol. I, *Praefatio*, pp. IV–V.

20 "Vacuum possibile ex solo examine idearum deducitur. Omne enim quod clare concipimus existere posse, possibile est. Quaestio ergo eo redit, an habeamus ideam extensionis non solidae. [...] In extensionis autem idea non continentur idea soliditatis, hanc non nisi ex contactu, illam vero sine illo acquirimus, et si quis nunquam corpus tetigisset, ei soliditas omnino ignota esset," 's Gravesande 1720–1721, vol. I, pp. 4–5.

21 "Corpus ulterius examinando videmus quasdam leges dari generales, secundum quas corpora moventur. *Corpus motum in motu continuare, actioni semper aequalem esse et contrariam reactionem* extra omne dubium est. Multaeque aliae similes circa corpus deteguntur leges, quae minime ex proprietatibus, quae ipsum corpus constituere dicuntur, deduci possunt. Cumque hae leges semper, id est, in omnibus occasionibus, et ubique obtineant, et omnia corpora iis subiciantur, pro generalibus naturae legibus habendae sint. Circa has in obscuro est, an ex materiae essentia fluant, an deducendae sint ex proprietatibus corporibus, ex quibus constat mundus, a Deo tributis, sed corpori minime essentialibus, tandem an non pendeant effectus, qui pro naturae legibus habentur, a causiis extraneis nobis nequidem ideis attingendis," 's Gravesande 1720–1721, vol. I, *Praefatio*, p. V.

22 "Omnis lex immediate a Dei voluntate pendet. Est etiam nostri respectu lex naturae, omnis effectus, qui in omnibus occasionibus, eodem modo producitur, cuius causa nobis est ignota, et quem videmus ex nulla lege nobis nota fluere posse. Nostri enim respectu non interest, an quid immediate a Dei voluntate pendeat, an vero mediante causa, cuius nullam ideam habemus, producatur. Leges naturae nisi ex examine phaenomenorum naturalium, non possunt elici," 's Gravesande 1720–1721, vol. I, p. 2.

Being merely concerned with phenomena, insofar as natural laws are universal effects, 's Gravesande assumes as a methodological, but also as a metaphysical criterion the unknowability of their causes, these being God himself or some other secondary causes.[23] Eventually, scientific discovery must follow the Newton's rules of philosophy and avoid any speculation on the causes of natural laws.

7.4 A first foundation: The survival axiom

The first formulation of the problem of the foundation of knowledge in natural philosophy as *scientia*, attained by following Newton's rules of philosophy, can be found in 's Gravesande's *De matheseos* and in his *Praefatio primae editionis*. With respect to mathematical statements, according to his *De matheseos* these are clear, indubitable, concern simple entities, and do not depend on the will of God, since He cannot violate the principle of contradiction ('s Gravesande 1717, pp. 7, 11). Hence, mathematical statements are necessarily true, i.e. are endowed with mathematical evidence. On the other hand, the knowledge of matters of fact has another kind of certainty, which relies on the use of testimony for history and of analogy for physics, allowing mathematical generalisations from the observation of phenomena ('s Gravesande 1717, pp. 11–12). Such certainty is firstly provided with a foundation in 's Gravesande's *Praefatio primae editionis*, where he addresses the legitimacy of the use of Newton's first three rules in physics, these being:

> Regula 1. Causas rerum naturalium non plures admitti debere quam quae et verae sint, et earum phenomenis explicandis sufficient.
>
> Regula 2. Effectuum naturalium eiusdem generis easdem esse causas.
>
> Regula 3. Qualitates corporum quae intendi et remitti nequeunt, quaeque corporibus omnibus competunt in quibus experimenta instituere licet, pro qualitatibus corporum universorum habenda sunt. ('s Gravesande 1720–1721, vol. I, p. 2)

Such rules concern matters of fact, whose existence is contingent, that is, their contrary is still possible ('s Gravesande 1720–1721, *Praefatio*, p. VII). One can grasp

---

23 "Satis ergo patet, quinam sit scopus physices, ex quibus naturae legibus phaenomena sint deducenda. Et quare, quando ad leges generales pervenimus, non ulterius in causarum inquisitione penetrare possimus," 's Gravesande 1720–1721, vol. I, *Praefatio*, p. IV. See Gori 1972, pp. 170–177.

their existence only through the senses: however, God himself provided us with some rules aimed at ensuring the truth of our knowledge of such matters, that is, Newton's *regulae philosophandi* ('s Gravesande 1720–1721, vol. I, p. 7). Whereas the first rule, according to 's Gravesande, is self-justified as it is the expression of a principle of economy, the other two rules require some premises as they determine the use of analogy in reasoning. In any case, 's Gravesande still does not explicitly provide a foundation of science on divine goodness as he would do in his later works. Instead, he grounds such rules on his well-known survival axiom, "pro vero habendum omne quod si negetur societas inter homines destruitur" ('s Gravesande 1720–1721, vol. I, p. 8). According to him, insofar as society cannot survive if men cease to reason on the basis of sense data and analogy, and given the fact that God himself has put us in the necessity of reasoning by analogy, Newton's second and third rules are given a foundation.[24] Such a foundation is theological since it appeals to the role of God in creating us as beings forced to use analogy. According to 's Gravesande, this argument leads to the necessary conclusion that reasoning by analogy will not deceive us. On the other hand, the conclusions reached by analogy are not as necessary as their foundation is, as one may fail in any particular reasoning.[25] However, 's Gravesande does not define what kinds of necessity are involved, nor does he appeal to divine goodness to ensure the truth of our statements. Given the roughness of his justification of analogical reasoning in the first edition of his *Elementa*, in his later works 's Gravesande would ground his foundation of science on more detailed definitions of the logical, metaphysical, and theological concepts entailed by this early argument.

7.5 Logic and metaphysics as the introduction to natural philosophy

A complete justification of the use of experience, analogical reasoning and testimony is provided by 's Gravesande in his *Oratio de evidentia*, given in 1724, and in a more comprehensive form in his *Introductio*. The core arguments of

---

24 "Quotidie, nequidem ad illud attendendo, sequentia ratiocinia unusquisque pro indubitatis habet, et clare videt horum conclusiones, sine praesentis rerum constitutionis destructione, in dubium vocari minime posse. Aedificium, hodie in omnibus partibus firmissimum, crastino die sponte non ruet [...] Haec omnia ratiocinia analogiam pro fundamento habent, et extra omne dubium est, nos a rerum Conditore in necessitate per analogiam ratiocinandi redactos esse, et hanc ideo ratiociniorum legitimum esse fundamentum," 's Gravesande 1720–1721, vol. I, pp. 8–9.
25 "Adde ex necessitate quidem generaliter deduci, ratiocinandi methodum esse legitimam, ratiocinia vero peculiaria ab hac necessitate non pendere," 's Gravesande 1720–1721, vol. I, p. 9.

his *Introductio* are the very contents of his earlier *De evidentia*, and testify to a substantial continuity in his philosophical position.

As stated above, 's Gravesande's *Introductio* is a logico-metaphysical treatise: it is divided into two books, devoted to metaphysics and to logic respectively. Logic, in fact, is the place of 's Gravesande's foundational arguments: as underlined by Paul Schuurman, it is a logic of ideas, devoted to the use of the mental faculties, ideas, and method (Schuurman 2004, pp. 133–134). However, this is introduced by a metaphysics as this concerns the most general concepts to be later used in logic, it helps in exercising our abstractive capacities, and introduces the reader to the problems concerning the classification of ideas, the objects of intelligence or *facultas percipiendi* ('s Gravesande 1736, pp. 1–2, 267). The first part of the book on metaphysics, thus, concerns the concepts of being, essence, substance and mode, relation, possibility and impossibility, necessity and contingence, time, identity, effect and cause. His considerations (which show some influence of Jean Le Clerc's *Ontologia*)[26] are consistent with the ontology entailed by his *Elementa*. In the section *De ente*, for instance, one can find an account of the notions of substance and modes: substances can be thinking – i.e., mind and God – and not thinking, namely body and space distinguished by 's Gravesande ('s Gravesande 1736, pp. 8–9). Moreover, 's Gravesande examines the different kinds of causes, yet without any commitment to the study of causes as something different from universal effects in physics ('s Gravesande 1736, pp. 29–37). Thirdly, he focuses on the notions of possibility, impossibility, necessity, and contingence, paving the way for his further evaluation of the knowledge of natural laws as morally evident, that is, as being as certain as mathematics but not acquired just by analysing ideas.

First of all, absolute impossibility characterises what contains in itself the reason for its non-existence, as "mons sine valle," and contradictory propositions in mathematics. Physical impossibility concerns a relation of two physical things: for instance, one cannot insert a cylinder into a hole smaller than its size – this kind of impossibility is determined by the geometrical features of a physical body. Finally, moral impossibility is a matter of probability, that is, its opposite has some degree of probability of its existence. Its consideration is postponed by 's Gravesande until the book on logic; however, he assumes that it concerns intelligent actions: for instance, it is morally impossible that a reasoning man wants to step into boiling water, as this would contradict his being rational.[27] With such

---

26 See Gori 1972, p. 135; Le Clerc 1692. Logic, ontology, and pneumatology roughly correspond to the three parts of 's Gravesande's foundation: logic, metaphysics, and rational theology. On Le Clerc's logic, see Schuurman 2003a, Schuurman 2004, pp. 70–109.

27 "Impossibilitas non semper ex eodem fonte fluit. Absolute impossibile dicitur, quod in se consideratum propriam impedit exsistentiam. Hoc revera nihil est, quamvis verbis exprimatur

notions 's Gravesande develops his views on necessity and contingence. He first defines absolute necessity as concerning those things whose contrary is absolutely impossible, even if sometimes it refers to what is physically impossible. Hypothetical necessity, on the other hand, concerns those things whose contrary is impossible in relation to some other thing. 's Gravesande broadly defines necessity as characterising those things whose contrary is impossible, no matter what the nature of this impossibility is. Absolute necessity, hence, concerns those things whose contrary is absolutely impossible, and physical or fatal necessity concerns those things whose contrary is physically impossible. Moral impossibility, finally, determines moral necessity: for instance, it is morally necessary that a rational man avoids poisoned food.[28] Contingency, on the other hand, characterises what can exist or not exist, i.e., what is undetermined according to its nature. According to a vulgar meaning, everything opposed to necessity is contingent;

---

quasi esset aliquid. Mons sine valle impossibilis est, et proprie loquendo nihil est. [...] Dantur variae impossibilitates [...] diversae. [...] Impossibilitas saepe tantum tribui debet relationi inter duas res: cylindrus foramen crassitie superans non potest intrudi propter relationem inter has magnitudines. [...] Praeter hasce impossibilitates, quas omnes physicas vocamus, aliam non debemus negligere, quam moralem vocabimus. Impossibilitas saepe moralis dicitur, quando oppositum exiguam, sed quandam tamen, habet probabilitatem. De tali impossibilitate nunc non agitur: ad materiam probabilitatis pertinet haec, et in logicis examinanda erit. Moralem in hisce vocamus impossibilitatem, ubi huius causa in intelligentia quaerenda est. Ex. gr. homo sana mente praeditus, sponte balneum aquae bullientis non intrabit, et impossibile hoc est, si intraret, non esset sana mente praeditus. Sed impossibilitas non ad ullam ex ante explicatis, praeter ultimam, referri potest: non physica est, sed intelligentiae soli tribui debet," 's Gravesande 1736, pp. 14–16. See also the chapter *De libertate*, where 's Gravesande points out that perfect human freedom consists in the absence of physical constraints (p. 43), and coincides with moral necessity itself (p. 45) and the chapter *De fato*, where he addresses Spinoza's views on freedom by pointing out that the mind is not subject to mechanical causes (p. 53). This is restated in the chapter *Examen diversarum sententiarum de libertate*, (p. 62).

28 "Inter omnes quidem convenit, illud necessarium esse, cuius contrarium impossibile est. Sed non intelligunt omnes impossibilitates de quibus egimus: saepe ad solam primam, id est, absolutam, attendunt [...]. Hypotheticam quidam vocant necessitatem, quando contrarium impossibile est, non sua natura, sed aliunde. Ut omnis confusio vitetur, iisdem vocibus, semper eaedem ideae exprimendae sunt. Generaliter ergo necesse vocabimus cuius contrarium impossibile est, quaecunque sit impossibilitatis causa. Absolutam vocabimus necessitatem cuius contrarium absolute impossibile est: id est, ubi non datur contrarium [...]. Hanc etiam physicam dicemus necessitatem, ut in omni alio casu, ubi impossibilitas contrarii physica est [...]. Omnemque physicam necessitatem etiam fatalem vocabimus, sed si impossibilitas moralis sit [...] moralis etiam necessitas. Talis est qua homo sana mente praeditus, inter venenum et cibum salubrem eligens, illud reiicit, hunc sumit: si aliter ageret, non esset sana mente praeditus," 's Gravesande 1736, pp. 17–18. See Mugnai 1990.

according to a more precise meaning, however, 'contingent' is what is ruled by moral necessity.[29] In fact, what is certain – such as that God foresees and creates – even if only contingent in itself, turns out to be necessary according to the general definition of necessity, even if not according to a fatal or absolute necessity.[30] This is the case of natural laws, which depend on God's free act of creation, entailing the highest degree of freedom, as He is governed only by himself.[31] Hence, natural laws are morally necessary and certain: their contrary is impossible (according to the broad definition of necessity), and they are the objects of our certain knowledge. Given these ontological assumptions, 's Gravesande can provide physics with the status of *scientia* as, like mathematics, this concerns necessary entities and is indubitable. Indeed, he does not admit that natural laws, even if morally necessary, are knowable through mere probability, as his account of moral impossibility may suggest. Rather, they are the objects of a knowledge as persuasive as mathematics, even if gained by experience, whose reliability in providing us with certain conclusions is maintained by 's Gravesande by stressing its being a gift of God (see *infra*, section 7.6). This apparent inconsistency, actually, goes back to one of the main philosophical problems of 's Gravesande, i.e., his need to maintain the universal or necessary status of the laws based on moral necessity and grasped by experience, and to avoid the absolute necessitarism of Spinoza on the constitution of the world.[32]

---

29 "Contingens dicitur, quod potest esse, aut non esse: id est, quod ex propria natura non determinatur. Confusionem autem non exiguam detegimus in usu huius vocis: nam multi contingentiam ita intelligunt, quasi omni necessitati opponeretur, sed minus vulgaris est haec significatio. Quotidie contingens vocatur quod morali necessitate adstringitur, quod cum contingentiae definitione congruit. Haec enim spectat rem, et moralis necessitas personam quae rem agit," 's Gravesande 1736, pp. 18–19.

30 "Inter illos qui dicunt nullum contingens esse necessarium, quidam distinguunt inter necessarium et certum. Sed illud quod certum est, aliter esse non potest, et quod aliter esse non potest, hoc ipsum quotidie necessarium dicitur, et hoc cum ipsa huius vocis definitione congruit [...], a qua si recedamus, confusio difficulter vitari poterit, sed distinguendum inter necessitates sua natura diversas. Hac de causa necessariam dicimus rem contingentem a Deo praevisam: con[tra]rium enim illius quod ita praevisum est, impossibile est. Sed cum rem contingens sit, non agitur de necessitate absoluta, aut alia quacunque fatali," 's Gravesande 1736, p. 19. 's Gravesande seems to address the distinction between certain and necessary expressed in Leibniz's *Discours de metaphysique* (1686), § 13 (see Leibniz 1923–, series VI, vol. IV/2, pp. 1546–1547.

31 On divine freedom, see 's Gravesande 1736, pp. 42, 56–57.

32 's Gravesande's identification of certainty with necessity would raise the criticisms contained in the anonymous *Lettre à monsieur G.J. S'Gravesande*: "il me semble, monsieur, que vôtre distinction entre necessité physique et necessité morale [...] n'est qu'une distinction faite à plaisir qui consiste seulement en paroles, n'y ayant au fond aucune difference réelle," Anonymous 1736, pp. 7–8. In his posthumously published *Essais de métaphysique*, 's Gravesande would point out

After having defined the ontological premises of his theory of physical laws, 's Gravesande devotes his further ontological considerations to mental faculties, which are introduced in the second section of the book on metaphysics, *De mente humana*. In this section he defines the relation of identity between consciousness and perception, which makes the perception of the relation between ideas unavoidable and indubitable:[33] this is the ground for the justification of mathematical evidence as this concerns mere ideas – no matter what their origin.[34] On the other hand, the problem of the foundation of science concerns ideas as these represent something different from themselves. In fact, a classification of ideas according to their properties is discussed in the first part of the second book of 's Gravesande's *Introductio*, i.e., his *Logica*, which aims to introduce the topic of the method of scientific discovery and its foundation by considering, first, the different sorts of our ideas ('s Gravesande 1736, pp. 103–104). 's Gravesande distinguishes between simple and composite ideas – as they are considered in themselves – and clear, obscure, adequate, inadequate, distinct, confused, abstract, concrete, singular,

---

that God did not create the world driven by a geometrical necessity – since a different constitution of the world is not contradictory in itself – but by a moral necessity, i.e. in accordance with his attributes. This is a Leibnitian solution: "pour ce qui regarde le Pouvoir physique, Dieu peut tout ce qui n'est pas contradictoire en soi [...]. Mais si nous faisons attention ou Pouvoir moral, il est clair qu'il est contradictoire que Dieu fasse autre chose que ce qu'il veut; il ne peut donc que ce qu'il veut. Mais il est contradictoire qu'il ne veuille pas ce qui est conforme à ses attributs, ou qu'il veuille autre chose; il est donc contradictoire que Dieu eusse une autre volonté que celle qu'il a, & par conséquent il est de même contradictoire qu'il fasse autre chose que ce qui'l fait, & dans le sens moral Dieu ne peut que ce qu'il fait," 's Gravesande 1774, vol. II, p. 208. See Ducheyne 2014a, pp. 46–47.

33 "A perceptione quacunque inseparabilis est conscientia ipsius perceptionis. Qui percipit conscius sibi est se percipere, et eo ipso propriae exsistentiae conscientiam habet," 's Gravesande 1736, p. 39.

34 The problem of the actual source of ideas is left in doubt by 's Gravesande. Recalling Locke's classification (for a detailed comparison, see Schuurman 2004, p. 135), in his *Metaphysica* 's Gravesande divides roughly ideas into three categories: those that the mind perceives in itself (i.e. are implied by self-perception, like the ideas of the affections of mind); those that the mind develops by comparing, judging, and reasoning about other ideas, and the ideas coming from the senses. However, whereas the ideas of the first category are undoubtedly innate, and the ideas of the second category rely on simpler ideas, the ideas of the third kind cannot be defined in their origin ('s Gravesande 1736, pp. 91–95), since both the solutions of Leibniz and Malebranche present some unavoidable difficulties in ascertaining whether all ideas are innate (pp. 95–101), and since the evidence of medicine and anatomy does not exclude an actual communication between substances (p. 84). In fact, 's Gravesande admits that the body as the instrument of the mind is required for the activity of the mind itself: however, we do not know to what extent the mind relies on the body to perform its function, because we do not have a complete knowledge of the nature of the mind (pp. 72–73).

particular, universal, absolute, and relative ideas – insofar as they represent something different from themselves ('s Gravesande 1736, part II, book I, chapters 1 to 6). Simple ideas are sensations themselves – which, strictly speaking, do not represent anything existent outside the mind – as well as the ideas of extension, motion, and mental acts, which seem to be innate ('s Gravesande 1736, pp. 102–104). 's Gravesande relies on the Lockean distinction between ideas of sensations and reflection as the building blocks of all our knowledge, or simple ideas.[35] Simple ideas are all clear, whereas composite ideas can be obscure: these, in fact, include the ideas of substances, which are all obscure since we can know only their modes ('s Gravesande 1736, p. 108). As mentioned above, in his *Elementa* 's Gravesande rejects the Cartesian clear and distinct notion of material substance by admitting the distinction of body and space. Moreover, every clear idea is adequate and distinct, but not every distinct idea is clear ('s Gravesande 1736, pp. 109–110). This is the case with the idea of the body, conceived as something extended and impenetrable, which can therefore be distinguished from other ideas but which we cannot grasp in its entirety. Eventually, in his *Introductio* this distinction is given with a logical justification. Even if body is extended and can be mathematically described, we cannot deduce all its properties and the natural laws from its essence. Hence, experience is our only means of grasping its properties, thus being a necessary foundation of its reliability. The main problem faced by 's Gravesande in his foundation of science, therefore, is whether we can know physical reality by experience, since the idea of a geometrical body does not entirely match the essence of the physical body, and the basic laws of nature are not attainable by means of deduction.

7.6 The theological foundation of moral evidence

The foundation of the scientific role of experience is addressed by 's Gravesande by considering the notions of judgment and truth. Judgment is a comparison of ideas implying a perception of their relation ('s Gravesande 1736, chapter 7 to 10),

---

35 "Ideam vocamus simplicem, in qua plures detegere non possumus. Compositam, quae ex pluribus simplicibus constat. Simplices ideae sunt omnes sensationes, ut colorum, odorum sonorum, et c. gaudii, doloris, et c. [...] Simplices etiam ideae extensionis, motus determinationis voluntatis, et similes," 's Gravesande 1736, p. 105. See Locke 1690, book II, chapter 1. In any case, 's Gravesande is not clear on the actual source of mathematical ideas, even if the simple idea of extension seems to be innate also according to his *Elementa*, where it is described as independent from the sense of touch. On the other hand, later in his *Introductio* 's Gravesande admits that by the senses we acquire the ideas of figures: still, mathematics deals with ideas alone, independently of their sources: 's Gravesande 1736, pp. 149–150.

whereas truth is the correspondence of ideas and things, and has two classes. First of all, it concerns ideas of mental actions and passions: the truth of judgments pertains to this class, because a judgment is true if it represents the relation between two ideas, which is a mental act. However, one needs to distinguish between the truth of the ideas involved in judgment and the truth of judgment itself. In this way, 's Gravesande can maintain that by judgments we may grasp something different from mere mental acts ('s Gravesande 1736, pp. 135–136). Indeed, the other kind of truth concerns those ideas acquired though an external cause. This kind of truth is provided by 's Gravesande with a foundation on moral evidence. Like Descartes, 's Gravesande recognises in evidence the criterion of truth: whenever we have an immediate perception of an idea, we are persuaded that this idea is true, or agrees with the thing it represents. Thus, evidence is the very immediacy of the perception of something. Therefore, all ideas of mental acts are evident and true, as well as those we may acquire without any means different from the mind itself.[36] For instance, in mathematics one can attain evidence as it deals only with ideas, no matter if they correspond to physical things – whose existence is only hypothetical insofar as it must be acquired by means different from the mere immediate perception of ideas: i.e., sense experience, testimony, and analogy.

These points are also considered in *De evidentia*, where 's Gravesande – like De Volder – stresses the importance of indubitability or forced assent as the mark of truth of judgments, i.e., of the perception of the relation of several ideas. The perception of two ideas entails the consciousness of their relation, which is thus indubitable: like the relation of the ideas of four and three, which is represented by the idea of seven ('s Gravesande 1736, pp. 139–140; see 's Gravesande 1734a, *De evidentia*, pp. 7–8). Moreover, evidence characterises all those disciplines concerning ideas grasped through a reflection on mental acts or affections: such as the ideas of being, spirits, soul, and God ('s Gravesande 1736, p. 140). Hence, mathematics, ontology, pneumatology, and rational theology are characterised by evidence. In his *De evidentia*, moreover, syllogistic logic is added to the list of

---

36 "Evidentiam vocamus immediatam perceptionem. Evidentia haec criterium est veri, pro omnibus ideis rerum quas immediate percipimus. Id est, haec ipsa est legitimum fundamentum persuasionis, et conclusionis huius, ideam quam acquirimus convenire cum re, quam immediate percipimus. Ipsa enim res cum huius immediata perceptione congruit. Dum cogito, cogitatio in mente distincta non est ab huius perceptione. Gaudium in mente mea, et huius perceptio, sunt unum et idem. Haec perceptio, ideo, mihi veram dat illius gaudii, quo mens nunc fruitur, ideam. Haec observatio ad res omnes, quas immediate percipimus, referri debet. Hae enim nisi cum ipsis ideis convenirent, immediate percipi non possent, cum mens nostra ideas tantum percipiat," 's Gravesande 1736, pp. 137–138.

such disciplines ('s Gravesande 1734a, *De evidentia*, p. 11). Such reflection grounds the mathematical evidence of pneumatology, as the notion of mind is revealed by any mental act, as well as that of rational theology. Indeed, 's Gravesande provides a Lockean demonstration of the existence of God. Since something exists – i.e., the mind – something eternal must exist: that is, God, defined as an unlimited intelligence or the source of limited intelligences. Thus, God has infinite wisdom from which one can deduce His infinite goodness.[37] Actually, 's Gravesande will base his foundation of moral evidence on this mathematical demonstration of the existence and goodness of God. From the acknowledgment of the existence and the attributes of God, moreover, one can deduce ethical rules, which are, as for Locke, capable of mathematical evidence ('s Gravesande 1734a, *De evidentia*, pp. 17–18; Locke 1690, book I, chapter 3). So all the disciplines dealing with mere ideas have a mathematical evidence, since our consciousness or perception of such ideas entails their existence. All the other disciplines, as they concern entities different from ideas, are provided with a moral certainty because one needs to assess their existence by means different from mere consciousness.

Given this notion of evidence as immediate perception – or indubitable perception of an idea – two problems arise: the explanation of the difficulties in reaching a consensus in those disciplines capable of a mathematical evidence, and the justification of the knowledge of things different from ideas themselves. Plainly, metaphysical considerations often gave rise to the most acrimonious dissensions among philosophers. Such dissensions are caused by the ignorance of the rules of reasoning, by the influence of the passions of the soul, and by the use of obscure terms, whereas in mathematics one always uses distinct terms and divides ideas into their simplest elements.[38] Hence, the immediacy of perception – or the only ground of mathematical evidence according to 's Gravesande's *Introductio* – seems not to be a sufficient condition for the attainment of truth,

---

37 "Si ad illam pneumatologiae partem nos convertamus in qua de Deo agitur et hanc in totum circa ideas versari videbimus, et ex talibus deduci, circa quas dubium nullum in mente haerere potest, quod ex ipsarum natura sequitur ideoque evidentia mathematica etiam niti, quae de intelligentia suprema et infinita disputantur. Aliquid nunc est, ergo aliquid ab aeterno fuit. Cogito ego, id est datur quid intelligens, inde deduco huius primum auctorem ab aeterno fuisse et infinitum intelligentia superare quam produxit intelligentiam [...]. Constat ergo Deum esse unicum, aeternum, immensa scientia praeditum, huiusque nullis terminis circumscribi potentiam. Quibus demonstratis ex his alia quae de Deo deteguntur profluunt. Bonitas ex. gr. in gradu supremo, ex infinita deducitur sapientia. [...] Illud ipsum quo probamus Deum esse, et sapientem esse, ex examine rerum deductum, argumentum mathematica concomitari evidentia defendimus," 's Gravesande 1734a, *De evidentia*, pp. 12–13. See Locke 1690, book III, chapter 10.
38 's Gravesande 1734a, *De evidentia*, pp. 14–17. See the second part of 's Gravesande's logic, concerning error. See also Schuurman 2004, pp. 148–149.

according to his *De evidentia*: the analysis or the distinction of ideas is also required. Such a distinction, eventually, helps in avoiding any arbitrariness in the definition of the evidence for an idea.[39]

The other main problem raised by his account of mathematical evidence is that of the justification of our knowledge of material things, or the application of mathematics to the study of natural phenomena (*mathesis mixta*). The knowledge of external things does not rely on ideas considered in themselves ('s Gravesande 1734a, *De evidentia*, pp. 17, 20; 's Gravesande 1736, pp. 143–144). This is the case with the knowledge pertaining to history and to physics ('s Gravesande 1734a, *De evidentia*, p. 17), where one can attain only moral evidence and a consequent moral certainty. The sources of moral evidence are different from those of mathematical evidence, namely sense experience, testimony, and analogy.[40] Moral evidence, indeed, characterises the perception of things as these are something different from ideas: their knowledge, thus, is not immediate and must rely on these three sources of knowledge, which are the sole means to assess the correspondence of ideas and things.[41] Therefore, 's Gravesande devotes several paragraphs in chapters 14 to 17 of the second book of his logic to the rules of the right use of the senses, testimony, and analogy. The senses, plainly, are our means to know phenomena and the properties and laws of matter. In order to gain certain knowledge by means of experience, one must rely on more than one sense, the senses must not be affected by any disease, and, in case of doubt about the constitution of one body, one must experience other bodies. Finally, the senses

---

39 's Gravesande 1734a, *De evidentia*, p. 15. The differences between the two treatises are to be evaluated in the light of his *Introductio* being a preliminary discourse to Newtonian physics: thus, it is more focused on moral evidence, whereas his *De evidentia* – again published in the third edition of his *Elementa* – has a more general character.

40 The source of 's Gravesande's positions on moral evidence seems to have been Humphry Ditton's *Discourse on the Resurrection of Jesus Christ* (Ditton 1712) apparently reviewed by 's Gravesande: see *Journal littéraire* I (1713): 391–435. See Gori 1972, pp. 218, 229, 231, 232, 247, 249.

41 "Res aliae, extra mentem positae, non immediate percipiuntur, neque ad se ipsam attendendo, mens harum acquirere potest notitiam. Nunquam ergo, sine auxilio extraneo, cognosci hae possunt. Aliud, ergo, veri criterium, ab evidentia diversum, nobis in hisce quaerendum est. Ut tamen eodem nomine veri criterium, in omni casu, exprimeretur, evidentiam moralem dixere illud, quo veritate idearum, de quibus in hisce agitur, determinamus. Et, ad omnem confusionem vitanda, simplici evidentiae, de qua praecedenti capite egimus, nomen evidentiae mathematicae dedere. Rerum extranearum, id est, extra mentem positarum, tribus mediis homines acquirunt ideas: sensibus, testimonio, et analogia, et tria haec dantur evidentiae moralis fundamenta. Nullum ex his, per se, id est, sua natura, est veri criterium, et eo respectu, evidentia moralis differt a mathematica. Conveniunt tamen, respectu persuasionis, quae utramque sequitur," 's Gravesande 1736, pp. 144–145. See also 's Gravesande 1720–1721, vol. I, *Praefatio*, pp. VII–VIII, 's Gravesande 1734a, *De evidentia*, p. 19.

should not be employed in quantitative analyses ('s Gravesande 1736, pp. 149–163). Testimony, on the other hand, proves to be crucial in collective scientific research. As a result of the propagandistic needs of his early edition of the Elementa, in which 's Gravesande presents all his experiments as if these were performed by Newton alone, testimony is not mentioned in his 1720 *Praefatio*. In his *Introductio*, on the other hand, 's Gravesande addresses the use of testimony and defines the criteria to accept others's witnesses, to be traced back to his studies of law. Its use must be controlled and obey three conditions: a witness must not have been deceived, must not want to deceive, and must express his thoughts in the clearest way. Such conditions, to be fulfilled, must respect nine rules, which are borrowed from the conduct of trials ('s Gravesande 1736, pp. 164–171). Finally, analogy or the generalisation of sense observations and testimony is intended to ground the inductive reasoning entailed by Newton's second and third rule. Still, attention is to be paid to the use of analogy, insofar as it may concern composite entities, which are to be analysed in all their parts and circumstances before arriving at a generalisation of their properties.[42] Eventually, the combined, right use of such means leads to evidence that provides us with certainty as persuasive as mathematical evidence, even if by different means, i.e. not resulting from ideas themselves but from the divine will.[43] Physics can then be defined as characterised by *scientia*, and its reliability is provided with a rational-theological foundation. Set forth in his early *Praefatio*, this solution is developed in his *De evidentia* and Introductio.

First of all, in his *De evidentia* 's Gravesande points out that the existence of bodies is not only morally evident, since we can perceive that some of our ideas depend on something external to the mind. Still, he does not openly declare whether the existence of bodies is subject to mathematical evidence ('s Gravesande 1734a, *De evidentia*, p. 19). In any case, the actual features and laws of matter are subject to moral evidence: according to his *Praefatio*, indeed, these rely on the will of God, being thus morally necessary. Moreover, according to his *De evidentia* and *Introductio*, the reliability of senses, testimony, and analogy is guaranteed by their relying on the moral necessity governing His acts, hence, they provide us with moral evidence. The core of 's Gravesande's arguments is that it is not contradictory that through the senses, testimony, and analogy one can err: however, it is contradictory – in a moral sense, according to the aforementioned considerations

---

42 's Gravesande 1736, pp. 171–174. An example of the misuse of analogy in reasoning is Huygens's conclusion that other planets are inhabited in his Κοσμοθεωρος: see 's Gravesande 1736, p. 173.
43 "Vidimus toto coelo differre evidentiam mathematicam a morali. Prima per se [...], secunda, ex Dei voluntate [...], id est, ex institutione, est criterium veri. Cum autem utriusque fundamentum sit firmum, plena etiam est persuasio quae moralem evidentiam sequitur," 's Gravesande 1736, pp. 175–176.

on necessity – that such sources of persuasion were provided to us by God as deceptive means, since they are essential to a long and happy life, given the material goods we may collect through them. The demonstration of their reliability is mathematically, i.e. demonstratively clear.[44] Therefore, sense perception, when used with due attention, provides us with moral evidence and certainty; testimony, even though provided by other men, may be evaluated by reason, which is a gift of God ('s Gravesande 1736, p. 148). The use of analogy, finally, justifies the very existence of universal laws, insofar as God does not deceive us in the use of generalisations. If there were no fixed rule, we would be deceived in our analogical reasoning.[45] Eventually, this is the ultimate foundation of Newton's second

[44] "Moralis autem evidentia non sua natura, sed ex Dei voluntate, persuasionis est fundamentum. Non, si rem in se consideremus, contradictionem involvit, sensus, testimonium, analogiam, adhibitis cautelis quibuscumque, nos in errorem inducere, sed contradictionem involvit, Deum voluisse haec esse persuasionis fundamenta, et haec, adhibitis legitimis cautelis, nos ad veritatem non conducere. Deum autem voluisse sensus, testimonium, et analogiam, talia esse fundamenta, et illum non frustra hoc voluisse, non erit demonstratu difficile, argumentis mathematice perspicuis. Talibus constat argumentis Deum esse, huncque esse bonum, et quidem in summo gradu. Hinc deducimus illum voluisse, ut homines iis utantur commodis quae ipsis largitus est. Iis autem rebus, quae ad vitam in superficie telluris ducendam, ubi Deus ipse homines collocavit, necessariae sunt, uti non posse demonstrabimus, nisi memorata admittamus criteria veri, unde patebit haec talia esse. Suprema sapientia sibi ipsi fuisset contraria, si datis ipsis rebus, facultatem de hisce diiudicandi denegasset. Quod tamen non excludit legitimas adhibendas esse cautelas," 's Gravesande 1734a, *De evidentia*, pp. 21–22. "Deus bonus hominibus magnam rerum ubertatem concessit, voluitque ipsos his uti, dum in superficie telluris vivunt, remotis autem sensibus, homines harum rerum cognitionem nullam omnino habere possent, et commodis ex his ipsis profluentibus privarentur. Unde manifestum est, universi moderatorem hominibus sensus dedisse, ut his in examine rerum uterentur, et ipsis fidem haberent. Sibi ipsi contraria esset suprema sapientia, si, concessis rebus, datisque mediis quibus cognoscantur, haec homines in errorem inducerent. [...] concludimus ex his omnibus, sensus, testimonia, analogiam, esse valida evidentiae moralis fundamenta," 's Gravesande 1736, pp. 146–148. Given the use of such terminology, he was accused of embracing a Spinozistic standpoint, i.e. to see divine actions as governed by a mathematical or absolute necessity. These arguments were rejected by means of Leibnizian arguments: see *supra*, section 7.5.

[45] "Infelices homines, qui singulis diebus in dubio haererent, utrum veneno an utili cibo vescerentur! [...] Nos summi numinis liberavit benignitas, nobis concessit observationes nostra ad non observata applicare, quo ad vitam necessaria a noxiis separamus, et futura saepe determinamus. [...] Non timeo aedificium firmum sponte casurum. Ex analogia ergo in rebus physicis mihi est ratiocinandum, et omnipotentem rerum Conditorem illud voluisse quis dubitabit, qui dum Conditorem bonum novit, ad rerum constitutione attendit. Sed dum Deus hoc voluit, et illa quae ut talibus ratiociniis vis communicetur necessario requiruntur etiam voluit: id est, fixis et immutatus rerum congeriem adstrinxit legibus. Positis enim his firmo stabilitur fundamento analogia, iisdem sublatis omnia sunt incerta in rebus physicis, et brevi genus integrum peribit humanum," 's Gravesande 1734a, *De evidentia*, p. 24. "Ratiocinia, quae analogiam

and third rule of philosophy, which in 's Gravesande's *Praefatio* was based on the survival axiom and which is provided in his *Introductio* and *De evidentia* with a more comprehensive theological justification.

Whenever the three means of moral evidence do not find complete application one must follow the hypothetical method, which leads to a probable knowledge or, according to a vulgar way of speaking – rejected by 's Gravesande – to moral certainty ('s Gravesande 1736, pp. 175–176). Aiming at providing Newtonian empirical physics with the status of *scientia*, that is, of indubitable knowledge, 's Gravesande stresses that experience, testimony, and analogy can lead us to an indubitable certainty. Thus, in his *Introductio* he softens the distinction between certainty and necessity, while maintaining a difference between absolute, physical, and moral necessity. In fact, he clearly distinguishes between *scientia* and probable knowledge. The latter, moreover, admits of several degrees and may finally acquire the status of *scientia*, or be provided with moral evidence. Its degrees are uncertainty, doubtfulness, probability, and certainty, all entailing the demonstration of possibility.[46] The first means in arguing for probable conclusions is the use of hypotheses, which is openly recognised and systematised by 's Gravesande in the section *De methodo* of his *Introductio*, after having assessed the rules of the analytical and the synthetic method, on which I have already focused. Hypotheses are used to provisionally explain those facts which are unexplainable otherwise; therefore they must be verified in order to lead to scientific, morally evident knowledge ('s Gravesande 1736, p. 292; see 's Gravesande 1742, vol. I, *Prefatio tertiae editionis*, p. X). In any case, their use must be subject to some rules, defined by 's Gravesande and applied to cryptography in his *Introductio* ('s Gravesande 1736, pp. 292–314). In fact, 's Gravesande's use of hypotheses is borrowed from Huygens's *Traité de la lumiere* (1690) ('s Gravesande 1736, pp. 295–296; see Gori 1972, pp. 271–272), and has its primary use in understanding the intentions of men: that is, the proper field of moral impossibility and necessity as defined in his book on metaphysics. Moreover, hypothetical

---

pro fundamento habent, nos ad veram rerum cognitionem ducere vidimus [...]. Circa hanc nunc observamus, ipsam admodum late patet, et hoc simplici principio niti. Rerum universitatem legibus immutatis regi. Nisi hanc admittamus propositionem, nullam omnino analogiam dari quis non videt? Huius firmitatem ex Dei voluntate deduximus. Manifestum idcirco est, hunc voluisse rerum materialium congeriem fixis adstringere legibus, et indubitatae erunt conclusiones quas ex principio deducemus," 's Gravesande 1736, p. 171. This is also stated in an interpolation in the *Praefatio primae editionis* printed in the second and third edition of his *Elementa* (1725, 1742): cf. 's Gravesande 1720–1721, vol. I, *Praefatio*, p. IX with 's Gravesande 1742, p. IX. This interpolation makes the contents of his *Praefatio* more consistent with his *De evidentia*, published in 1724.

46 's Gravesande 1736, pp. 179–180. Moreover, there are composed and opposing probabilities, which are mathematically dealt with by 's Gravesande: see pp. 181–211.

reasoning may concern the investigation of natural phenomena as these are ruled by laws rooted in the divine will. In both cases, its use is justified as both human and natural phenomena do not follow only mechanical or geometrical reasons.

7.7 's Gravesande's Newtonian philosophy

In basing his criterion of truth on the survival axiom in his *Praefatio primae editionis*, 's Gravesande recognises the moral function of the exercise of experience, testimony, and analogical reasoning. The right use of mental faculties allows the establishment of human society as God's wish; hence, such means are the right means in attaining the truth, in accordance with the survival axiom. Moreover, insofar as the establishment of society is enabled by our capacity to foresee events – such as that houses do not collapse, or that some food will not poison us – God has established fixed laws corresponding to our ideas in order to take care of men as His privileged creatures. This is determined by His goodness and wisdom, the keystone of 's Gravesande's foundational arguments in his *Introductio*. Not surprisingly, in presenting his views on the purpose of philosophy in his *Oratio de vera philosophia* (1734) 's Gravesande declares the coincidence of the 'true philosophy' with moral philosophy, aimed at the good life. Astronomy, mathematics, physics or mixed mathematics, as well as mechanics, optics, and hydraulics – or the scope of his academic teaching – relate to true philosophy but do not constitute philosophy as such.[47] This is the search for wisdom (*sapientia*) enabling men to fulfil the end God destined for them,[48] i.e., the attainment of the happiness or beatitude for which God provided them with reason, made possible as all men act for the good of other men.[49] To social life, indeed, men are driven by the right use

---

47 "Haec omnia utilia sunt, haec philosophiae debentur: sed non constituunt philosophiam," 's Gravesande 1734a, *De vera philosophia*, pp. 27–28.

48 "Hominem non fortuito natum, sed scopo cuidam peculiari destinatum esse, dum vitam in telluris superficie cum reliquis hominibus agit, illi tantum negant, qui ei quod evidentissime demonstrari potest assensum dare recusant. Summa hominis sapientia est huic scopo satisfacere, et ille merito philosophus vocatur, qui ut eo perveniat omnem operam impendit, neque alter hocce nomen meretur," 's Gravesande 1734a, *De vera philosophia*, p. 30.

49 "Beatitudinis ergo capax est homo, et hac sola de causa quia intelligentia praeditus est. Hanc autem amat, hanc optat beatitudinem, propriamque potius ipse destrueret naturam, quam hunc extingueret affectum, quod ab illa non potest separari perceptione, qua sibi constat se esse. Quaerit ideo homo, omne quod felicitatem, augere potest, et cum hanc tantum possideat quia est intelligens, etiam cognoscendi facultatem extendere cupit, haecque dum profluunt ex ipsa hominis natura, omnibus hominibus innata sunt. Homo, cum aliis hominibus in telluris superficie aetatem degens, singulis momentis aliorum indiget auxilio, quod ab his sperare non poterit,

of those faculties God provided them with.[50] In sum, the use of our faculties as given by God, the acknowledgment of His ends and the establishment of society are unmistakably related in 's Gravesande's view. In accordance with this view, he can set forth his system of philosophy, in which logic as the art of reasoning is the first part and is followed by metaphysics as the study of the human faculties and of being – according to an order later reversed in his *Introductio*. Physics finds a place in philosophy as it aims to grasp the order of nature.[51] Finally, philosophy includes rational theology, or the acknowledgment of the providential order of creation and the moral duties of men.[52] Recalling his studies in law, 's Gravesande praises *iurisperiti* as those able to embrace the true philosophy, underlining the difficulties in keeping law and theology detached from philosophy, since these higher arts concern the same topics as philosophy ('s Gravesande 1734a, *De vera philosophia*, pp. 43–45). As emphasised by Gori, the establishment of society through the right use of the mental faculties as the end of Providence is a concept unmistakably characterising the views on natural law of Grotius and Pufendorf, with which 's Gravesande's came to be acquainted under the influence of Gerardus Noodt, professor at Leiden University, addressing in his *De religione ab imperio iure gentium libera* (1706) the correspondence of nature and society, both following divine obligations.[53] In fact, 's Gravesande's arguments evidently recall the views of Grotius and Pufendorf – as interpreted by Barbeyrac – on our being

---

nisi et ipse alios adiuverit, et ab hoc mutuo officiorum commercio magis extenso, augmentum illius, quam dum in vivis est sperare potest, felicitatis pendere detegit. Unde hanc deducimus conclusione, tunc esse hominum felicitatem in tellure maximam, quando omnibus bonum aliorum cordi est. Ubi quisque hanc officiorum primam ponit regulam, unumquemque quantum potest alios adiuvare debere, quaerere ut ipsis utilis sit, nihil humani a se alienum putare," 's Gravesande 1734a, *De vera philosophia*, pp. 33–35.

50 "Ita etiam res disposuit, ut homo homini prodesse possit, aut potius inter omnes homines societatem esse voluit; hocque in ipsa rerum constitutione manifeste declaravit," 's Gravesande 1734a, *De vera philosophia*, p. 36.

51 's Gravesande 1734a, *De vera philosophia*, pp. 38–41. According to his *Praefatio primae editionis*, physics has the main purpose of disclosing the power and wisdom of God (see 's Gravesande 1720–1721, vol. I, *Praefatio*, p. VI). Still, 's Gravesande does not embrace any cosmological argument and rejects any metaphysical implications of his physics – while still maintaining its theological foundation: see Gori 1972, pp. 48–63, and Ducheyne 2014a, pp. 38–40.

52 "Officiorum tamen doctrina philosopho scopus erit, huic omnes animi vires applicabit et omnium primum investigabit quid debeat illi, a quo omnia accepit a quo omnia sperat. Dei perfectiones meditabitur, et ex his officia erga ipsum deducet," 's Gravesande 1734a, *De vera philosophia*, pp. 40–41. In fact, rational theology is present in logic also, as to the foundation of science.

53 See Gori 1972, pp. 66–67. Noodt's *Dissertatio de religione ab imperio iure gentium libera* (Noodt 1708), is referred to by 's Gravesande in his *De vera philosophia*. See 's Gravesande 1774, vol. II, p. 364 (n.).

provided with means enabling the establishment of human society as one of the ends of God,[54] and on the possibility of a demonstrative morality (Pufendorf 1740, book I, chapter 2, §§ 9–11; see Gori 1972, pp. 82, 184). However, although inscribed in a broader theological and moral perspective (in fact, the latter also characterised Descartes's plan of philosophy), for 's Gravesande philosophy is functional to physics, as his actual theories testify. In conclusion, 's Gravesande provides a foundation of science by arguments to be traced back to logic, metaphysics, and rational theology, as in the case of his Cartesian predecessors. This foundation is both an introduction of scholars to the new science and its justification as a means to understand natural phenomena. Such a foundation has a metaphysical nature since metaphysics provides the basic ontology for physics, whilst not being any physical principle.[55] Moreover, it is carried out by logical means, as logic defines the method of natural philosophy and the limits of science. Accordingly, it is within logic that 's Gravesande provides a rational-theological foundation of moral evidence. His foundation, therefore, entails both a de-metaphysicalisation of physics and a reflection on its limits.

The premises of such arguments mainly rely on the theories of Descartes, Malebranche, and Locke, used by 's Gravesande to support his own metaphysical and logical considerations. On the other hand, the core of his theological foundation is evidently influenced by Pufendorf's views on divine law. The recourse to natural law arguments can actually be explained by taking into account that 's Gravesande aimed at prevailing over Descartes's differentiation between moral certainty, which serves life and is open to doubt, and mathematical certainty or *scientia*. If according to Descartes empirical knowledge only serves the good life, the vindication of the 'scientific' status of such knowledge rests on its definition as the means given to us by God in order to lead a good life as the accomplishment of His providential order. For this reason, 's Gravesande borrowed his arguments from the natural law theories of Pufendorf and Grotius. However, this foundation of science not merely served to support the validity of the new physics in defiance of Cartesian philosophy, but resulted from a broader philosophical view also found in his *Lettre sur le mensonge*, where he justifies the use of lying as a means to preserve society, in accordance with right reason and the plan of God. The same

---

54 See Pufendorf 1740, book I, chapter 1, §§ 3–4, chapter 3, §§ 3, 5 (especially p. 71); Pufendorf 1723, book I, chapter 3, §§ 7–11; Grotius 1768, book I, chapter 1, § 10, comment 5. See also the text of 's Gravesande's teacher Philipp Reinhard Vitrarius, *Institutiones iuris naturae et gentium* (Vitrarius 1745), p. 9. See Gori 1972, pp. 130–133, 236–238, 256.

55 The 'anti-metaphysical' attitude of 's Gravesande was noticed for the first time by Ernst Cassirer: see Cassirer 1951, pp. 61–64. On this topic, see also Schliesser 2011 and Ducheyne 2014a.

view had been expressed by Pufendorf and his commentator Barbeyrac.[56] Thus, 's Gravesande's foundation of science was determined by two main factors: first, by the need to defend the reliability of empirical knowledge as *scientia*, since Newtonian physics had to be defended in order to allow its inclusion among the philosophical disciplines. Secondly, it served to show the truth of such knowledge as one of the means enabling the achievement of the providential plan of God. Therefore, 's Gravesande's foundation had a further end other than demonstrating the status of *scientia* of Newtonian physics and making it a part of the academic curriculum. Yet, following the Cartesian revolution in philosophy he pulled the final transformation of logic, metaphysics, and rational theology into foundational theories, embodying a philosophy of science as a justification and a reflection on the methods and concepts of mathematical-experimental physics.

---

56 See 's Gravesande 1774, vol. II, p. 258, Pufendorf 1740, vol. II, p. 172, n. 1 Gori 1972, pp. 82–85.

8 Conclusion: From *ancilla theologiae* to philosophy of science: a systematic assessment

Through a consideration of the philosophical debates occurring in the Dutch and Dutch-related intellectual framework in the early modern period, in the present study some alternatives in the foundation of philosophy and science have been highlighted and analysed. In conclusion, it is time to assess them in a more systematic manner. Each alternative entails a different view on foundational arguments, which may be grouped into theological, metaphysical, and logical ones. This research reveals the essential features of a philosophical milieu created by Descartes and constituting the framework for the dissemination of Newtonian science in Europe, leading to the birth of a philosophy of science as the study of the foundations, assumptions, methods, and limits of the study of natural realms. In fact, one can recognise several reasons for such foundation: on a general level, the need of providing philosophy with a secure foundation against the 'sceptical crisis' of the sixteenth and seventeenth centuries highlighted by Richard Popkin. More specifically, however, the foundation of Cartesian and Newtonian philosophy and science served to defend the conceptual premises of new ways of thinking in an academic context, demonstrating the validity of such new ways and rejecting the commonsensical and non-explanatory assumptions of Scholastic philosophy, and, in the case of 's Gravesande, the speculative character of Descartes's physics itself.

The first, main philosopher here taken into account is Henricus Regius, who was interested in the development of a Cartesian physics aimed at providing medicine with a basis – as elaborated upon by Descartes with his metaphor of the tree – but with no rational foundation, since Regius rejected Descartes's metaphysics. This rejection was based on metaphysical reasons, that is, by Regius's assumption of an empiricist standpoint with respect to the sources of knowledge and of an ontological lack of concern about the nature of the mind. Accordingly, he rejected any purely rational solution to philosophical problems, such as the formulation of an explanatory model of the constitution of the world. Still concerned with the problem of a foundation, however, he appealed to the Bible as the only solution to the metaphysical problems raised by Descartes and as the guarantee for the reliability of the use of the mental faculties, the use of which, however, can lead us only to a hypothetical or moral certainty with regard to our knowledge of natural laws.

The second figure who shaped Cartesian foundationalism was Johannes Clauberg, who had a view on the function of philosophy broader than Regius's.

https://doi.org/10.1515/9783110569698-009

Indeed, Clauberg aimed to replace the whole corpus of Scholastic thought with Descartes's philosophy, providing a basis for medicine, law, and theology. In other words, Clauberg aimed at developing a Cartesian scholasticism. For this sake, he developed a metaphysical foundation, with relevant implications for logic and *ontosophia*. According to him, metaphysics – which embodies rational-theological arguments on the role of God as guarantor of the truth of our knowledge – is the first discipline in the corpus of the sciences as it introduces students to a radically new way of thinking, *via* Cartesian doubt. Hence, metaphysics has the function of granting the reliability of the use of the mental faculties through theological arguments and of outlining the basic concepts of philosophy. Moreover, metaphysics finds its methodological counterpart in logic, aimed at guiding the mind in the formation of ideas and in their expression in words, but still including foundational arguments, as Clauberg analyses the degrees of certainty of metaphysical arguments though logical considerations. These, in turn, include concepts expressly dealt with in *ontosophia* as the last part of philosophy. So there is a threefold foundation of philosophy in Clauberg's case: logic, metaphysics, and *ontosophia*, with metaphysics as the major part. His solution shows that the problem of the foundation of philosophy as a comprehensive corpus of diverse disciplines required a metaphysical foundation embodying natural theology as the guarantee of the truth of philosophical arguments. Moreover, it required the development of a comprehensive methodology to be explained through a logical introduction.

Arnold Geulincx, on the other hand, adopted a more theologically-oriented foundation. Mainly interested in the development of an ethics responding to both the demands of Reformed theology and the philosophical standard of Leiden University, Geulincx developed a philosophical ethics conceived as the keystone of his system of philosophy. The foundation of philosophy intended as a rational ethics has a more consistent theological character since it is aimed at the attainment of salvation and beatitude. According to him, logic is the basement, metaphysics is the column, physics is the floor, and ethics is the roof of the House of Philosophy. In fact, rational theology plays an essential role in Geulincx's metaphysics as it is through an analysis of the role of God that he defines the status of physics as a hypothetical science (since natural laws rely on the will of God and can be grasped only through experience), and he defines the basics of his ethics, whose acknowledgment begins with the awareness of the dependence on God of our actions. On the other hand, logic has no foundational role as it neither delivers the basic concepts of the sciences nor guarantees the right functioning of the mind. Rather, it has a mere instrumental value as it helps to exercise the mind in demonstrative reasoning.

Johannes de Raey rejected both Regius's and Descartes's solutions and proposed more straightforward foundational arguments. His case shows that the foundation of philosophy as a purely rational enterprise – in response to the problem of the use of philosophy in practical disciplines raised by Regius, Sylvius, Meijer, and Spinoza – led to a foundation examining the basic rules of philosophical reasoning, their metaphysical presuppositions, and a rational theology aimed at guaranteeing their reliability. This foundation has a logical nature because it primarily shows how we may deal with philosophical notions and because it entails metaphysics, since by considering concepts and words it takes into account their reference. According to De Raey, logic has the function of providing physics with a foundation; physics, in turn, is the main part of philosophy and is to be distinguished from all the other academic disciplines and from any kind of practical art. Indeed, physics is aimed only at the formulation of a theoretical model of the constitution of the world on a purely rational basis. It has to be based on logic, since logic teaches us to use the four rules of Descartes's method. It provides the demonstration, via rational theology, of the existence of God as guarantor of our knowledge, and it consists in the analysis of the main notions and principles of philosophy. Thus, logic provides both the guarantee of the right functioning of the mind and discloses the basic notions of science. Moreover, since language is the vehicle of concepts, De Raey provides his analysis of the main notions of philosophy through a Cartesian interpretation of language. This is carried out by paying attention to the errors arising from a rejection of Descartes's metaphysics and from the application of the Cartesian paradigm of knowledge to the empirical disciplines and everyday practice. These errors, prompted by Regius's misuse of Cartesian philosophy, lead to a materialist standpoint on the references of ordinary and philosophical language, and accordingly, to the collapse of communication among men and of the formulation of philosophical arguments. De Raey thus developed a foundation of philosophy that was, at the same time, a reflection on the actual methods of philosophers and practitioners.

The fifth alternative is the metaphysical foundation of physics by Burchard de Volder, who was mainly interested in teaching natural philosophy by means of experiments confirming a Cartesian worldview. Moreover, he showed openness towards an empirical method of discovery inspired by Galileo, Huygens, and Newton, that is, the early figures of the modern experimental-mathematical science. Attracted by the successes of English science, De Volder's appreciation of an empirical methodology in the discovery of natural laws goes along with his unconcern with the deduction of the first principles of science from the attributes of God, carrying out a 'de-metaphysicalisation' of natural philosophy that had already commenced with Geulincx's hypothetical physics. Moreover, his interest in the practice of teaching which prevailed over the development of a comprehensive theory led him to neglect the formulation of a clear scientific methodology. Therefore, his foundation of physics does not entail a deduction of natural laws from the attributes of God, and mainly consists in the justification of the validity of the basic notions of mechanicism as our only means to formulate explanatory hypotheses on the causes of phenomena. Moreover, according to De Volder nothing but mere consciousness shows us that such principles are valid, which is only confirmed by the demonstration of the existence of God. Also, their actual causal role with regard to phenomena cannot be demonstrated, since they concern only one possible explanatory model. Therefore, the development of an empirical physics, having hypothetical certainty, goes along with a metaphysical foundation including some theological arguments, being mainly focused on the assessment of the main concepts to be dealt with experimentally.

Finally, the foundation of Newtonian physics of 's Gravesande provides a confirmation of the categorsation of foundational arguments, since a threefold classification is assumed by 's Gravesande in his foundation of Newtonian physics. The development of a systematic, experimental-mathematical physics, aimed at gaining a necessary knowledge in an academic framework dominated by a Cartesian stance on *scientia* as evident, purely intellectual knowledge, required the development of a comprehensive foundation. This concerns both metaphysical and logical (i.e. methodological) aspects; moreover, it includes a rational theology as the only basis for a justification of the reliability of sense perception. The core of the justification of the use of Newton's *regulae philosophandi* is theological, and bears witness to the influence of Pufendorf's and Grotius's views on the role of God in the establishment of human society. Indeed, this is made possible by the goodness of God, which makes the use of the senses, testimony, and analogy reliable in discovering the constant laws of nature. 's Gravesande's theological foundation, however, is introduced by an overview of the metaphysical assumptions underlying Newtonian physics, and serves as the guarantee of the ways the mind may become acquainted with the truth, that is, the methodological rules of analysis, synthesis, and hypothetical reasoning, which are considered from a logical point of view. By means of this threefold foundation of science 's Gravesande aims at providing Newtonian experimental-mathematical physics with the status of *scientia*, that is, with a certainty as persuasive as that of mathematics, even if provided by different means. While avoiding the development of a 'metaphysical physics', as Descartes did, 's Gravesande still provided physics with the highest degree of certainty.

These six main figures prove that philosophers paid attention to a defence of the reliability of Cartesian and Newtonian philosophy and science, and to their systematisation according to the needs of academia. One may object, however, that the case of Regius is a counter-proof to the claim that a foundation of philosophy and science was required by academic needs: indeed, he did not provide either a philosophical or a consistent foundation of philosophy, relying on a few biblical quotations as the only guarantee of the reliability of the mental faculties. This may prove Regius's lack of interest in metaphysical issues and the foundation of physics itself, but may also have been a particular strategy to avoid accusations of enthusiasm for the new philosophy, as these occurred during the Utrecht and Leiden crises. In this manner, Regius facilitated the introduction of the new philosophy in the academy by avoiding providing it with a foundation on Cartesian metaphysics: rather, he adopted a foundation on revealed theology. Yet, he could not demonstrate the status of *scientia* – i.e. as indubitable knowledge – of physics itself, which was vulnerable to the arguments of the sceptics. The next cases, which do entail a philosophical foundation of science, may indeed be interpreted as a reaction to Regius's solution, and at the same time as a response to the demands of academia. Clauberg developed a metaphysics serving as an introduction for students to the new ways in philosophy, and a logic teaching them how to conduct reason in every academic discipline. On the other hand, De Raey developed a foundational theory defining the very limits of Cartesian philosophy, making it consistent with the use of a commonsensical approach in medicine, law, theology, and the practical arts. Both Clauberg and De Raey, however, provided the new philosophy with a foundation aimed at justifying the function of the new philosophy in academic culture. Similarly, Geulincx was deeply concerned with the integration of the new philosophy into the academic context. He developed a philosophical ethics based on a coherent system of disciplines replacing the traditional matter of teaching. On the other hand, Spinoza, who was not concerned with academic demands, did not provide his ethics with a proper foundational theory, as he starts with a list of definitions and develops his theory of substance, mind and passions through a demonstrative reasoning which is neither justified nor clarified in its methodological implications.[1] In fact, he developed a metaphysics without a foundation, since the use of metaphysical notions is aimed at developing an ethics but is not justified in its reliability. In turn, De Volder was mainly concerned with the practice of academic teaching by means of experiments and with a method of discovery roughly inspired by that of Galileo, Huygens, and Newton, even though he maintained a Cartesian cosmological model. Therefore, he developed a Cartesian foundation of the principles of mechanicism for the benefit of students while avoiding deducing from these all the physical explanations. As in the case of Geulincx's hypothetical

---

[1] As pointed out by Paul Schuurman, "neither the *Ethica* nor the *Tractatus de intellectus emendatione* [...] devotes much attention to the two stages of the logic of ideas. Spinoza is interested primarily in establishing how we can obtain the clear and distinct ideas by which we can overcome our passions," Schuurman 2004, p. 65. See Spinoza's *Ethica*, part V, *Praefatio*.

physics, based on a set of notions dealt with by metaphysics (*somatologia*) but then developed by means of experience and generalisations, one can recognise a progressive de-metaphysicalisation of physics, as foundational theories were progressively kept detached from the discovery of new truths. Finally, 's Gravesande's *Introductio ad philosophiam* embodies all the solutions mentioned to the problem of a foundation, fitting the needs of the introduction, justification, and teaching of the basics of a new paradigm in the university. 's Gravesande's solution confirms that metaphysics, logic, and rational theology assumed a specific function in the introduction of new philosophies in early modernity, aimed at defending and clarifying their methodological and conceptual assumptions. So foundational theories led to the emergence of a philosophy of science, but also to a radical shift in the function of philosophy and in the very system of academic teaching, since philosophy progressively lost its character as the handmaiden of the higher faculties, which can still be recognised in Regius's and Clauberg's theories but which disappears in the next generations of Cartesian and Newtonian philosophers. Provided with increasing autonomy, justified through its foundational theories, philosophy found in itself the problems it aimed to solve, rather than being a mere preparation of scholars for law, medicine, and theology. Furthermore, this transformation was the prelude to a more substantial change in the organisation of knowledge in the modern age: that is, to the differentiation of philosophy and science. As natural philosophy progressively diversified its branches into independent sciences, logic and metaphysics as philosophical disciplines would increasingly assume the function of philosophy of science.

Bibliography

Primary sources

Andala, Ruardus (1701): *Oratio inauguralis de physicae praestantia, utilitate et iucunditate*. Franeker: apud Franciscum Halmam.

Andala, Ruardus (1709): *Exercitationes academicae in philosophiam primam et naturalem in quibus philosophia Renati Des-Cartes clare et perspicue explicatur, valide confirmatur nec non solide vindicatur*. Franeker: apud Wibium Bleck bibliopolam.

Andala, Ruardus (1711): *Syntagma theologico-physico-metaphysicum complectens Compendium theologiae naturalis, Paraphrasin in Principia philosophiae Renati Des-Cartes, ut et Dissertationum philosophicarum heptada*. Franeker: apud Wibium Bleck bibliopolam.

Andala, Ruardus (1716): *Examen Ethicae Clar. Geulingii sive Dissertationum philosophicarum, in quibus praemissa introductione sententiae quaedam paradoxae ex Ethica Clar. Geulingii examinantur, pentas*. Franeker: apud Wibium Bleck bibliopolam.

Andala, Ruardus (1718): *Apologia pro vera et saniore philosophia*. Franeker: excudit Henricus Halma.

Andreae, Tobias (1653): *Brevis replicatio reposita Brevi explicationi mentis humanae, sive animae rationalis D. Henrici Regii*. Amsterdam: typis Ludovici Elzevirii.

Andreae, Tobias (1653–1654): *Methodi cartesianae assertio opposita Jacobi Revii praefatae methodi cartesianae Considerationi theologicae quam vocat*. Groningen: typis Joannis Cöelleni.

Anonymous (1736): *Lettre à monsieur G.J. S'Gravesande, professeur en philosophie à Leide, sur son Introduction à la philosophie et particulièrement sur la nature de la liberté*. Amsterdam: chez J.F. Bernard.

Arbuthnot, John (1710): "An Argument for Divine Providence, Taken from the Constant Regularity Observed in the Births of Both Sexes". In: *Philosophical Transactions* 27, pp. 186–190.

Arnauld, Antoine/Lancelot, Claude (1660): *Grammaire générale et raisonnée contenant les fondemens de l'art de parler, expliqués d'une manière claire et naturelle*. Paris: chez Pierre le Petit.

Arnauld, Antoine/Nicole, Pierre (1683): *La logique ou l'Art de penser, contenant, outre les règles communes, plusieurs observations nouvelles, propres à former le jugement*. Paris: chez Guillaume Desprez. (1st ed. 1662).

Augustine (1970): "Soliloquiorum libri duo". In: Augustine: *Opera omnia*. Vol. III-1. Rome: Città Nuova Editrice.

Bacon, Francis (1605): *The Two Bookes of the Proficience and Advancement of Learning Divine and Humane*. London: printed for Henrie Tomes.

Bacon, Francis (1620): "Novum organum scientiarum sive indicia vera De interpretatione naturae". In: Bacon, Francis: *Instauratio magna*. London: apud Ioannem Billium, pp. 35–360.

Bacon, Francis (1623): *De augmentis scientiarum libri IX*. London: in officina Joannis Haviland.

Bacon, Francis (1653): "De principiis atque originibus secundum fabulas Cupidinis et Coeli: sive Parmenidis et Telesii et praecipue Democriti Philosophia". In: Bacon, Francis: *Scripta in naturali et universali philosophia*. Amsterdam: apud Ludovicum Elzevirum, pp. 208–285.

Bellini, Lorenzo (1663): *De urinis et pulsibus, de missione sanguinis, de febribus, de morbis capitis et pectoris*. Bologna: Ex typographia Antonii Pisarii.

https://doi.org/10.1515/9783110569698-010

Bellini, Lorenzo (1695): *Opuscula aliquot ad Archibaldum Pitcarnium*. Pistoia: ex nova officina Stephani Gatti.

Boerhaave, Herman (1703): *De usu ratiocinii mechanici in medicina*. Leiden: apud Joann. Verbessel.

Boerhaave, Herman (1715): *Sermo academicus de comparando certo in physicis*. Leiden: apud Petrum vander Aa.

Boerhaave, Herman (1718): *Sermo academicus de chemia suos errores expurgante*. Leiden: sumptibus Petri vander Aa.

Bontekoe, Cornelis/Geulincx, Arnold (1688): *Metaphysica, et liber singularis De motu, nec non eiusdem Oeconomia animalis, opera posthuma: quibus accedit Arnoldi Geulincx [...] Physica vera, opus posthumum*. Leiden: apud Johannem de Vivie, et Fredericum Haaring.

Borelli, Giovanni Alfonso (1680): *De motu animalium*. Rome: ex typographia Angeli Bernabò.

Boyle, Robert (1660): *New Experiments Physico-Mechanicall, touching the Spring of the Air, and its Effects*. Oxford: Printed by H. Hall.

De Bruyn, Johannes (1670): *Defensio doctrinae cartesianae de dubitatione et dubitandi modo*. Amsterdam: apud Danielem Elzevirium.

Burgersdijk, Franco (1626): *Institutionum logicarum libri duo ex Aristotelis, Keckermanni, aliorum praecipuorum logicorum praeceptis recensitis*. Leiden: apud Abraham Commelinum.

Burgersdijk, Franco (1640): *Institutionum metaphysicarum libri duo*. Leiden: apud Hieronymum de Vogel.

Burgersdijk, Franco (1645): *Institutionum logicarum synopsis, sive Rudimenta logica*. Leiden: apud Abrahamum Commelinum.

Burgersdijk, Franco (1649): *Institutionum logicarum synopsis, sive Rudimenta logica*. Amsterdam: apud Aegidium Valckenier.

Burgersdijk, Franco (1660): *Institutionum logicarum libri editio novissima*. Amsterdam: apud Aegidium Valckenier, et Casparum Commelinum.

Clauberg, Johannes (1647): *Elementa philosophiae sive Ontosophia*. Groningen: ex officina J. Nicolai.

Clauberg, Johannes (1652): *Defensio cartesiana, adversus Iacobum Revium [...] et Cyriacum Lentulum [...]: pars prior exoterica, in qua Renati Cartesii Dissertatio de methodo vindicatur, simul illustria cartesianae logicae et philosophiae specimina exhibentur*. Amsterdam: apud Ludovicum Elzevirium.

Clauberg, Johannes (1655): *Initiatio philosophi, sive Dubitatio cartesiana, ad metaphysicam certitudinem viam aperiens*. Leiden: ex officina A. Wyngaerden.

Clauberg, Johannes (1656): *De cognitione Dei et nostri exercitationes centum*. Duisburg: ex officina Wyngaerden.

Clauberg, Johannes (1657): *Unterschied zwischen der cartesianischer und der sonst in Schulen gebraeuchlicher Philosophie*. Duisburg: bey Adryan Wyngarten.

Clauberg, Johannes (1658): *Logica vetus et nova modum inveniendae ac tradendae veritatis, in genesi simul et analysi, facili methodo exhibens, editio secunda*. Amsterdam: ex officina Elzeviriana. (1st ed. 1654).

Clauberg, Johannes (1660): *Ontosophia nova, quae vulgo metaphysica, theologiae, iurisprudentiae et philologiae, praesertim germanicae studiosis accomodata. Accessit Logica contracta, et quae ex ea demonstratur orthographia germanica*. Duisburg: typis Adriani Wyngaerden.

Clauberg, Johannes (1664a): *Physica, quibus rerum corporearum vis et natura, mentis ad corpus relatæ proprietates, denique corporis ac mentis arcta et admirabilis in homine coniunctio explicantur*. Amsterdam: apud Danielem Elzevirium.

Clauberg, Johannes (1664b): *Metaphysica de ente, quae rectius Ontosophia*. Amsterdam: apud Danielem Elzevirium.

Clauberg, Johannes (1680): *Differentia inter cartesianam et in scholis vulgo usitatam philosophiam*. Berlin: apud Rupertum Völckern.

Clauberg, Johannes (1691): *Opera omnia philosophica*. Amsterdam: ex typographia P. et T. Blaev.

Le Clerc, Jean (1692): *Logica, sive, Ars ratiocinandi. Ontologia, sive De ente in genere. Pneumatologia seu De spiritibus*. London: impensis Awnsham et Johan. Churchill.

Le Clerc, Jean (1709): "Éloge de feu Mr. de Volder". In: *Bibliothèque choisie* 18 (Amsterdam: chez Henri Schelte), pp. 346–354.

De Cordemoy, Géraud (1666): *Le discernement du corps et de l'âme, en six discours, pour servir à l'éclaircissement de la physique*. Paris: chez Florentin Lambert.

De Cordemoy, Géraud (1668): *Discours physique de la parole*. Paris: chez Florentin Lambert.

Craanen Theodor (1686): *Lumen rationale medicum, hoc est Praxis medica reformata sive Annotationes in Praxin Henrici Regii*. Middelburg: apud Johannem de Reede.

Craanen, Theodor (1689): *Tractatus physico-medicus de homine, in quo status eius tam naturali, quam praeternaturalis, quoad theoriam rationalem mechanice demonstratur*. Leiden: apud Petrum vander Aa.

Denzinger, Heinrich Joseph Dominicus/Schönmetzer, Adolf (1991): *Enchiridion symbolorum definitionum et declarationum de rebus fidei et morum*. Freiburg, Basel, Rome, Vienna: Herder. (37th ed.) (1st ed. 1854).

Descartes, René (1641): Meditationes de prima philosophia, in qua Dei existentia et animae immortalitas demonstratur. Paris: apud Michaelem Soly.

Descartes, René (1648): *Notae in Programma quoddam*. s.l [Amsterdam]: s.n.

Descartes, René (1657): *Meditationes de prima philosophia: of Bedenkingen van d'eerste wysbegeerte*. Jan Hendrik Glazemaker (Trans.). Amsterdam: voor Jan Rieuwertsz.

Descartes, René (1662): *De homine figuris et latinitate donatus*. Florentius Schuyl (Ed. and Trans.). Leiden: apud Petrum Leffen et Franciscum Moyardum.

Descartes, René (1664a): *Le monde […] ou le traité de la lumière*. Paris: chez Michel Bobin.

Descartes, René (1664b): *L'homme […] et un Traité de la formation du foetus*. Claude Clerselier (Ed.) Paris: Chez Charles Angot.

Descartes, René (1897–1913): *Oeuvres*. 11 vols. Adam, Charles/Tannery, Paul (Eds.). Paris: L. Cerf.

Descartes, René/Schuyl, Florentius (1662): *De homine, figuris, et latinitate donatus*. Leiden: apud Petrum Leffen et Franciscum Moayardum.

Desmarets, Samuel (1645): *Collegium theologicum sive Systema breve universae theologiae*. Groningen: ex officina J. Nicolai.

Desmarets, Samuel (1649): *Collegium theologicum, sive Breve systema universae theologiae*. Groningen: apud Joannem Nicolaum. (1st ed. 1645).

Digby, Kenelm (1645): *Two Treatises*. London: printed for John Williams.

Ditton, Humphry (1712): *Discourse on the Resurrection of Jesus Christ*. London: printed by J. Darby.

Eustache de Saint-Paul (1620): *Summa philosophiae quadripartita, de rebus dialecticis, moralibus et metaphysicis*. Paris: apud Claudium Obert. (1st ed. 1609).

Fernel, Jean (1567): *Universa medicina*. Paris: apud Andream Wechelum.

Galen (1821–1833): *Opera omnia*. 20 vols. Kühn, Carl G. (Ed.). Leipzig: C. Cnobloch.

Galilei, Galileo (1632): *Dialogo [...] sopra i due massimi sistemi del mondo, tolemaico e copernicano*. Florence: per Gio. Batista Landini.

Gassendi, Pierre (1644): *Disquisitio metaphysica seu Dubitationes et instantiae adversus Renati Cartesii Metaphysicam et Responsa*. Amsterdam: apud Iohannem Blaeu.

Geulincx, Arnold (1653): *Questiones quodlibeticae in utramque partem disputatae*. Antwerp: bij de We Cnobbaert.

Geulincx, Arnold (1662): *Logica fundamentis suis restituta*. Leiden: apud Henricum Verbiest.

Geulincx, Arnold (1663): *Methodus inveniendi argumenta, quae solertia quibusdam dicitur* Leiden: apud Isaacum De Waal Verbiest.

Geulincx, Arnold (1665a): *Saturnalia, seu (ut passim vocantur) Quaestiones quodlibeticae in utramque partem disputatae. Editio secunda ab auctore recognita et aucta*. Leiden: ex officina Henrici Verbiest.

Geulincx, Arnold (1665b): *De virtute et primis eius proprietatibus, quae vulgo virtutes cardinales vocantur tractatus ethicus primus*. Leiden: apud Philippum de Croy.

Geulincx, Arnold (1667): *Van de hooft-deuchden. De eerste tucht-verhandeling*. Leiden: bij Philips de Croy.

Geulincx, Arnold (1675): Γνῶϑι σεαυτόν, *sive Ethica*. Leiden: apud Adrianum Severini.

Geulincx, Arnold (1691a): *Metaphysica vera et ad mentem peripateticam*. Amsterdam: apud Joannem Wolters.

Geulincx, Arnold (1691b): *Annotata maiora in Principia philosophiae Renati des Cartes*. Dordrecht: ex officina T. Goris.

Geulincx, Arnold (1891–1893): *Opera philosophica*. 3 vols. Land, Jan Pieter Nicolaas (Ed.). The Hague: Martinus Nijhoff.

Geulincx, Arnold (2006): *Ethics. With Samuel Beckett's Notes*.Van Ruler, Han/Uhlmann, Anthony/Wilson, Martin (Eds. and Trans.). Leiden: Brill.

Goclenius, Rudolph (1597): *Problemata logica*. 5 vols. Marburg: typis Pauli Egenolphi.

Goclenius, Rudolph (1613): *Lexicon philosophicum*. Frankfurt: typis viduae Matthiae Beckerii.

's Gravesande, Willem Jacob (1711): *Essai de perspective*. The Hague: chez la veuve d'Abraham Troyel.

's Gravesande, Willem Jacob (1714): "Remarques sur la construction des machines pneumatiques & sur les dimensions qu'il faut leur donner". In: *Journal literaire* 4, pp. 182–208.

's Gravesande, Willem Jacob (1717): *Oratio inauguralis de matheseos in omnibus scientiis, praecipue in physicis, usu, nec non de astronomiae perfectione ex physica haurienda*. Leiden: apud Samuelem Luchtmans.

's Gravesande, Willem Jacob (1720–1721): *Physices elementa mathematica, experimentis confirmata. Sive Introductio ad philosophiam newtonianam*. Leiden: apud Petrum van der Aa.

's Gravesande, Willem Jacob (1721): "Examen des raisons de Mr. Bernard contre le mensonge officieux". *Journal literaire* 11, pp. 344–366.

's Gravesande, Willem Jacob (1723): *Philosophiae newtonianae institutiones, in usus academicos*. Leiden: apud Petrum vander Aa.

's Gravesande, Willem Jacob (1725): *Physices elementa mathematica, experimentis confirmata. Sive Introductio ad philosophiam newtonianam. Editio secunda, auctior et emendatior*. Leiden: apud Petrum vander Aa.

's Gravesande, Willem Jacob (1727): *Matheseos universalis elementa, quibus accedunt, Specimen commentarii in arithmeticam universalem Newtoni: ut et De determinanda forma seriei infinitae adsumtae regula nova*. Leiden: apud Samuelem Luchtmans.

's Gravesande, Willem Jacob (1734a): *Orationes tres. De matheseos in omnibus scientiis, praecipue in physicis, usu, nec non de astronomiae perfectione ex physica haurienda. Altera, De evidentia. Tertia De vera et nunquam vituperata philosophia*. Leiden: apud Samuelem Luchtmans.

's Gravesande, Willem Jacob (1734b): *Orationes duae. Prima De vera, et nunquam vituperata, philosophia, altera De evidentia*. Leiden: apud Samuelem Luchtmans.

's Gravesande, Willem Jacob (1736): *Introductio ad philosophiam, metaphysicam et logicam continens*. Leiden: s.n.

's Gravesande, Willem Jacob (1737): *Introductio ad philosophiam: metaphysicam et logicam continens. Editio altera*. Leiden: apud Joh. et Herm. Verbeek.

's Gravesande, Willem Jacob (1742): *Physices elementa mathematica, experimentis confirmata. Sive Introductio ad philosophiam newtonianam. Editio tertia duplo auctior*. Leiden: apud Johannem Arnoldum Langerak, Johannem et Hermannum Verbeek.

's Gravesande, Willem Jacob (1774): *Oeuvres philosophiques et mathématiques*. 2 vols. Allamand, Jean Nicolas Sébastien (Ed.). Amsterdam: chez Marc Michel Rey.

Gronovius, Johannes (1709): *Burcheri de Volder laudatio*. Leiden: apud Cornelium Boutestein.

Grotius, Hugo (1768): *Le droit de la guerre et de la paix*. Barbeyrac, Jean/Gronovius, Johannes (Trans. and Comm.). Leipzig: s.n. (1st ed. 1724).

Heereboord, Adriaan (1650): Ἑρμηνεία *logica, seu Explicatio Synopseos logicae Burgersdicianae*. Leiden: apud Severinum Matthæi, et Davidem a Lodensteyn.

Heereboord, Adriaan (1654): *Meletemata philosophica, maximam partem, metaphysica*. Leiden: ex officina Francisci Moyardi.

Heidanus, Abraham (1645): *De causa Dei: dat is De sake Godts verdedight tegen den mensche, ofte Wederlegginge van de Antwoorde van M. Simon Episcopius*. Leiden: gedruckt by Paulus Aertsz van Ravesteyn.

Heidanus, Abraham (1669): *Advijs van de Theologische Faculteyt tot Leyden*. S.l: Voor Jan Recht-zinnigh.

Heidanus, Abraham, *et al.* (1676): *Consideratien, over eenige saecken onlanghs voorgevallen in de Universiteyt binnen Leyden*. Leiden: bij Aernout Doude.

Heidanus, Abraham, *et al.* (1678): *Considerationes ad res quasdam nuper gestas in Academia Lugduno-Batava*. Hamburg: apud Petrum Grooten.

Hobbes, Thomas (1640): *Elements of Law, Natural and Politic*. (Manuscript).

Hobbes, Thomas (1642): *Elementa philosophica De cive*. Paris: s.n.

Hobbes, Thomas (1647): *Elementa philosophica De cive*. Amsterdam: apud Danielem Elzevirium.

Hobbes, Thomas (1649): *Elements philosophiques du citoyen*. Samuel Sorbière (Trans). Amsterdam: de l'Imprimerie de Iean Blaeu.

Hobbes, Thomas (1650): *De corpore politico, or, The Elements of Law, Moral & Politick*. London: printed for J. Martin, and J. Ridley.

Hobbes, Thomas (1655): *Elementorum philosophiae sectio prima De corpore*. London: Crook.

Hobbes, Thomas (1651): *Leviathan or The Matter, Forme and Power of a Common-Wealth Ecclesiasticall and Civil*. London: printed for Andrew Crooke.

Hobbes, Thomas (1657): *Elementa philosophica De cive*. Amsterdam: apud Ludovicum et Danielem Elzevirios.

Hobbes, Thomas (1667*): Leviathan: of van de stoffe, gedaente, ende magt vande kerckelycke ende wereltlycke regeeringe*. Abraham van Berckel (Trans.) Amsterdam: By Jacobus Wagenaar. (2nd ed. 1672).

Hobbes, Thomas (1668): *Opera philosophica quae latine scripsit, omnia*. Amsterdam: Apud Ioannem Blaeu.

Hobbes, Thomas (1669): *Elementa philosophica De cive*. Amsterdam: apud Danielem Elzevirium.

Hobbes, Thomas (1670): *Leviathan, sive de materia, forma et potestate civitatis ecclesiasticae et civilis*. Amsterdam: Apud Ioannem Blaeu.

Hobbes, Thomas (1675): *De eerste beginselen van een burger-staat*. Amsterdam: s.n.

Van Hogelande, Cornelis (1646): *Cogitationes, quibus Dei existentia, item animae spiritalitas, et possibilis cum corpore unio, demonstrantur, nec non, brevis historia oeconomiae corporis animalis, proponitur, atque mechanice explicatur*. Amsterdam: apud Ludovicum Elzevirium.

Holwarda, Johannes Phocylides (1651): *Philosophia naturalis, seu Physica vetus-nova*. Franeker: excudit Idzardus Alberti.

Huet, Pierre-Daniel (1689): *Censura philosophiae cartesianae*. Paris: apud Danielem Horthemels.

Huygens, Christiaan (1690): *Traité de la lumière [...]. Avec un Discours de la cause de la pesanteur*. Leiden: Chez Pierre vander Aa.

Huygens, Christiaan (1698): *Κοσμοθεωρός: sive De terris coelestibus, earumque ornatu, coniecturae, ad Constantinum Hugenium fratrem*. Burchard de Volder, Bernhard Fullenius (Eds.). The Hague: apud Adrianum Moetjens, bibliopolam.

Huygens, Christiaan (1703): *Opuscula posthuma*. Burchard de Volder, Bernhard Fullenius (Eds.). Leiden: apud Cornelium Boutesteyn.

Huygens, Christiaan (1888–1950): *Oeuvres complètes*. 22 vols. Bierens de Haan, David/ Bosscha, Johannes/Korteweg, Diederik J./Vollgraff, Johan A. (Eds.). The Hague: Martinus Nijhoff.

Irenaeus, Philalethius [Heidanus, Abraham] (1656a): *Bedenkingen op den Staat des geschils* over de *Cartesiaensche philosophie en op de Nader openinghe over eenige stucken de theologie raeckende*. Rotterdam: by Johannes Benting.

Irenaeus, Philalethius [Heidanus, Abraham] (1656b): *De overtuigde quaetwilligheidt van Svetonius Tranquillus*. Leiden: by Adrianus Wijngaerden.

Jansenius, Cornelius (1640): *Augustinus seu doctrina Sancti Augustini de humanae naturae sanitate, aegritudine medicina aduersus Pelagianos et Massilienses*. Leuven: Typis Iacobi Zegeri.

Keckermann, Bartholomäus (1600): *Systema logicae, tribus libris adornatum, pleniore praeceptorum methodo, et commentariis scriptis ad praeceptorum illustrationem*. Hannover: apud Guilielmum Antonium.

Keckermann, Bartholomäus (1611): *Scientiae metaphysicae compendiosum systema*. Hannover: apud Guilielmum Antonium. (1st ed. 1609).

Keckermann, Bartholomäus (1613a): "Systema logicae, tribus libris adornatum, pleniore praeceptorum methodo, et commentariis scriptis ad praeceptorum illustrationem". In: Keckermann, Bartholomäus: *Systema systematum*. Vol. I. Hannover: apud Guilielmum Antonium, pp. 67–766.

Keckermann, Bartholomäus (1613b): "Praecognitorum logicorum tractatus III". In: Keckermann, Bartholomäus: *Systema systematum*. Vol. I. Hannover: apud Guilielmum Antonium, pp. 1–66.

Keckermann, Bartholomäus (1623): *Systema physicum*. Hannover: impensis Ioannis Stockelii. (1st ed. 1610).

Keill, John (1702): *Introductio ad veram physicam; seu Lectiones physicae habitae in Schola naturalis philosophiae Academiae Oxoniensis*. Oxford: impensis T. Bennet.

Van Lamzweerde, Jan Baptist (1674): *Respirationis Swammerdammianae exspiratio, una cum anatomia neologices Joannis de Raei*. Amsterdam: ex officina Joannis à Someren.

Leibniz, Gottfried Wilhelm von (1875–1890): *Die philosophischen Schriften*. Carl Immanuel Gerhardt (Ed.). 7 vols. Berlin: Weidmannsche Buchhandlung.

Leibniz, Gottfried Wilhelm von (1923): *Sämtliche Schriften und Briefe*. Darmstadt, Leipzig, Berlin: Akademie Verlag.

Lentulus, Cyriacus (1651): *Nova Renati Des Cartes sapientia: faciliori quam antehac methodo detecta*. Herborn: s.n.

Lentulus, Cyriacus (1653): *Cartesius triumphatus et nova sapientia ineptiarum et blasphemiae convicta*. Frankfurt: typis et sumptibus Johannis Friderici Weiss.

Locke, John (1690) [1689]: *An Essay Concerning Human Understanding*. London: printed for Tho. Basset.

Meijer, Lodewijk (1666): *Philosophia S. Scripturae interpres, exercitatio paradoxa, in qua, veram philosophiam infallibilem S. Literas interpretandi normam esse, apodictice demonstratur*. s.l.: s.n.

Molhuysen, Philip Christiaan (1913–1924): *Bronnen tot de geschiedenis der Leidsche Universiteit*. 7 vols. The Hague: Martinus Nijhoff.

Newton, Isaac (1687): *Philosophiae naturalis principia mathematica*. London: Iussu Societatis Regiae ac Typis Josephi Streater.

Newton, Isaac (1704): *Opticks: Or, a Treatise on the Reflections, Refractions, Inflections and Colours of Light*. London: printed for Sam. Smith, and Benj. Walford.

Newton, Isaac (1726): *Philosophiae naturalis principia mathematica. Editio tertia aucta et emendata*. London: apud Guil. et Joh. Innys. (1st ed. 1687).

Noodt, Gerard (1708): *Dissertatio de religione ab imperio iure gentium libera*. Leiden: apud Fredericum Haaring.

Pitcairne, Archibald (1692): *Oratio, qua ostenditur, medicinam ab omni philosophorum secta esse liberam*. Leiden: apud Abrahamum Elzevier.

Pitcairne, Archibald (1717): *Elementa medicinae physico-mathematica, libris duobus: quorum prior theoriam, posterior praxim exhibet*. London: Impensis Gulielmi Innys.

Pufendorf, Samuel (1723): *Les devoirs de l'homme, et du citoien, tels qu'ils lui sont prescrits par la loi naturelle*. Barbeyrac, Jean (Trans. and Comm.). Amsterdam: chez Pierre de Coup.

Pufendorf, Samuel (1740): *Le droit de la nature et des gens, ou Système général des principes les plus importans de la morale, de la jurisprudence, et de la politique*. Barbeyrac, Jean (Trans. and Comm.). London: chez Jean Nours, 1740.

De Raey, Johannes (1651–1652): *Disputationum physicarum ad Problemata Aristotelis prima-quinta*. Leiden: e typographeo Francisci Hackii, et ex officina Joannis Maire.

De Raey, Johannes (1651): *Oratio inauguralis de gradibus et vitiis notitiae vulgaris circa contemplat. naturae et officio philosophi circa eandem*. Leiden: e typographeo Francisci Hackii.

De Raey, Johannes (1654): *Clavis philosophiae naturalis, seu Introductio ad naturae contemplationem, aristotelico-cartesiana*. Leiden: ex officina Joannis et Danielis Elsevier.

De Raey, Johannes (1661): *Disputatio philosophica de mundi systemate et elementis prima*. Leiden: apud Johannem Elsevirium.

De Raey, Johannes (1663): *Disputatio philosophica qua quaeritur quo pacto anima humana in corpore moveat et sentiat*. Leiden: apud viduam et haeredes Johannis Elsevirii.

De Raey, Johannes (1665–1667): *Disputatio philosophica, de forma substantiali et anima hominis, prima-tertia*. Leiden: apud viduam et haeredes Johannis Elsevirii.

De Raey, Johannes (1668a): *Disputatio philosophica de constitutione logicae*. Leiden: apud viduam et haeredes Johannis Elsevirii.

De Raey, Johannes (1668b): *Disputatio philosophica de constitutione physicae*. Leiden: apud viduam et haeredes Johannis Elsevirii.

De Raey, Johannes (1669): *Dissertatio philosophica de sapientia veterum*. Amsterdam: apud Joannem Ravesteinium.

De Raey, Johannes (1677): *Clavis philosophiae naturalis aristotelico-cartesiana. Editio secunda, aucta opuscolis philosophicis varii argumenti*. Amsterdam: apud Danielem Elsevirium.

De Raey, Johannes (1685): *Miscellanea philosophica*. Amsterdam: apud Joannem Riwuwetsz.

De Raey, Johannes (1692): *Cogitata de interpretatione, quibus natura humani sermonis et illius rectus usus, tum in communi vita et disciplinis ad vitae usum sptectantibus, tum in philosophia, ab huius seculi errore et confusione vindicantur. Accedunt Notae recentes ad partem primam generalem. Cum Appendice ex olim scriptis, propter cognationem*. Amsterdam: apud Henricum Wetstenium.

De Raey, Johannes/Wolzogen, Ludwig/Van Leeuwen, G. (1689): *Copie van de acte van de heeren professoren der Illustre Schoole tot Amsterdam, J. de Raei, L. Wolzogue, en G. van Leeuwen [...] tegen het misbruyk der philosophie*. Amsterdam: s.n.

Ramus, Petrus (1569): *Scholae in liberales artes*. Basel: per Eusebium Episcopium.

Ramus, Petrus (1543a): *Dialecticae partitiones*. Paris: excudebat Iacopus Bogardus.

Ramus, Petrus (1543b): *Dialecticae institutiones, ad celeberrimam, et illustrissimam Parisiorum Academiam*. Paris: excudebat Iacobus Bogardus.

Ramus, Petrus (1543c): *Aristotelicae animadversiones*. Paris: excudebat Iacopus Bogardus.

Regius, Henricus (1640a): *Disputatio medico-physiologica pro sanguinis circulatione*. Utrecht: ex officina Aegidii Roman.

Regius, Henricus (1641a): *Physiologia sive cognitio sanitatis. Tribus disputationibus in Academia Ultraiectina publice proposita*. Utrecht: ex officina Aegidii Roman.

Regius, Henricus (1641b): *Disputatio medica prima [-tertia] de illustribus aliquot quaestionibus physiologicis*. Utrecht: ex officina Aegidii Roman.

Regius, Henricus (1642): *Responsio sive Notae in Appendicem ad Corollaria [...] Voetii*. Utrecht: apud Ioannem a Doorn.

Regius, Henricus (1646): *Fundamenta physices*. Amsterdam: apud Lodovicum Elzevirium.

Regius, Henricus (1647a): *Fundamenta medica*. Utrecht: apud Theodorum Ackersdycium.

Regius, Henricus (1647b): *Medicatio viri cachexia leucophlegmatica affecti*. Utrecht: Johannes a Noortdyck.

Regius, Henricus (1648): *Brevis explicatio mentis humane*. Utrecht: ex officina Theodori Ackersdicii.

Regius, Henricus (1654): *Philosophia naturalis*. Amsterdam: apud Lodovicum Elzevirium.

Regius, Henricus (1657a): *Medicinae libri quatuor*. Utrecht: typis Theodori ab Ackersdijck, et Gisberti a Zijll.

Regius, Henricus (1657b): Praxis medica. Utrecht: typis Theodori ab Ackersdijck, et Gisberti a Zijl.

Regius, Henricus (1661): *Philosophia naturalis, in qua tota rerum universitas, per clara et facilia principia, explanantur*. Amsterdam: apud Lodovicum et Danielem Elzevirios.

Revius, Jacob (1647): *Analectorum theologicorum disputatio XXIII. De cognitione Dei, tertia*. Leiden: apud haeredes Johannis Nicolai a Dorp.

Revius, Jacob (1648): *Methodi cartesianae consideratio theologica*. Leiden: apud Hieronymum de Vogel.

Revius, Jacob (1650): *Statera philosophiae cartesianae, qua principiorum eius falsitas et dogmatum impuritas expenditur ac castigatur*. Leiden: ex officina Petri Leffen.

Revius, Jacob (1653): *Thekel, hoc est levitas Defensionis Cartesianae, quam Johannes Claubergius Considerationi et Staterae Jacobi Revii opposuit*. Brielle: excudebat Michael Feermans.

Revius, Jacob (1654): *Psychotheomachia. Hoc est Appendix Tobiae Andreae, qua I. Animae immortalitatem obscurat, II. Dei veracitatem negat, armis e Scriptura et ratione petitis profligata*. Leiden: s.n.

Revius, Jacobus (1654): *Kartēsiomania, hoc est, Furiosum nugamentum, quod Tobias Andreae, sub titulo Assertionis methodi cartesianae, orbi literato obtrusit, succincte ac solide confutatum*. Leiden: apud Hieronymum de Vogel.

Revius, Jacobus (1655): *Kartēsiomanias pars altera, qua ad secundam partem rabiosae Assertionis Tobiae Andreae responditur*. Leiden: apud Hieronymum de Vogel.

Rohault, Jacques (1671): *Traité de physique*. Paris: Chez la veuve de Charles Savreux.

Scaliger, Julius Caesar (1557): *Exotericarum exercitationum libri XV De subtilitate ad Hieronymum Cardanum*. Paris: ex officina typographica Michaelis Vascosani.

Schoock, Martin (1643): *Admiranda methodus novae philosophiae Renati des Cartes*. Utrecht: ex officina Joannis van Waesberge.

Schuyl, Florentius (1672): *De veritate scientiarum et artium academicarum*. Leiden: apud Johannis Hogenackerum.

Senguerd, Wolferd (1681): *Philosophia naturalis: quatuor partibus primarias corporum species affectiones, differentias, productiones mutationes et interitus exhibens*. Leiden: apud Danielem a Gaesbeeck.

Senguerd, Wolferd (1685): *Philosophia naturalis: quatuor partibus primarias corporum species affectiones, differentias, productiones mutationes et interitus exhibens. Editio secunda, priore auctior*. Leiden: apud Danielem a Gaesbeeck.

Senguerd, Wolferd [*praes.*]/Boerhaave, Herman [*resp.*] (1687): *Disputatio de cohaesione corporum*. Leiden: apud Abraham Elzevier.

Senguerd, Wolferd [*praes.*]/Boerhaave, Herman [*resp.* and author] (1687–1688): *Disputatio pneumatica de mente humana prima-tertia*. Leiden: apud Abraham Elzevier.

Senguerd, Wolferd (1699): *Inquisitiones experimentales, quibus praeter particularia nonnulla phaenomena, atmosphaerici aëris natura explicatius traditur*. Leiden: apud Cornelium Boutesteyn.

Senguerd, Wolferd (1715): *Rationis atque experientiae connubium continens experimentorum physicorum, mechanicorum, hydrostaticourm, barometricorum, thermometricorum, aliorumque compendiosam narrationem*. Rotterdam: apud Bernardum Bos.

Spinoza, Baruch de (1663): *Renati des Cartes Principiorum philosophiae pars I, et II, more geometrico demonstratae. Accesserunt eiusdem Cogitata metaphysica*. Amsterdam: Apud Johannem Rieuwertsz.

Anonymous [Spinoza, Baruch de] (1670): *Tractatus theologico-politicus, continens dissertationes aliquot, quibus ostenditur libertatem philosophandi non tantum salva pietate, et reipublicae pace posse concedi, sed eandem nisi cum pace reipublicae, ipsaque pietate tolli non posse*. Hamburg: Apud Henricum Künrath.

Stuart, Adam (1647): *Thesium metaphysicarum de Deo disputatio prima-secunda*. Leiden: B. et A. Elzevier.

Svetonius Tranquillus [pseudonym] (1656a): *Staat des geschils, over de Cartesiaansche philosophie*. Utrecht: by Johannes van Waesberge.

Svetonius Tranquillus [pseudonym] (1656b): *Nader openinge van eenighe stucken in de Cartesiaensche philosophie raeckende de [...] theologie*. Leiden: voor Cornelis Banheining.

Svetonius Tranquillus [pseudonym] (1656c): *Den overtuyghden cartesiaen ofte Claere aenwysinge uyt de Bedenckingen van Irenaeus Philalethius [...]*. Leiden: gedruckt door Cornelis Banheinning.

Sylvius, Franciscus (1658): *Oratio inauguralis de hominis cognitione*. Leiden: ex officina Joannis Elsevier.

Sylvius, Franciscus (1663): *Disputationum medicarum pars prima: primarias corporis humani functiones ex anatomicis, practicis et chymicis experimentis deductas complectens*. Amsterdam: apud Johannem van den Bergh.

Triglandius, Jacob (1647): *Disputationum theologicarum in confessionem et apologiam remonstrantium quinquagesima*. Leiden: ex officina Bonaventurae et Abrahami Elsevir.

Van Velthuysen, Lambert (1651a): *Epistolica dissertatio de principiis iusti et decori, continens apologiam pro tractatu clarissimi Hobbaei De Cive*. Amsterdam: apud Ludovicum Elzevirium.

Van Velthuysen, Lambert (1651b): *Disputatio de finito et infinito, in qua defenditur sententia clarissimi Cartesii, de motu, spatio et corpore*. Amsterdam: apud Ludovicum Elzevirium.

Van Velthuysen, Lambert (1680): *Opera omnia [...] quibus accessere duo tractatus novi, hactenus inediti: prior est De articulis fidei fundamentalibus, alter De cultu naturali, oppositus Tractatui theologico-politico et Operi posthumo Benedicti de Spinoza*. Rotterdam: Typis Reineri Leers.

Van Velthuysen, Lambertus (2013): *A Letter on the Principles of Justness and Decency, Containing a Defence of the Treatise De Cive of the Learned Mr Hobbes*. De Mowbray, Malcom (Ed. and Trans.)/Secretan, Catherine (Intr.). Leiden, Boston: Brill.

De Versé, Noël Aubert (1684): *L'impie convaincu ou Dissertation contre Spinosa, dans laquelle l'on réfute les fondements de son atheisme. L'on trouvera dans cét ouvrage non seulement la réfutation des maximes impies de Spinosa, mais aussi celle des principales hypotheses du cartesianisme, que l'on fait voir être l'origine du spinosisme*. Amsterdam: chez Jean Crelle.

De Versé, Noël Aubert (2015): *L'impie convaincu ou Dissertation contre Spinoza*. Benigni, Fiormichele (Ed.). Rome: Edizioni di Storia e Letteratura.

Vinson, Carolus (1676–1677): *Experimenta philosophica naturalia auctore Magistro de Kaldo [sic]: 1676, 1677*. In Sloane 1292, ms., British Library, ff. 78–141.

Vitrarius, Philipp Reinhard (1745): *Institutiones iuris naturae et gentium*. Lausanne: apud Antonium Chapuis.

Voetius, Gysbertus *et al.* (1643): *Testimonium Academiae Ultraiectinae, et Narratio historica*. Utrecht: ex typographia Wilhelmi Strickii.

De Volder, Burchard (1664): *Disputatio medica inauguralis de natura*. Leiden: apud Severinum Matthiae.

De Volder, Burchard (1674–1676): *Disputatio philosophica de rerum naturalium principiis prima-quintadecima*. Leiden: apud viduam et haeredes J. Elsevirii.

De Volder, Burchard (1676–1678): *Disputatio philosophica de aëris gravitate prima-quinta*. Leiden: apud viduam et haeredes Joannis Elsevirii.

De Volder, Burchard (1680–1681): *Disputatio philosophica contra atheos prima-quarta*. Leiden: apud viduam et heredes Johannis Elsevirii.

De Volder, Burchard (1681a): *Quaestiones academicae de aëris gravitate*. Middelburg: typis viduae Remigii Schreverii.

De Volder, Burchard (1681b): *Disputationes philosophicae sive Cogitationes rationales de rerum naturalium principiis*. Middelburg: typis Remigii Schreverii.

De Volder, Burchard (1681c): *Disputationes philosophicae sive Cogitationes rationales de rerum naturalium principiis, ut et De aëris gravitate*. Leiden: apud Jacobum Moukee.

De Volder, Burchard (1682): *Oratio de coniugendis philosophicis et mathematicis disciplinis, cum philosophicae professioni adiunctam mathematicam rite auspicaretur*. Leiden: apud Jacobum Voorn.

De Volder, Burchard (1685): *Disputationes philosophicae omnes contra atheos*. Middelburg: apud Joannem Lateranum.

De Volder, Burchard (1689): *Disputatio philosophica de certitudine clarae et distinctae perceptionis*. Leiden: apud Abrahamum Elzevier.

De Volder, Burchard (1691–1693): *Exercitationum philosophicarum adversum Censuram prima-vicesima octava*. Leiden: apud Abrahamum Elzevier.

De Volder, Burchard (1694): *Disputatio philosophica quae est de mundi systemate*. Leiden: apud Abrahamum Elzevier.

De Volder, Burchard (1695): *Exercitationes academicae quibus Ren. Cartesii philosophia defenditur adversus Petri Danielis Huetii [...] Censuram philosophiae cartesianae*. Amsterdam: apud Arnoldum van Ravestein.

De Volder, Burchard (1698): *Oratio de rationis viribus, et usu in scientiis, dicta publice cum rectoris Academiae Lugd. Bat. munere abiret*. Leiden: apud Fredericum Haringium.

Wittich, Christopher (1659): *Consensus veritatis in Scriptura divina et infallibili revelatae cum veritate philosophica a Renato Descartes detecta*. Nijmegen: apud Adrianum Wyngaerden.

Wittich, Christopher (1690): *Anti-Spinoza, sive Examen Ethices Benedicti de Spinoza, et Commentarius de Deo et eius attributis*. Amsterdam: apud Joannem Wolters.

Wolzogen, Ludwig von (1668): *De Scripturarum interprete, adversus exercitatorem paradoxum libri duo*. Utrecht: Apud Johannem Ribbium.

Zabarella, Jacopo (1597): "De methodis". In: Zabarella, Jacopo: *Opera logica, editio tertia*. Cologne: impensis Lazari Zetzneri. (1st ed. 1578).

Secondary sources

Aalderink, Mark (2009): *Philosophy, Scientific Knowledge, and Concept Formation in Geulincx and Descartes*. Utrecht: Utrecht University, Publications of the Department of Philosophy.

Ablondi, Fred (2005): *Gerauld de Cordemoy: Atomist, Occasionalist, Cartesian*. Milwaukee: Marquette University Press.

Achinstein, Peter (1968): *Concepts of Science: A Philosophical Analysis*. Baltimore: John Hopkins Press.

Alanen, Lili (2003): *Descartes's Concept of Mind*. Cambridge, London: Harvard University Press.

Alexander, Werner (1993): Hermeneutica generalis. *Zur Konzeption und Entwicklung der allgemeinen Verstehenslehre im 17. und 18. Jahrhundert*. Stuttgart: M&P.

Alexandrescu, Vlad (2013): "Regius and Gassendi on Human Soul". In: *Intellectual History Review* 23/2, pp. 1–20.

Allamand, Jean (1774): "Histoire de la vie et des ouvrages du Mr. 's Gravesande". In: *Oeuvres philosophiques et mathématiques de Mr. G.J. 's Gravesande*. 2 vols. Allamand, Jean Nicolas Sébastien (Ed.). Amsterdam: chez Marc Michel Rey, vol. I, pp. ix–lix.

Althaus, Paul (1914): *Die Prinzipien der deutschen reformierten Dogmatik im Zeitalter der aristotelischen Scholastik.* Leipzig: Deichert.

Angelini, Annarita (2008): *Metodo ed enciclopedia nel cinquecento francese.* Vol. I: *Il pensiero di Pietro Ramo all'origine dell'enciclopedismo moderno*; Vol. II: *Tableaux di Savigny.* Florence: Olschki.

Angyal, A. (1941): "Kleiner Beitrag zur Literaturgeschichte des deutschen Cartesianismus". In: *Germanisch-romanische Monatsschrift* 29, pp. 69–73.

Anstey, Peter (2005): "Experimental versus Speculative Natural Philosophy". In: Anstey, Peter/ Schuster, John (Eds.): *The Science of Nature in the Seventeenth Century: Patterns of Change in Early Modern Natural Philosophy.* Dordrecht: Kluwer-Springer, pp. 215–242.

Anstey, Peter/Jalobeanu, Dana (Eds.) (2011): *Vanishing Matter and the Laws of Motion: Descartes and Beyond.* London: Routledge.

Antoine-Mahut, Delphine/Gaukroger, Stephen (Eds.) (2016): *Descartes'* Treatise on Man *and Its Reception.* Dordrecht: Springer.

Ariew, Roger (1999): *Descartes and the Last Scholastics.* Ithaca: Cornell University Press.

Ariew, Roger (2011): *Descartes among the Scholastics.* Leiden, Boston: Brill.

Ariew, Roger (2013): "Censorship, Condemnations, and the Spread of Cartesianism". In: Dobre, Mihnea/Nyden, Tammy (Eds.): *Cartesian Empiricisms.* Dordrecht, Heidelberg, New York, London: Springer, pp. 25–46.

Ariew, Roger (2014): *Descartes and the First Cartesians.* Oxford: Oxford University Press.

Armogathe, Jean-Robert (2005): "Cartesian Physics and the Eucharist in the Documents of the Holy Office and the Roman Index (1671–6)". In: Schmaltz, Tad (Ed.): *Receptions of Descartes. Cartesianism and Anti-Cartesianism in Early Modern Europe.* London, New York: Routledge, pp. 157–182.

Ashworth, Earline Jennifer (1974): *Language and Logic in the Post-Medieval Period.* Dordrecht, Boston: Reidel.

Ashworth, Earline Jennifer (1981): "Do Words Signify Ideas or Things?". In: *Journal of the History of Philosophy* 19, pp. 299–326.

Ashworth, Earline Jennifer (1988): "Changes in Logic Textbooks from 1500 to 1650: The New Aristotelianism". In: Kessler, Eckhart/Lohr, Charles H./Sparn, Walter (Eds.): *Aristotelismus und Renaissance. In Memoriam Charles B. Schmitt.* Wiesbaden: Harrossowitz, pp. 75–87.

Van Asselt, Willem J. (2001): *The Federal Theology of Johannes Cocceius (1603–1669).* Leiden: Brill.

Ast, Friedrich (1807): *Grundriß einer Geschichte der Philosophie.* Landshut: J. Thomann.

Atherton, Margaret (2005): "Descartes among the British: The Case of the Theory of Vision". In: Schmaltz, Tad (Ed.): *Receptions of Descartes. Cartesianism and Anti-Cartesianism in Early Modern Europe.* London, New York: Routledge, pp. 200–214.

Aucante, Vincent (2006): *La philosophie médicale de Descartes.* Paris: Presses Universitaires de France.

Ayers, Michael (2008) "Theories of Knowledge and Belief". In: Garber, Daniel/Ayers, Michael (Eds.): *The Cambridge History of Seventeenth-Century Philosophy.* Cambridge: Cambridge University Press, vol. II, pp. 1003–1061.

Barth, Christian (2017): *Intentionalität und Bewusstsein in der frühen Neuzeit. Die Philosophie des Geistes von René Descartes und Gottfried Wilhelm Leibniz.* Frankfurt: Klostermann.

Battail, J.-F. (1993): "Arnold Geulincx". In: Schobinger, Jean-Pierre (Ed.): *Grundriss der Geschichte der Philosophie.* Basel: Schwabe, vol. II, pp. 375–397.

Beck, Andreas J. (2001): "Gisbertus Voetius (1589–1676): Basic Features of His Doctrine of God". In: Van Asselt, Willem/Dekker, Eef (Eds.): *Reformation and Scholasticism. An Ecumenical Enterprise*. Grand Rapids: Baker Book House, pp. 205–226.

Beck, Andreas J. (2007): *Gisbertus Voetius (1589–1676): Sein Theologieverständnis und seine Gotteslehre*. Göttingen: Vandenhoeck & Ruprecht.

Belgioioso, Giulia (Ed.) (1985): *Paolo Mattia Doria fra rinnovamento e tradizione*. Galatina: Congredo.

Belgioioso, Giulia (1992): *Cultura a Napoli e cartesianismo. Scritti su G. Gimma, P.M. Doria, C. Cominale*. Galatina: Congredo.

Belgioioso, Giulia (1999): *La variata immagine di Descartes. Percorsi della metafisica da Parigi a Napoli*. Lecce: Milella.

Belgioioso, Giulia (2005): "Images of Descartes in Italy". In: Schmaltz, Tad (Ed.): *Receptions of Descartes. Cartesianism and Anti-Cartesianism in Early Modern Europe*. London, New York: Routledge, pp. 171–196.

Bellis, Delphine (2013): "Empiricism without Metaphysics: Regius' Cartesian Natural Philosophy". In: Dobre, Mihnea/Nyden, Tammy (Eds.): *Cartesian Empiricisms*. Dordrecht, Heidelberg, New York, London: Springer, pp. 169–172.

Ben-Yami, Hanoch (2015): *Descartes' Philosophical Revolution: A Reassessment*. Basingstoke: Palgrave Macmillan.

Van Berkel, Klaas (1985): "Franeker als centrum van ramisme". In: Jensma, Goffe T./Smit, Franck/Westra, Frans (Eds.): *Universiteit te Franeker, 1585–1811: bijdragen tot de geschiedenis van de Friese hogeschool*. Leeuwarden: Fryske Akademy-Walburg Pers., pp. 424–437.

Berti, Enrico (1983): "Differenza tra il metodo risolutivo degli aristotelici e la 'resolutio' dei matematici". In: Olivieri, Luigi (Ed.): *Aristotelismo veneto e scienza moderna*. Padua: Antenore, pp. 435–457.

Van Besouw, Jip (2016): "The Impeccable Credentials of an Untrained Philosopher: Willem Jacob 's Gravesande's Career before his Leiden Professorship, 1688–1717". In: *Notes and Records: The Royal Society Journal of the History of Science* 70/3, pp. 231–249.

Van Besouw, Jip (2017a): *Out of Newton's Shadow: An Examination of Willem Jacob 's Gravesande's Scientific Methodology*. Ph.D. dissertation, Brussels: Vrije Universiteit Brussel.

Van Besouw, Jip (2017b): "'s Gravesande on the Application of Mathematics in Physics and Philosophy". In: *Noctua* 4/1–2, pp. 17–55.

Beukers, Harm (1999): "Acid Spirits and Alkaline Salts: The Iatrochemistry of Franciscus dele Boë, Sylvius". In: *Sartoniana* 11, pp. 39–58.

Bitbol-Hespériès, Annie (1993): "Descartes et Regius: leur pensée médicale". In: Verbeek, Theo (Ed.): *Descartes et Regius. Autour de l'Explication de l'esprit humain*. Amsterdam: Rodopi, pp. 47–68.

Bizer, Ernst (1958): "Die reformierte Orthodoxie und der Cartesianismus". In: *Zeitschrift für Theologie und Kirche* 55, pp. 306–372.

Bizer, Ernst (1965): "Reformed Orthodoxy and Cartesianism". In: *Journal for Theology and the Church* 2, pp. 20–82.

Blum, Paul Richard (1998): *Philosophenphilosophie und Schulphilosophie. Typen des Philosophierens in der Neuzeit*. Stuttgart: Franz Steiner.

Bodeüs, Richard (1991): "Leibniz, Jean de Raey et la Physique Reformée". In: *Studia Leibnitiana* 23, pp. 103–110.

Bohatec, Josef (1912): *Die cartesianische Scholastik in der Philosophie und reformierten Dogmatik des 17. Jahrhunderts*. Leipzig: A. Deichert.

Bordoli, Roberto (1997): *Ragione e Scrittura tra Descartes e Spinoza. Saggio sulla 'Philosophia S. Scripturae Interpres' di Lodewijk Meyer e sulla sua recezione*. Milan: Franco Angeli.

Bordoli, Roberto (2009): *Dio ragione verità. Le polemiche su Descartes e su Spinoza presso l'Università di Franeker*. Macerata: Quodlibet.

Borghero, Carlo (2011): *Les Cartésiens face à Newton. Philosophie, science et religion dans la première moitié du XVIIIe siècle*. Turnhout: Brepols.

Borghero, Carlo/Del Prete, Antonella (Eds.) (2011): *Immagini filosofiche e interpretazioni storiografiche del cartesianismo*. Florence: Le Lettere.

Bos, Egbert P./Krop, Henri A. (Eds.) (1993): *Franco Burgersdijk (1590–1635): Neo-Aristotelianism in Leiden*. Amsterdam: Rodopi.

Bos, Erik-Jan (2002): *The Correspondence between Descartes and Henricus Regius*. Utrecht: Utrecht University, Publications of the Department of Philosophy.

Bos, Erik-Jan (2013): "Henricus Regius et les limites de la philosophie cartésienne". In: Kolesnik-Antoine, Delphine (Ed.): *Qu'est-ce qu'être cartésien?* Paris: Editions de l'ENS de Lyon, pp. 53–68.

Bottin, Francesco (1972): "La teoria del *regressus* in Giacomo Zabarella". In: Giacon, Carlo (Ed.): *Saggi e ricerche su Aristotele, S. Bernardo, Zabarella*. Padua: Antenore, pp. 49–70.

Bouillier, Francisque (1854): *Histoire de la Philosophie Cartésienne*. Paris: Durand, Lyon: Brun.

Boyle, Deborah (2009): *Descartes on Innate Ideas*. London, New York: Continuum.

Brockliss, Lawrence W.B. (1981): "Aristotle, Descartes and the New Science: Natural philosophy at the University of Paris, 1600–1740". In: *Annals of Science* 38/1, pp. 33–69.

Brockliss, Laurence W.B. (1987): *French Higher Education in the Seventeenth and Eighteenth Centuries: A Cultural History*. Oxford: Clarendon.

Brockliss, Laurence W.B. (1995): "Descartes, Gassendi, and the Reception of the Mechanical Philosophy at the French Collèges de Plein Exercice, 1640–1730". In: *Perspectives on Science* 3/4, pp. 450–479.

Brosch, Pius (1926): *Die Ontologie des Johannes Clauberg. Eine historische Würdigung und eine Analyse ihrer Probleme*. Greifswald: Hartmann.

Broughton, Janet (2002): *Descartes's Method of Doubt*. Princeton: Princeton University Press.

Brown, Harcourt (1934): *Scientific Organizations in Seventeenth Century France (1620–1680)*. Baltimore: Williams and Wilkins.

Brulez, Lucien (1926): *Holländische Philosophie*. Breslau: Hirt.

Brunet, Pierre (1926): *Les physiciens hollandais et la méthode expérimentale en France au XVIIIe siècle*. Paris: Albert Blanchard.

Bruyère, Nelly (1984): *Méthode et dialectique dans l'oeuvre de La Ramée: Renaissance et Age classique*. Paris: Vrin.

Buchdahl, Gerd (1969): *Metaphysics and the Philosophy of Science: The Classical Origins, Descartes to Kant*. Cambridge, MA: MIT Press.

Van Bunge, Wiep (2001): *From Stevin to Spinoza. An Essay on Philosophy in the Seventeenth-Century Dutch Republic*. Leiden, Boston, Cologne: Brill.

Van Bunge, Wiep (2013): "Dutch Cartesian Empiricism and the Advent of Newtonianism". In: Dobre, Mihnea/Nyden, Tammy (Eds.): *Cartesian Empiricisms*. Dordrecht, Heidelberg, New York, London: Springer, pp. 89–104.

Van Bunge, Wiep (2017): "The Waning of the Radical Enlightenment in the Dutch Republic". In: Ducheyne, Steffen (Ed.): *Reassessing the Radical Enlightenment*. London: Routledge, pp. 178–193.

Buning, Robin (2013): *Henricus Reneri (1593–1639), Descartes' Quartermaster in Aristotelian Territory*. Utrecht: Utrecht University, Publications of the Department of Philosophy.

Burtt, Edwin Arthur (1932): *The Metaphysical Foundations of Modern Physical Science: A Historical and Critical Essay*. London: Routledge and Kegan Paul.

Buys, Ruben (2011): "Between Actor and Spectator: Arnout Geulincx and the Stoics". In: *British Journal for the History of Philosophy* 18, pp. 741–761.

Caps, Géraldine (2010): *Les "médecins cartésiens". Héritage et diffusion de la représentation mécaniste du corps humain (1646–1696)*. Hildesheim, Zürich, New York: Olms.

Caroti, Stefano (2014): "Ruardus Andala e la nuova filosofia". In: Caroti, Stefano/Siclari, Alberto (Eds.): *Filosofia e religione. Studi in onore di Fabio Rossi*. Parma: E-Theca On Line Open Access Edizioni, pp. 236–257.

Carraud, Vincent (1999): "L'ontologie peut-elle être cartésienne? L'exemple de L'*Ontosophia* de Clauberg, de 1647 à 1664: de l'*ens* à la *mens*". In: Verbeek, Theo (Ed.): *Johannes Clauberg (1622–1665) and Cartesian Philosophy in the Seventeenth Century*. Dordrecht: Kluwer, pp. 13–38.

Cassirer, Ernst (1951): *The Philosophy of the Enlightenment*. Pettegrove, James P./Koelin, Fritz C. (Trans.). Princeton: Princeton University Press.

Cerrato, Francesco (1999): "The Influence of Pierre de la Ramée at Leiden University and on the Intellectual Formation of the Young Spinoza". In: *Studia Spinozana* 15, pp. 15–33.

Des Chene, Dennis (2002): "Cartesian Science: Régis and Rohault". In: Nadler, Steven (Ed.): *A Companion to Early Modern Philosophy*. Malden: Blackwell, pp. 183–196.

Chomsky, Noam (1966): *Cartesian Linguistics: A Chapter in the History of Rationalist Thought*. New York: Harper & Row.

Clarke, Desmond M. (1989): *Occult Powers and Hypotheses. Cartesian Natural Philosophy under Louis XIV*. Oxford: Clarendon.

Clarke, Desmond M. (1982): *Descartes' Philosophy of Science*. Manchester: Manchester University Press.

Clarke, Desmond M. (2005): *Descartes's Theory of Mind*. Oxford: Clarendon.

Clarke, Desmond M. (2006): *Descartes. A Biography*. New York: Cambridge University Press.

De Clercq, Peter (1997a): *The Leiden Cabinet of Physics. A Descriptive Catalogue*. Leiden: Museum Boerhaave.

De Clercq, Peter (1997b): *At the Sign of the Oriental Lamp. The Musschenbroek Workshop in Leiden, 1660–1750*. Rotterdam: Erasmus.

Condren, Conal/Gaukroger, Stephen/Hunter, Ian (Eds.) (2006): *The Philosopher in Early Modern Europe*. Cambridge: Cambridge University Press.

Cooney, Brian (1978): "Arnold Geulincx: A Cartesian Idealist". In: *Journal of the History of Philosophy* 16, pp. 167–180.

Corneanu, Sorana (2011): *Regimens of the Mind. Boyle, Locke and the Early Modern* Cultura Animi *Tradition*. Chicago: University of Chicago Press.

Cottingham, John (2007): "The Role of God in Descartes's Philosophy". In: Broughton, Janet/ Carriero, John (Eds.): *A Companion to Descartes*. Malden (MA), Oxford: Blackwell, pp. 288–301.

Cottingham, John (2008a): *Cartesian Reflections. Essays on Descartes's Philosophy*. Oxford: Oxford University Press.

Cottingham, John (2008b): "The only sure sign …' Thought and Language in Descartes". In: Cottingham, John: *Cartesian Reflections*. Oxford: Oxford University Press, pp. 107–128.

Cousin, Victor (1866): *Fragments Philosophiques pour servir à l'histoire de la philosophie*. Paris: Didier.

Cramer, Jan Anthony (1889): *Abraham Heidanus en zijn Cartesianisme*. Utrecht: Stoom-boek- en steendrukkerij "De Industrie", J. van Druten.

Cunning, David (2010): *Argument and Persuasion in Descartes' Meditations*. Oxford: Oxford University Press.

Cunning, David (2014): "The First Meditation: Divine Omnipotence, Necessary Truths and the Possibility of Radical Deception". In: Cunning, David (Ed.): *The Cambridge Companion to Descartes' Meditations*. Cambridge: Cambridge University Press, pp. 68–87.

Cunningham, Andrew (1990): "Medicine to Calm the Mind. Boerhaave's Medical System and Why It Was Adopted in Edinburgh". In: Cunningham, Andrew/French, Roger (Eds.): *The Medical Enlightenment of the Eighteenth Century*. Cambridge: Cambridge University Press, pp. 40–66.

Curley, Edwin (1993): "Certainty: Psychological, Logical and Metaphysical". In: Voss, Stephen (Ed.): *Essays on the Philosophy and Science of René Descartes*. New York: Oxford University Press, pp. 11–30.

Curley, Edwin (2008): "The *Cogito* and the Foundations of Knowledge". In: Gaukroger, Stephen (Ed.): *The Blackwell Guide to Descartes' Meditations*. Malden, Oxford, Carlton: Blackwell, pp. 30–47.

Damiron, Jean-Philibert (1846): *Essai sur l'histoire de la philosophie en France, au XVIIe siècle*. Paris: Hachette.

Danneberg, Lutz (1998): "Logik und Hermeneutik: Die *analysis logica* in den ramistischen Dialektiken". In: Scheffler, Uwe/Wuttich, Klaus (Eds.): *Terminigebrauch und Folgebeziehung*. Berlin: Logos, pp. 129–157.

Danneberg, Lutz (2001): "Logik und Hermeneutik im 17. Jahrhundert". In: Schröder, Jan (Ed.): *Theorie der Interpretation vom Humanismus bis zur Romantik – Rechtswissenschaft, Philosophie, Theologie*. Stuttgart: Franz Steiner, pp. 75–131.

Danneberg, Lutz (2005): "Vom *grammaticus* und *logicus* über den *analyticus* zum *hermeneuticus*". In: Schönert, Jörg/Vollhardt, Friedrich (Ed.): *Geschichte der Hermeneutik und die Methodik der textinterpretierenden Disziplinen*. Berlin, New York: De Gruyter, pp. 255–337.

Danneberg, Lutz (2006): "Logique et herméneutique au XVIIe siècle". In: Gens, Jean-Claude (Ed.): *La logique herméneutique du XVIIe siècle – J.-C. Dannhauer et J. Clauberg*. Argenteil: Vrin, pp. 15–65. (French translation of Danneberg 2001).

Datson, Lorraine (1998): "Probability and Evidence". In: Garber, Daniel/Ayers, Michael (Eds.): *The Cambridge History of Seventeenth-Century Philosophy*. Cambridge: Cambridge University Press, vol. II, pp. 1108–1144.

Dawson, Hannah (2011): *Locke, Language and Early-Modern Philosophy*. Cambridge: Cambridge University Press.

Dear, Peter (1998): "Method and the Study of Nature". In: Garber, Daniel/Ayers, Michael (Eds.): *The Cambridge History of Seventeenth-Century Philosophy*. Cambridge: Cambridge University Press, vol. I, pp. 147–177.

Dear, Peter (2012): "Philosophy of Science and its Historical Reconstruction". In: Mauskopf, Seymur/Schmaltz, Tad (Eds.): *Integrating History and Philosophy of Science. Problems and Prospects*. Dordrecht: Springer, pp. 67–82.

Dechange, Klaus (1966): *Die frühe Naturphilosophie des Henricus Regius (Utrecht 1641)*. Ph.D. dissertation, Münster: University of Münster.

Della Rocca, Michael (2005): "Descartes, the Cartesian Circle, and Epistemology without God". In: *Philosophy and Phenomenological Research* 70/1, pp. 1–33.

Dessì, Paola/Lotti, Brunello (2011): *Eredità cartesiane nella cultura britannica*. Florence: Le Lettere 2011.

Dibon, Paul (1954): *La Philosophie néerlandaise au siècle d'or. Tome I. L'enseignement philosophique dans les Universités à l'époque pre-cartésienne (1575–1650)*. Paris, Amsterdam, London, New York: Elsevier.

Dibon, Paul (1990): *Regards sur la Hollande du siècle d'or*. Naples: Vivarium.

Dijksterhuis, Fokko J. (Ed.) (1950): *Descartes et le cartésianisme hollandais – Etudes et documents*. Paris: Presses Universitaires de France.

De Dijn, Herman (1996): *Spinoza: The Way to Wisdom*. West Lafayette: Purdue University Press.

Dobre, Mihnea (2010): *Metaphysics and Physics in Cartesian Natural Philosophy: Descartes and Early French Cartesians on the Foundation of Natural Philosophy*. Ph.D. dissertation, Radboud University, Nijmegen.

Dobre Mihnea (2013a): "Knowledge and Certainty in the Foundation of Cartesian Natural Philosophy". In: *Revue Roumaine de philosophie* 57/1, pp. 95–110.

Dobre, Mihnea (2013b): "Rohault's Cartesian Physics". In: Dobre, Mihnea/Nyden, Tammy (Eds.): *Cartesian Empiricisms*. Dordrecht, Heidelberg, New York, London: Springer, pp. 203–226.

Dobre, Mihnea (2017): *Descartes and Early French Cartesianism: Between Metaphysics and Physics*. Bucharest: Zeta Books.

Dobre, Mihnea/Nyden, Tammy (Eds.) (2013a): *Cartesian Empiricisms*. Dordrecht, Heidelberg, New York, London: Springer.

Dobre, Mihnea/Nyden, Tammy (2013b): "Introduction". In: Dobre, Mihnea/Nyden, Tammy (Eds.): *Cartesian Empiricisms*. Dordrecht, Heidelberg, New York, London: Springer, pp. 1–21.

Donati, Silvia (1988): "La dottrina delle dimensioni indeterminate in Egidio Romano". In: *Medioevo* 14, pp. 149–233.

Douglas, Alexander X. (2014): "Christopher Wittich's Anti-Spinoza". In: *Intellectual History Review* 24/2, pp. 153–166.

Douglas, Alexander X. (2015): *Spinoza & Dutch Cartesianism: Philosophy and Theology*. Oxford: Oxford University Press.

Ducheyne, Steffen (2014a): "'s Gravesande's Appropriation of Newton's Natural Philosophy, Part I: Epistemological and Theological Issues". In: *Centaurus* 56/1, pp. 31–55.

Ducheyne, Steffen (2014b): "'s Gravesande's Appropriation of Newton's Natural Philosophy, Part II: Methodological Issues". In: *Centaurus* 56/2, pp. 97–120.

Dürr, K. (1939–1940): "Die mathematische Logik des Arnold Geulincx". In: *The Journal of Unified Science* 8, pp. 361–368.

Dürr, Karl (1965): "Arnold Geulincx und die klassische Logik des 17. Jahrhunderts". In: *Studium Generale* 18, pp. 520–541.

Easton, Patricia (Ed.) (1997): *Logic and the Workings of the Mind: The Logic of Ideas and Faculty Psychology in Early Modern Philosophy*. Atascadero: Ridgeview.

Easton, Patricia/Gholamnejad, Melissa (2016): "Louis de la Forge and the Development of Cartesian Medical Philosophy". In: Distelzweig, Peter/Goldberg, Benjamin/Ragland, Evan (Eds.): *Early Modern Medicine and Natural Philosophy*. Dordrecht, Heidelberg, New York, London: Springer, pp. 207–225.

Farina, Paolo (1975): "Sulla formazione scientifica di Henricus Regius: Santorio Santorii e il *De statica medicina*". In: *Rivista critica di storia della filosofia* 30, pp. 363–399.

Farina, Paolo (1977): "Il corpuscolarismo di Henricus Regius: materialismo e medicina in un cartesiano olandese del seicento". In: Baldini, Ugo/Farina, Paolo/Trevisani, Francesco/Zanier, Giancarlo (Eds.) (1977): *Ricerche sull'atomismo del Seicento*. Florence: La Nuova Italia, pp. 119–178.

Feingold, Mordechai (2001): "English Ramism: A Reinterpretation". In: Feingold, Mordechai/Freedman, Joseph S./Rother, Wolfgang (Eds.) (2001): *The Influence of Petrus Ramus: Studies in Sixteenth and Seventeenth Century Philosophy and Sciences*. Basel: Schwabe, pp. 127–176.

Fichant, Michel (1998): *Science et mètaphysique dans Descartes et Leibniz*. Paris: Presses Universitaires de France.

Fix, Andrew C. (1999): *Fallen Angels: Balthasar Bekker, Spirit Belief and Confessionalism in the Seventeenth-Century Dutch Republic*. Dordrecht: Kluwer.

Flage, Daniel E./Bonnen, Clarence A. (1999): *Descartes and Method. A Search for a Method in Meditations*. London, New York: Routledge.

Fouke, Daniel (1997): *The Enthusiastical Concerns of Dr. Henry More: Religious Meaning and the Psychology of Delusion*. Leiden, New York, Cologne: Brill.

Fowler, Colin F. (1999): *Descartes on the Human Soul. Philosophy and the Demands of Christian Doctrine*. Dordrecht: Kluwer.

Frank, Günter (2012): "Petrus Ramus als Interpret der aristotelischen Metaphysik – Anmerkungen zum Theologie-Kapitel in Metaphysik XII, cap. 6, 7 und 9". In: Selderhuis, Herman J./Frank, Günter (Eds.): *Philosophie der Reformierten*. Stuttgart, Bad Cannstatt: Frommann-Holzboog, pp. 93–112.

Freedman, Joseph S. (1984): *Deutsche Schulphilosophie im Reformationszeitalter (1500–1650): Ein Handbuch für den Hochschulunterricht*. Münster: MAKS.

Freedman, Joseph S. (1999): *Philosophy and the Arts in Central Europe, 1500–1700. Teaching and Texts at Schools and Universities*. Aldershot: Ashgate.

Frijhoff, Willem/Spies, Marijke (2004): *Dutch Culture in a European Perspective I: 1650: Hard-won Unity*. Assen: Van Gorcum.

Gabbey, Alan (1982): "*Philosophia Cartesiana Triumphata*: Henry More (1646–1671)". In: Lennon, Thomas M./Nicholas, John M./Davis, John W. (Eds.): *Problems of Cartesianism*. Kingston, Montreal: McGill-Queens University Press, pp. 171–250.

Garber, Daniel (1992): *Descartes' Metaphysical Physics*. Chicago: University of Chicago Press.

Garber, Daniel (2001): *Descartes Embodied. Reading Descartes's Philosophy through Cartesian Science*. Cambridge: Cambridge University Press.

Garber, Daniel (2006): "Physics and Foundations". In: Park, Katharine/Daston, Lorraine (Eds.): *Cambridge History of Science*. Vol. III.: *Early Modern Science*. New York: Cambridge University Press, pp. 21–69.

Garber, Daniel/Roux, Sophie (Eds.) (2013): *The Mechanization of Natural Philosophy*. New York: Kluwer.

Gariepy, Thomas (1990): *Mechanism without Metaphysics. Henricus Regius and the Establishment of Cartesian Medicine*. Ph.D. dissertation, New Haven: Yale University.

Gaukroger, Stephen (1989): *Cartesian Logic. An Essay on Descartes's Conception of Inference*. Oxford: Oxford University Press.

Gaukroger, Stephen (1995): *Descartes: An Intellectual Biography*. Oxford: Oxford University Press.

Gaukroger, Stephen (2002): *Descartes' System of Natural Philosophy*. Cambridge: Cambridge University Press.

Gaukroger, Stephen (2008): "Knowledge, Evidence and Method". In: Rutherford, Donald (Ed.): *The Cambridge Companion to Early Modern Philosophy*. Cambridge: Cambridge University Press, pp. 39–66.

Gaukroger, Stephen/Schuster, John/Sutton, John (Eds.) (2000): *Descartes' Natural Philosophy*. London: Routledge.

Gilbert, Neal Ward (1960): *Renaissance Concepts of Method*. New York: Columbia University Press.

Goldstein, Catherine (2008): "Écrire l'expérience mathématique au XVIIe siècle; la méthode selon Bernard Frenicle de Bessy". In: Dubourg-Glattigny, Pascal/Vérin, Hélène (Eds.): *Réduire en art. La technologie de la Renaissance aux Lumières*. Paris: Éditions de la Maison des Sciences de l'Homme, pp. 213–224.

Gori, Giambattista (1972): *La fondazione dell'esperienza in 's Gravesande*. Florence: La Nuova Italia.

Goudriaan, Aza (1999): *Philosophische Gotteserkenntnis bei Suarez und Descartes im Zusammenhang mit der niederländischen reformierten Theologie und Philosophie des 17. Jahrhunderts*. Leiden: Brill.

Goudriaan, Aza (2002): *Jacobus Revius: A Theological Examination of Cartesian Philosophy: Early Criticisms (1647)*. Leiden: Brill.

Goudriaan, Aza (2006): *Reformed Orthodoxy and Philosophy, 1625–1750: Gisbertus Voetius, Petrus van Mastricht, and Anthonius Driessen*. Leiden, Boston: Brill.

Goudriaan, Aza/Van Lieburg, Fred (Eds.) (2011): *Revisiting the Synod of Dordt (1618–1619)*. Leiden: Brill.

Goulding, Robert (2010): *Defending Hypatia. Ramus, Savile, and the Renaissance Rediscovery of Mathematical History*. Dordrecht, Heidelberg, London, New York: Springer.

Grene, Marjorie (1999): "Descartes and Skepticism". In: *The Review of Metaphysics* 52/3, pp. 553–571.

Guéroult, Martial (1953): *Descartes selon l'ordre des raisons*. Paris, Aubier.

Guerrini, Anita (1987): "Archibald Pitcairne and Newtonian Medicine". In: *Medical History* 31/1, pp. 70–83.

De Haan, A.A.M. (1993): "Geschiedenis van het wijsgerig onderwijs te Deventer". In: Blom, H.W./Krop, Henri A./Wielema, Michiel R. (Eds.) (1993): *Deventer denkers, de geschiedenis van het wijsgerig onderwijs te Deventer*. Hilversum: Uitgeverij Verloren, pp. 29–122.

Hahn, Roger (1971): *The Anatomy of a Scientific Institution. The Paris Academy of Sciences, 1666–1803*. Berkeley, Los Angeles, London: University of California Press.

Hall, A. Rupert (1982): "Further Newton Correspondence". In: *Notes and Records of the Royal Society of London* 37/1, pp. 7–34.

Hartog, Jan (1876): "Het collegie der scavanten te Utrecht". In: *De Gids* 40/2, pp. 77–114.

Hasso Jæger, H.-E. (1974): "Studien zur Frühgeschichte der Hermeneutik". In: *Archiv für Begriffgeschichte* 18:1, pp. 35–84.

Hatfield, Gary (1990): "Metaphysics and the New Science". In: Lindberg, David/Westman, Robert (Eds.): *Reappraisals of the Scientific Revolution*. Cambridge: Cambridge University Press, pp. 93–166.

Hatfield, Gary (2013): "The Cartesian Psychology of Antoine Le Grand". In: Dobre, Mihnea/ Nyden, Tammy (Eds.): *Cartesian Empiricisms*. Dordrecht, Heidelberg, New York, London: Springer, pp. 251–274.

Heyd, Michael (1995): *"Be Sober and Reasonable": The Critique of Enthusiasm in the Seventeenth and Early Eighteenth Centuries*. Leiden: Brill.

Hirschfield, John Milton (1981): *The Académie royale des sciences. 1666–1683*. New York: Arno Press.

De Hoog, Adriaan C. (1974): *Some Currents of Thought in Dutch Natural Philosophy: 1675–1720*. Ph.D. dissertation, Oxford: Oxford University.

Hooykas, Reijer (1968): *Humanisme, Science et Réforme. Pierre de la Ramée (1515–1572)*. Leiden: Brill.

Hotson, Howard (1994): *Philosophical pedagogy in reformed central Europe between Ramus and Comenius*: A survey of the continental background of the 'Three Foreigners'. In Samuel Hartlib and universal reformation: Studies in intellectual communication, ed. Mark Greengrass, Michael Leslie, and Timothy Raylor, 27–50. Cambridge: Cambridge University Press.

Hotson, Howard (2007): *Commonplace Learning: Ramism and Its German Ramifications, 1543–1630*. Oxford: Oxford University Press.

Hubert, Christiane (1994): *Les premières réfutations de Spinoza: Aubert de Versé, Wittich, Lamy*. Paris: Presses Paris Sorbonne.

Hutton, Sarah (Ed.) (1990): *Henry More (1614–1687). Tercentenary Studies*. Dordrecht: Kluwer.

Huussen, Henk jr. (Ed.) (2013): *Onderwijs en onderzoek. Studie en wetenschap aan de academie van Groningen in de 17e en 18e eeuw*. Groningen: Hilversum.

Jalobeanu, Dana (2011): "The Cartesians of the Royal Society. The Debate Over Collisions and the Nature of Body (1668–1670)". In: Jalobeanu, Dana/Anstey, Peter (Eds.): *Vanishing Matter and the Laws of Motion. Descartes and Beyond*. New York, London: Routledge, pp. 103–129.

Janiak, Andrew/Schliesser, Eric (Eds.) (2012): *Interpreting Newton. Critical Essays*. Cambridge: Cambridge University Press.

Jesseph, Douglas (2005): "Mechanism, Skepticism, and Witchcraft: More and Glanvill on the Failures of the Cartesian Philosophy". In: Schmaltz, Tad (Ed.): *Receptions of Descartes. Cartesianism and Anti-Cartesianism in Early Modern Europe*. London, New York: Routledge, pp. 183–199.

Jorink, Erik/Maas, Ad (Eds.) (2012): *Newton and the Netherlands. How Isaac Newton was Fashioned in the Dutch Republic*. Leiden: Leiden University Press.

Karsksen, Michael (1993): "Subject, Object and Substance in Burgersdijk's Logic". In: Bos, Egbert P./Krop, Henri A. (Eds.): *Franco Burgersdijk (1590–1635): Neo-Aristotelianism in Leiden*. Amsterdam: Rodopi, pp. 29–36.

Klever, Wim (1988): "Burchardus de Volder (1643–1709): A Crypto-Spinozist on a Leiden Cathedra". In: *Lias. Journal of Early Modern Intellectual Culture and its Sources* 15, pp. 191–241.

Knoeff, Rina (2002): *Herman Boerhaave (1668–1738): Calvinist Chemist and Physician*. Amsterdam: Koninklijke Nederlandse Akademie van Wetenschappen.

Knoeff, Rina (2003): "Boerhaave, Herman (1668–1738)". In: Van Bunge, Wiep/Krop, Henry A./Leeuwenburgh, Bart/Van Ruler, Han/Schuurman, Paul/Wielema, Michiel (Eds.): *The Dictionary of Seventeenth and Eighteenth-Century Dutch Philosophers*. Bristol: Thoemmes Press, vol. I, pp. 118–124.

Kolesnik-Antoine, Delphine (Ed.) (2013a): *Qu'est-ce qu'être cartésien?* Paris: Editions de l'ENS de Lyon.

Kolesnik-Antoine, Delphine (2013b): "Le rôle des expériences dans la physiologie d'Henricus Regius: les 'pierres lydiennes' du cartésianisme". In: *Journal of Early Modern Studies* 2/1, pp. 125–145.

Kors, Alan Charles (2016): *Naturalism and Unbelief in France, 1650–1729*. Cambridge: Cambridge University Press.

Krop, Henri A. (1996): "Radical Cartesianism in Holland: Spinoza and Deurhoff". In: Van Bunge, Wiep (Ed.): *Disguised and Overt Spinozism around 1700*. Leiden, New York, Cologne: Brill, pp. 55–81.

Krop, Henri A. (1999): "Spinoza and the Calvinistic Cartesianism of Lambertus van Velthuysen". In: *Studia Spinozana* 15, pp. 107–132.

Krop, Henri A. (2003a): "Burgersdijk, Franco (1590–1635)". In: Van Bunge, Wiep/Krop, Henry A./Leeuwenburgh, Bart/Van Ruler, Han/Schuurman, Paul/Wielema, Michiel (Eds.): *The Dictionary of Seventeenth and Eighteenth-Century Dutch Philosophers*. Bristol: Thoemmes Press, vol. I, pp. 181–190.

Krop, Henri A. (2003b) "Medicine and Philosophy in Leiden around 1700: Continuity or Rupture?". In: Van Bunge, Wiep (Ed.): *The Early Enlightenment in the Dutch Republic, 1650–1750*. Leiden, Boston: Brill, pp. 173–196.

Kuiper, Ernst Jan (1958): *De Hollandse "Schoolordre" van 1625: Een studie over het onderwijs op de Latijnse scholen in Nederland in de 17de en 18de eeuw*. Groningen: J.B. Wolters.

Laerke Mogens/Smith, Justin E.H./Schliesser, Eric (Eds.): *Philosophy and Its History. Aims and Methods in the Study of Early Modern Philosophy*. New York: Oxford University Press.

Lagrée, Jaqueline (2002): "Sens et vérité chez Clauberg et Spinoza". In: *Philosophiques* 29/1, pp. 121–138.

Land, Jan Pieter Nicolaas (1887): "Arnold Geulincx te Leiden (1658–1669)". In: *Verslagen en Mededeelingen der Koninklijke Akademie van Wetenschappen, Afdeeling Letterkunde* 3, pp. 277–327.

Land, Jan Pieter Nicolaas (1891): "Arnold Geulincx and His Works". In: *Mind* 16, pp. 223–242.

Land, Jan Pieter Nicolaas (1895): *Arnold Geulincx und seine Philosophie*. The Hague: Martinus Nijhoff.

De Lattre, Alain (1967): *L'occasionalisme d'Arnold Geulincx*. Paris: Editions de Minuit.

Leinsle, Ulrich G. (1985): *Das Ding und die Methode. Methodische Konstitution und Gegenstand der frühen protestantischen Metaphysik*. 2 vols. Augsburg: Maro.

Leinsle, Ulrich G. (1999): "Comenius in der Metaphysik des jungen Clauberg". In: Verbeek, Theo (Ed.): *Johannes Clauberg (1622–1665) and Cartesian Philosophy in the Seventeenth Century*. Dordrecht: Kluwer, pp. 1–12.

Lennon, Thomas M. (1993): *The Battle of the Gods and Giants: The Legacies of Descartes and Gassendi 1655–1715*. Princeton: Princeton University Press.

Lennon, Thomas M. (Ed.) (2003): *Cartesian Views: Papers Presented to Richard A. Watson*. Leiden: Brill.

Lennon, Thomas M. (2008): *The Plain Truth. Descartes, Huet, and Skepticism*. Leiden, Boston: Brill.

Lennon, Thomas M. (2014): "The Fourth Meditation: Descartes's Theodicy *avant la lettre*". In: Cunning, David (Ed.): *The Cambridge Companion to Descartes' Meditations*. Cambridge: Cambridge University Press, pp. 168–185.

Lennon, Thomas M./Nicholas, John M./Davis, John W. (Eds.) (1982): *Problems of Cartesianism*. Kingston, Montreal: McGill-Queens University Press.

Lenz, Martin/Waldow, Anik (Eds.) (2013): *Contemporary Perspectives on Early Modern Philosophy*. Boston: Springer.

Lindeboom, Gerrit A. (1968): *Herman Boerhaave: The Man and his Work*. London: Methuen.

Lindeboom, Gerrit A. (1978): *Descartes and Medicine*. Amsterdam: Rodopi.

Lodge, Paul (1998): "The Failure of Leibniz's Correspondence with De Volder". In: *Leibniz Society Review* 8, pp. 47–67.

Lodge, Paul (2001): "The Debate over Extended Substance in Leibniz's Correspondence with De Volder". In: *International Studies in the Philosophy of Science* 15, pp. 155–166.

Lodge, Paul (2005): "Burchard de Volder: Crypto-Spinozist or Disenchanted Cartesian?". In: Schmaltz, Tad (Ed.): *Receptions of Descartes. Cartesianism and Anti-Cartesianism in Early Modern* Europe. London, New York: Routledge, pp. 128–145.

Lodge, Paul (Ed.) (2013): *The Leibniz-De Volder Correspondence: With Selections from the Correspondence Between Leibniz and Johann Bernoulli.* New Haven: Yale University Press.

Losonsky, Michael (2006): *Linguistic Turns in Modern Philosophy.* Cambridge: Cambridge University Press.

Lüthy, Christoph H. (2006): "Where Logical Necessity Becomes Visual Persuasion: Descartes's Clear and Distinct Illustrations". In: Kusukawa, Sachiko/Maclean, Ian (Eds.) (2006): *Transmitting Knowledge. Words, Images, and Instruments in Early Modern Europe.* Oxford: Oxford University Press, pp. 97–133.

Luyendijk-Elshout, Antonie M. (1975): "*Oeconomia Animalis*, Pores and Particles: The Rise and Fall of the Mechanical Philosophical School of Theodoor Craanen (1621–1690)". In: Lunsingh Scheurleer, Theodor H./Posthumus Meyjes, Guillaume H.M. (Eds.): *Leiden University in the Seventeenth Century: An Exchange of Learning.* Leiden: Brill, pp. 294–307.

Mancini, Italo (1960): "L'*ontosophia* di Johannes Clauberg e i primi tentativi di sostituzione cartesiana". In: Glockner, Hermann *et al.* (Ed.): *Festschrift H.J. De Vleeschauwer.* Pretoria: University of South Africa, pp. 66–83.

Marchand, Prosper (1758–1759): "'s Gravesande". In: Marchand, Prosper: *Dictionnaire historique ou mémoires critiques et littéraires.* La Haye: chez Pierre de Hondt, vol. II, pp. 214–242.

Marion, Jean-Luc (1993): *Sur l'ontologie grise de Descartes. Science cartésienne et savoir aristotélicien dans les "Regulae".* Paris: Vrin.

Mauskopf, Seymur/Schmaltz, Tad (2012): "Introduction". In: Mauskopf, Seymur/Schmaltz, Tad (Eds.): *Integrating History and Philosophy of Science. Problems and Prospects.* Dordrecht: Springer, pp. 1–10.

McClaughlin, Trevor (1977): "Le concept de science chez Jacques Rohault". In: *Revue d'histoire des sciences* 30/3, pp. 225–240.

McClaughlin, Trevor (1979): "Censorship and defenders of the Cartesian faith in mid-seventeenth century France". In: *Journal of the History of Ideas* 40/4, pp. 563–581.

McClaughlin, Trevor (1996): "Was There an Empirical movement in Mid-seventeenth Century France? Experiments in Jacques Rohault's *Traité de physique*". In: *Revue d'histoire des sciences* 49/4, pp. 459–481.

McClaughlin, Trevor (2000): "Descartes, Experiments, and a First Generation Cartesian, Jacques Rohault". In: Gaukroger, Stephan/Schuster, John/Sutton, John (Eds.) (2000): *Descartes' Natural Philosophy.* London, New York: Routledge, pp. 330–345.

McGahagan, Arthur (1976): *Cartesianism in the Netherlands. 1639–1676. The New Science and the Calvinist Counter-Reformation.* Ph.D. dissertation, Philadelphia: University of Pennsylvania.

Meerhoff, Kees/Magnien, Michel (Eds.) (2004): *Ramus et l'université.* Paris: Éditions Rue d'Ulm.

Menk, Gerhard (1980): "Kalvinismus und Pädagogik, Matthias Martinius (1572–1630) und der Einfluß der Herborner Hohen Schule auf Johann Amos Comenius". In: *Nassausiche Annalen* 91, pp. 77–104.

Menk, Gerhard (1981): *Die Hohe Schule Herborn in ihrer Frühzeit (1584–1660). Ein Beitrag zum Hochschulwesen des deutschen Kalvinismus im Zeitalter der Gegenreformation.* Wiesbaden: Veröffentlichungen der Historischen Kommission für Nassau.

Van Miert, Dirk (2009): *Humanism in an Age of Science: The Amsterdam Athenaeum in the Golden Age, 1632–1704*. Leiden, Boston: Brill.

Mignini, Filippo (1984): "Sur la genèse du Court Traité: l'hypothèse d'une dictée originaire est-elle fondée?". In: *Cahiers Spinoza* 5 (Winter 1984–1985). Paris: Éditions Réplique, pp. 147–165.

Mignini, Filippo (1985): "Nuovi contributi per la datazione e l'interpretazione del *Tractatus de Intellectus Emendatione*". In: Giancotti, Emilia (Ed.): *Spinoza nel 350o anniversario della nascita. Atti del congresso internazionale (Urbino 4–8 ottobre 1982)*. Naples: Bibliopolis, pp. 515–525.

Mijnhardt, Wijnand W. (2003): "The Construction of Silence: Religious and Political Radicalism in Dutch History". In: Van Bunge, Wiep (Ed.): *The Early Enlightenment in the Dutch Republic, 1650–1750*. Leiden, Boston: Brill, pp. 231–262.

Mikkeli, Heikki (1992): *An Aristotelian Response to Renaissance Humanism. Jacopo Zabarella on the Nature of Arts and Sciences*. Helsinki: Societas Historica Finlandiae.

Moltmann, Jürgen (1957): "Zur Bedeutung des Petrus Ramus für Philosophie und Theologie im Calvinismus". In: *Zeitschrift für Kirchengeschichte* 68, pp. 295–318.

Monchamp, Georges (1886): *Histoire du Cartésianisme en Belgique*. Brussels: F. Hayez.

Morman, Thomas (2010): "History of Philosophy of Science as Philosophy of Science by Other Means? Comment on Thomas Uebel". In: Stadler, Friedrich (Ed.): *The Present Situation in the Philosophy of Science*. Dordrecht, Heidelberg, New York, London: Springer, pp. 29–40.

Mouy, Paul (1934): *Le développement de la physique cartésienne: 1646–1712*. Paris: Vrin.

Mueller, Hermann (1891): *Johannes Clauberg und seine Stellung im Cartesianismus mit besonderer Berücksichtigung seines Verhältnisses zu der occasionalistischen Theorie*. Jena: Frommann.

Mugnai, Massimo (1990): "'Necessità *ex hypothesi*' e analisi infinita in Leibniz". In: Lamarra, Antonio (Ed.): *L'infinito in Leibniz. Problemi e terminologia. Das Unendliche bei Leibniz. Problem und Terminologie*. Rome: Edizioni l'Ateneo, pp. 143–155.

Mulsow, Martin (2002): *Moderne aus dem Untergrund. Radikale Frühaufklärung in Deutschland 1680–1720*. Hamburg: Meiner.

Mulsow, Martin (Ed.) (2009): *Spätrenaissance-Philosophie in Deutschland 1570–1650: Entwürfe zwischen Humanismus und Konfessionalisierung, okkulten Traditionen und Schulmetaphysik*. Tübingen: Niemeyer.

Mulsow, Martin (2015): *Enlightenment Underground: Radical Germany, 1680–1720*. Charlottesville: University of Virginia Press.

Mulsow, Martin/Rohls, Jan (Eds.) (2005): *Socinianism and Arminianism: Antitrinitarians, Calvinists, and Cultural Exchange in Seventeenth-Century Europe*. Leiden: Brill.

Murray, Gemma/Harper, William/Wilson, Curtis (2011): "Huygens, Wren, Wallis, and Newton on Rules of Impact and Reflection". In: Jalobeanu, Dana/Anstey, Peter (Eds.): *Vanishing Matter and the Laws of Motion. Descartes and Beyond*. New York, London: Routledge, pp. 153–191.

Nadler, Steven (1999): "Knowledge, Volitional Agency and Causation in Malebranche and Geulincx". In: *British Journal for the History of Philosophy* 7, pp. 263–274.

Nadler, Steven (2011): *Occasionalism: Causation among the Cartesians*. Oxford, New York: Oxford University Press.

Nagel, Karl (1930): *Das Substanzproblem bei Arnold Geulincx*. Cologne: s.n.

Nauta, Lodi (2009): *In Defense of Common Sense. Lorenzo Valla's Humanist Critique of Scholastic Philosophy*. Cambridge (MA), London: Harvard University Press.

Neele, Adriaan C. (2009): *Petrus van Mastricht (1630–1706). Reformed Orthodoxy: Method and Piety*. Leiden: Brill.

Nuchelmans, Gabriel (1988): *Geulincx' Containment Theory of Logic*. Amsterdam: Koninklijke Nederlandse Akademie van Wetenschappen.

Nuchelmans, Gabriel (1998): "Deductive Reasoning". In: Garber, Daniel/Ayers, Michael (Eds.): *The Cambridge History of Seventeenth-Century Philosophy*. Cambridge, Cambridge University Press, vol. I, pp. 132–146.

Nyden, Tammy (2007): *Spinoza's Radical Cartesian Mind*. New York: Continuum.

Nyden, Tammy (2013): "De Volder's Cartesian Physics and Experimental Pedagogy". In: Dobre, Mihnea/Nyden, Tammy (Eds.): *Cartesian Empiricisms*. Dordrecht, Heidelberg, New York, London: Springer, pp. 227–249.

Nyden, Tammy (2014): "Living Force at Leiden: De Volder, 's Gravesande and the Reception of Newtonianism". In: Biener, Zvi/Schliesser, Eric (Eds.): *Newton and Newtonianism*. Oxford: Oxford University Press, pp. 207–222.

Olivo, Gilles (1993): "L'homme en personne". In: Verbeek, Theo (Ed.) (1993): *Descartes et Regius. Autour de l'Explication de l'esprit humain*. Rodopi: Amsterdam, pp. 69–91.

Omodeo, Pietro Daniel (2017): "Lodewijk de Bils' and Tobias Andreae's Cartesian Bodies: Embalmment Experiments, Medical Controversies and Mechanical Philosophy". In: *Early Science and Medicine* 22/4, pp. 301–332.

Ong, Walter, J. (1958): *Ramus, Method, and the Decay of Dialogue. From the Art of Discourse to the Art of Reason*. Cambridge (MA): Harvard University Press.

Otterspeer, Willem (2001): "The University of Leiden: An Eclectic Institution". In: *Early Science and Medicine* 6/4, pp. 324–333.

Pagel, Walter (2002): *Joan Baptista van Helmont: Reformer of Science and Medicine*. Cambridge: Cambridge University Press.

Papuli, Giovanni (1983): "La teoria del 'regressus' come metodo scientifico negli autori della scuola di Padova". In: Luigi, Olivieri (Ed.): *Aristotelismo veneto e scienza moderna*. Padua: Antenore, pp. 221–277.

De Pater, Cornelis (1975): "Experimental Physics". In: Lunsingh Scheurleer, Theodor H./ Posthumus Meyjes, Guillaume H.M. (Eds.): *Leiden University in the Seventeenth Century: An Exchange of Learning*. Leiden: Brill, pp. 309–327.

De Pater, Cornelis (Ed.) (1988): *Willem Jacob 's Gravesande. Welzijn, wijsbegeerte en wetenschap*. Baarn: Ambo.

De Pater, Cornelis (1989): "The Textbooks of 's Gravesande and Van Musschenbroek in Italy". In: Maffioli, Cesare Sergio/Palm, Lodewijk C. (Eds.) (1989): *Italian Scientists in the Low Countries in the XVIIth and XVIIIth Centuries*. Amsterdam: Rodopi, pp. 231–241.

De Pater, Cornelis (1994): "Willem Jacob 's Gravesande (1688–1742) and Newton's 'Regulae philosophandi'". In: *Lias* 21/2, pp. 257–294.

De Pater, Cornelis (1995): "'s Gravesande on Moral evidence". In: Fresco, Marcel F./Geeraedts, Loek/Hammacher, Klaus (Eds.): *Frans Hemsterhuis (1721–1790). Quellen, Philosophie und Rezeption*. Münster, Hamburg: LIT, pp. 221–242.

Patterson, Sarah (2008): "Clear and Distinct Perception". In: Broughton, Janet/Carriero, John (Eds.): *A Companion to Descartes*. Malden (MA), Oxford: Blackwell, pp. 216–234.

Perler, Dominik (1996): *Repräsentation bei Descartes*. Frankfurt: Klostermann.

Perler, Dominik (1998): *René Descartes*. Munich: Beck.

Pesce, Mauro (1992): "Il 'Consensus veritatis' di Christopher Wittich e la distinzione tra verità scientifica e verità biblica". In: *Annali di storia dell'esegesi* 9/1, pp. 53–76.

Petrus, Klaus (1997): *Genese und Analyse: Logik, Rhetorik und Hermeneutik im 17. und 18. Jahrhundert*. Berlin: De Gruyter.

Petry, Michael J. (1984): "Hobbes and the Early Dutch Spinozists". In: De Deugd, Cornelis (Ed.): *Spinoza's Political and Theological Thought*. Amsterdam: North-Holland, pp. 150–170.

Popkin, Richard (1960): *The History of Scepticism From Erasmus to Descartes*. Assen: Van Gorcum.

Popkin, Richard (2003): *The History of Scepticism from Savonarola to Bayle*. New York: Oxford University Press.

Poppi, Antonino (1969): "Pietro Pomponazzi tra averroismo e galenismo sul problema del 'regressus'". In: *Rivista critica di storia della filosofia* 24, pp. 243–267.

Poppi, Antonino (1970): *Introduzione all'aristotelismo padovano*. Padua: Antenore.

Poppi, Antonino (1972): *La dottrina della scienza in Giacomo Zabarella*. Padua: Antenore.

Pozzo, Riccardo (2001): "Ramus' Metaphysics and its Criticism by the Helmstedt Aristotelians". In: Feingold, Mordechai/Freedman, Joseph S./Rother, Wolfgang (Eds.): *The Influence of Petrus Ramus*. Basel: Schwabe, pp. 92–106.

Pozzo, Riccardo (2002): "Ramus and Other Renaissance Philosophers on Subjectivity". In: *Topoi* 22/1, pp. 5–13.

Pozzo, Riccardo (2004): "Logic and Metaphysics in German Philosophy from Melanchton to Hegel". In: Sweet, William (Ed.): *Approaches to Metaphysics*. Dordrecht: Kluwer, pp. 57–66.

Pozzo, Riccardo (2012): Adversus Ramistas. *Kontroversen über die Natur der Logik am Ende der Renaissance*. Basel: Schwabe.

Del Prete, Antonella (2001): "Tra Galileo e Descartes: l'esegesi biblica filocopernicana di Christopher Wittich". In: Montesinos, José/Solis, Carlo (Eds.): *Largo campo di filosofare. Eurosymposium Galileo 2001*. La Orotava: Fundación Canaria, pp. 719–729.

Del Prete, Antonella (2002): "Ermeneutica cartesiana: il contributo di Christopher Wittich". In: Marcialis, Maria Teresa/Crasta, Francesca Maria (Eds.): *Descartes e l'eredità cartesiana nell'Europa sei-settecentesca*. Lecce: Conte.

Del Prete, Antonella (2013): "Y-a-t-il une interprétation cartésienne de la Bible? Le cas de Christopher Wittich". In: Kolesnik-Antoine, Delphine (Ed.): *Qu'est-ce qu'être cartésien?* Paris: Editions de l'ENS de Lyon, pp. 117–142.

Del Prete, Antonella (forthcoming): "*Duplex intellectus et sermo duplex*: Method and the Separation of Disciplines in Johannes De Raey". In: Roux, Sophie (Ed.): *Physique et métaphysique*.

Ragland, Evan (2016): "Mechanism, the Senses, and Reason: Franciscus Sylvius and Leiden Debates Over Anatomical Knowledge After Harvey and Descartes". In: Distelzweig, Peter/Goldberg, Benjamin/Ragland, Evan R. (Eds.): *Early Modern Medicine and Natural Philosophy*. Dordrecht, Heidelberg, New York, London: Springer, pp. 173–206.

Reid, Steven J./Wilson, Emma Annette (Eds.) (2011): *Ramus, Pedagogy and the Liberal Arts: Ramism in Britain and the Wider World*. Farnham: Ashgate.

Reiss, Timothy J. (2000): "Neo-Aristotle and Method: Between Zabarella and Descartes". In: Gaukroger, Stephen/Schuster, John/Sutton, John (Eds.) (2000): *Descartes's Natural Philosophy*. London, New York: Routledge, pp. 195–227.

Rey, Anne-Lise (2009a): *L'ambivalence de la notion d'action dans la Dynamique de Leibniz. La correspondance entre Leibniz et De Volder (Iere Partie)*. Studia Leibnitiana, vol. 41, n. 1, 2009a, pp. 47–66.

Rey, Anne-Lise (2009b): *L'ambivalence de la notion d'action dans la Dynamique de Leibniz. La correspondance entre Leibniz et De Volder (IIe Partie)*. Studia Leibnitiana, vol. 41, n. 2, 2009b, pp. 157–182.

Rey, Anne-Lise (Ed.) (2016): *Correspondance - Gottfried Wilhelm Leibniz, Burcher De Volder*. *Paris*: Vrin, 2016.

Van Rijen, J.B.M. (1993): "Burgersdijk, Logician or Textbook Writer?". In: Bos, Egbert P./ Krop, Henri A. (Eds.): *Franco Burgersdijk (1590–1635): Neo-Aristotelianism in Leiden*. Amsterdam: Rodopi, pp. 9–28.

Risse, Wilhelm (1964): *Die Logik der Neuzeit*. Vol. 1: *1500–1640*. Stuttgart: Friedrich Frommann.

Risse, Wilhelm (1970): *Die Logik der Neuzeit*. Vol. 2: *1640–1780*. Stuttgart: Friedrich Frommann.

Risse, Wilhelm (1983): "La dottrina del metodo di Zabarella". In: Olivieri, Luigi (Ed.): *Aristotelismo veneto e scienza moderna*. Padua: Antenore, pp. 173–186.

Robinet, André (1996): *Aux sources de l'esprit cartésien, l'axe La Ramée-Descartes de la Dialectique de 1555 aux* Regulae. Paris: Vrin.

Rodis-Lewis, Geneviève (1993) "Problèmes discutés entre Descartes et Regius: l'ame et le corps". In: Verbeek, Theo (Ed.) (1993): *Descartes et Regius. Autour de l'Explication de l'esprit humain*. Rodopi: Amsterdam, pp. 35–46.

Rothschuh, Karl E. (1953): *Geschichte der Physiologie*. Berlin, Göttingen, Heidelberg: Springer.

Rothschuh, Karl E. (1968): "Henricus Regius und Descartes. Neue Einblicke in die frühe Physiologie (1640–1641) des Regius". In: *Archives internationales d'histoire des sciences* 21, pp. 39–66.

Rousset, Bernard (1999): *Geulincx entre Descartes et Spinoza*. Paris: Vrin.

Roux, Sophie (1998): "Le scepticisme et les hypothèses de la physique". In: *Revue de synthèse* 119/2–3, pp. 211–255.

Roux, Sophie (2006): "La philosophie naturelle à l'époque de Le Nôtre. Remarques sur la philosophie mécanique et sur le cartésianisme". In: Fahrat, Georges (Ed.): *André Le Nôtre, fragments d'un paysage culturel. Institutions, arts, sciences et techniques*. Sceaux: Musée de l'Île-de-France, pp. 98–111.

Roux, Sophie (2013a): "Was There a Cartesian Experimentalism in 1660s France?". In: Dobre, Mihnea/Nyden, Tammy (Eds.) (2013): *Cartesian Empiricisms*. Dordrecht, Heidelberg, New York, London: Springer, pp. 47–88.

Roux, Sophie (2013b): "An Empire Divided: French Natural Philosophy (1670–1690)". In: Garber, Daniel/Roux, Sophie (Eds.) (2013): *The Mechanization of Natural Philosophy*. New York: Kluwer, pp. 55–95.

Ruestow, Edgar G. (1973): *Physics at Seventeenth and Eighteenth-Century Leiden: Philosophy and the New Science in the University*. The Hague: Martinus Nijhoff.

Van Ruler, Han (1995): *The Crisis of Causality, Voetius and Descartes on God, Nature and Change*. Leiden, New York, Cologne: Brill.

Van Ruler, Han (1999a): "Geulincx and Spinoza: Books, Backgrounds and Biographies". In: *Studia Spinozana: An International and Interdisciplinary Series* 15, pp. 89–106.

Van Ruler, Han (1999b): "'Something, I know not what'. The Concept of Substance in Early Modern Thought". In: Nauta, Lodi/Vanderjagt, Arjo (Eds.) (1999): *Between Imagination and Demonstration. Essays in the History of Science and Philosophy Presented to John D. North*. Leiden: Brill, pp. 365–393.

Van Ruler, Han (2000): "Minds, Forms, and Spirits: The Nature of Cartesian Disenchantment". In: *Journal of the History of Ideas* 61, pp. 381–395.

Van Ruler, Han (2001): "Reason Spurred by Faith: Abraham Heidanus and Dutch Philosophy". In: *Geschiedenis van de Wijsbegeerte in Nederland* 12, pp. 21–28.

Van Ruler, Han (2003a): "Bontekoe, Cornelis (c. 1644–85)". In: Van Bunge, Wiep/Krop, Henry A./Leeuwenburgh, Bart/Van Ruler, Han/Schuurman, Paul/Wielema, Michiel (Eds.): *The Dictionary of Seventeenth and Eighteenth-Century Dutch Philosophers*. Bristol: Thoemmes Press, pp. 128–132.

Van Ruler, Han (2003b): "Heidanus, Abraham (1597–1678)". In: Van Bunge, Wiep/Krop, Henry A./Leeuwenburgh, Bart/Van Ruler, Han/Schuurman, Paul/Wielema, Michiel (Eds.): *The Dictionary of Seventeenth and Eighteenth-Century Dutch Philosophers*. Bristol: Thoemmes Press, vol. I, pp. 397–402.

Van Ruler, Han (2003c): "The Shipwreck of Belief and Eternal Bliss: Philosophy and Religion in Later Dutch Cartesianism". In: Van Bunge, Wiep (Ed.): *The Early Enlightenment in the Dutch Republic, 1650–1750*. Leiden, Boston: Brill, pp. 109–136.

Rutherford, Donald (2013): "Descartes' Ethics". In: *Stanford Encyclopedia of Philosophy*. https://plato.stanford.edu/entries/descartes-ethics/, visited on 10 June 2017.

Sassen, Ferdinand (1941): *Henricus Renerius, de eerste "Cartesiaanse" hoogleraar te Utrecht*. Amsterdam: Noord-Hollandse Uitgeversmaatschappij.

Savini, Massimiliano (2004): *Le développement de la méthode cartésienne dans les Provinces-Unies (1643–1665)*. Lecce: Conte.

Savini, Massimiliano (2006): "L'invention de la logique cartésienne: la *Logica vetus et nova* de Johannes Clauberg". In: *Revue de métaphysique et de morale* 1, pp. 73–88.

Savini, Massimiliano (2011a): *Johannes Clauberg:* Methodus cartesiana *et ontologie*. Paris: Vrin.

Savini, Massimiliano (2011b): "*Methodus cartesiana* e *pansophia*: i primi dibattiti intorno al metodo cartesiano e il progetto di riforma del sapere nelle Provincie Unite". In: Dessì, Paola/Lotti, Brunello (Eds.): *Eredità cartesiane nella cultura britannica*. Florence: Le Lettere, pp. 29–47.

Savini, Massimiliano (2012): "La 'Medicina mentis' de Ehrenfried Walther von Tschirnhaus en tant que Philosophie première". In: *Les Cahiers philosophiques de Strasbourg* 2, pp. 147–172.

Schliesser, Eric (2011): "Newton's Challenge to Philosophy: A Programmatic Essay". In: *HOPOS: The Journal of the International Society for the History of Philosophy of Science* 1/1, 101–128.

Schmaltz, Tad (2002): *Radical Cartesianism: The French Reception of Descartes*. Cambridge: Cambridge University Press.

Schmaltz, Tad (Ed.) (2005a): *Receptions of Descartes. Cartesianism and Anti-Cartesianism in Early Modern Europe*. London, New York: Routledge.

Schmaltz, Tad (2005b): "French Cartesianism in Context: The Paris Formulary and Regis's *Usage*". In: Tad Schmaltz (Ed.): *Receptions of Descartes. Cartesianism and Anti-Cartesianism in Early Modern Europe*. London, New York: Routledge, pp. 73–89.

Schmaltz, Tad (2008): *Descartes on Causation*. New York: Oxford University Press.

Schmaltz, Tad (2016a): *Early Modern Cartesianisms: Dutch and French Constructions*. New York: Oxford University Press.

Schmaltz, Tad (2016b): "The Early Dutch Reception of *L'Homme*". In: Antoine-Mahut, Delphine/Gaukroger, Stephen (Eds.): *Descartes'* Treatise on Man *and its Reception*. Dordrecht: Springer, pp. 71–90.

Schmidt, Andreas (2009): *Göttliche Gedanken: Zur Metaphysik der Erkenntnis bei Descartes, Malebranche, Spinoza und Leibniz*. Frankfurt: Klostermann.

Schmidt-Biggemann, Wilhelm (1983): *Topica Universalis*. Hamburg: Meiner.

Schoneveld, Cornelis W. (1983): *Intertraffic of the Mind. Studies in Seventeenth-Century Anglo-Dutch Translation*. Leiden: Brill.

Schuster, John (1993): "Whatever Should We Do with Cartesian Method? Reclaiming Descartes for the History of Science". In: Voss, Stephen (Ed.): *Essays on the Philosophy and Science of René Descartes*. New York: Oxford University Press, pp. 195–223.

Schuster, John (2012): *Descartes-Agonistes. Physico-mathematics, Method & Corpuscular-Mechanism 1618–33*. Dordrecht, Heidelberg, New York, London: Springer.

Schuurman, Paul (2001): "*Ex naturæ lumine et Aristotele*: Johannes de Raeys verdediging van de Cartesiaanse fysica". In: *Algemeen Nederlands Tijdschrift voor Wijsbegeerte* 23, pp. 237–254.

Schuurman, Paul (2003a): "The Empiricist Logic of Ideas of Jean Le Clerc". In: Van Bunge, Wiep (Ed.): *The Early Enlightenment in the Dutch Republic, 1650–1750*. Leiden, Boston: Brill, pp. 70–109.

Schuurman, Paul (2003b): "'s Gravesande, Willem Jacob (1688–1742)". In: Van Bunge, Wiep/Krop, Henry A./Leeuwenburgh, Bart/Van Ruler, Han/Schuurman, Paul/Wielema, Michiel (Eds.): *The Dictionary of Seventeenth and Eighteenth-Century Dutch Philosophers*. Bristol: Thoemmes Press, vol. II, pp. 865–872.

Schuurman, Paul (2003c): "De Raey, Johannes (1620–1702)". In: Van Bunge, Wiep/Krop, Henry A./Leeuwenburgh, Bart/Van Ruler, Han/Schuurman, Paul/Wielema, Michiel (Eds.): *The Dictionary of Seventeenth and Eighteenth-Century Dutch Philosophers*. Bristol: Thoemmes Press, vol. I, pp. 813–816.

Schuurman, Paul (2003d): "Willem Jacob 's Gravesande's Philosophical Defence of Newtonian Physics: On the Various Uses of Locke". In: Anstey, Peter (Ed.): *The Philosophy of John Locke: New Perspectives*. London, New York, Routledge, pp. 43–57.

Schuurman, Paul (2004): *Ideas, Mental Faculties and Method: The Logic of Ideas of Descartes and Locke and its Reception in the Dutch Republic, 1630–1750*. Leiden, Boston: Brill.

Sécretan, Catherine (1987): "La réception de Hobbes aux Pays Bas". In: *Studia Spinozana* 3, pp. 27–46.

Sgarbi, Marco (2012): *The Aristotelian Tradition and the Rise of British Empiricism. Logic and Epistemology in the British Isles (1570–1689)*. Dordrecht, Heidelberg, New York, London: Springer.

Sgarbi, Marco (2014): *The Italian Mind: Vernacular Logic in Renaissance Italy (1540–1551)*. Leiden: Brill.

Shank, J.B. (2008): *The Newton Wars and the Beginning of the French Enlightenment*. Chicago: University of Chicago Press.

Shelford, April (2007): *Transforming the Republic of Letters: Pierre-Daniel Huet and European Intellectual Life, 1650–1720*. Rochester: University Rochester Press.

Skalnik, James Veazie (2002): *Ramus and Reform: University and Church at the End of the Renaissance*. Kirksville: Truman State University Press.

Smit, F.R.H./Jensma, G.T. (1985): *Academisch Onderwijs in Franeker en Groningen 1585–1843*. Ijver en Wedijver. Groningen: Universiteitsmuseum Groningen.

Smith, Justin E.H. (2013): "Heat, Action, Perception: Models of Living Beings in German Medical Cartesianism". In: Dobre, Mihnea/Nyden, Tammy (Eds.): *Cartesian Empiricisms*. Dordrecht, Heidelberg, New York, London: Springer, pp. 105–123.

Sorell, Tom (Ed.) (1993): *The Rise of Modern Philosophy: The Tension between the New and Traditional Philosophies from Machiavelli to Leibniz*. Oxford: Clarendon.

Sorell, Tom (2005): *Descartes Reinvented*. Cambridge: Cambridge University Press.

Sorell, Tom (2010): "*Scientia* and Sciences in Descartes". In: Sorell, Tom/Rogers, G.A.J/ Kraye, Jill (Eds.): Scientia *in Early Modern Philosophy. Seventeenth-Century Thinkers on Demonstrative Knowledge from First Principles.* Dordrecht, Heidelberg, London, New York: Springer, pp. 71–82.

Sorell, Tom/Rogers, G.A.J/Kraye, Jill (Eds.) (2010): Scientia *in Early Modern Philosophy. Seventeenth-Century Thinkers on Demonstrative Knowledge from First Principles.* Dordrecht, Heidelberg, London, New York: Springer.

Spruit, Leen (1999): "Johannes Clauberg on Perceptual Knowledge". In: Verbeek, Theo (Ed.): *Johannes Clauberg (1622–1665) and Cartesian Philosophy in the Seventeenth Century.* Dordrecht: Kluwer, pp. 1–12.

Stadler Friedrich (2014): "History and Philosophy of Science: Between Description and Construction". In: Galavotti, Maria Carla/Dieks, Dennis/Gonzalez, Wenceslao J./Hartmann, Stephan/Uebel, Thomas/Weber, Marcel (Eds.): *New Directions in the Philosophy of Science.* Dordrecht: Springer, pp. 747–767.

Strazzoni, Andrea (2011): "La filosofia aristotelico-cartesiana di Johannes De Raey". In: *Giornale critico della filosofia italiana* 7/1, pp. 107–132.

Strazzoni, Andrea (2012): "The Dutch Fates of Bacon's Philosophy: *Libertas Philosophandi*, Cartesian Logic and Newtonianism". In: *Annali della Scuola Normale Superiore di Pisa – Classe di Lettere e Filosofia* 4/1, pp. 251–281.

Strazzoni, Andrea (2013): "A Logic to End Controversies: The Genesis of Clauberg's *Logica Vetus et Nova*". In: *Journal of Early Modern Studies* 2/2, pp. 123–149.

Strazzoni, Andrea (2014a): "The Crypto-Dualism of Henricus Regius". In: Caroti, Stefano/ Spallanzani, Mariafranca (Eds.): *Individuazione, individualità, identità personale.* Florence: Le Lettere, pp. 133–151.

Strazzoni, Andrea (2014b): "On Three Unpublished Letters of Johannes de Raey to Johannes Clauberg". In: *Noctua* 1/1, pp. 68–106.

Strazzoni, Andrea (2015): "The Cartesian Philosophy of Language of Johannes de Raey". In: *Lias. Journal of Early Modern Intellectual Culture and its Sources* 42/2, pp. 89–120.

Strazzoni, Andrea (forthcoming a): "The Hidden Presence of Ramism in Early Modern Dutch Philosophy". In: Burton, Simon/Choptiany, Michał (Eds.): *The Tree of Knowledge: Theories of Sciences and Arts in Central Europe, 1400–1700.* Oxford: Routledge.

Strazzoni, Andrea (forthcoming b): "'How did Regius become Regius?'". In: *Early Science and Medicine* 23/4.

Strazzoni, Andrea (forthcoming c): "The Medical Cartesianism of Henricus Regius. Disciplinary Partitions, Mechanical Reductionism and Methodological Aspects". In: *Galilæana. Studies in Renaissance and Early Modern Science*, 15.

Suitner, Riccarda (2016): *Die philosophischen Totengespräche der Frühaufklärung.* Hamburg: Meiner.

Terraillon, Eugène (1912): *La morale de Geulincx dans ses rapports avec la philosophie de Descartes.* Paris: Alcan.

Theis, Robert/Ferrari, Jean/Ruffing, Margit/Vollet, Matthias/Guenancia, Pierre (Eds.) (2009): *Descartes und Deutschland – Descartes et l'Allemagne.* Hildesheim: Olms.

Thijssen-Schoute, C. Louise (1954): *Nederlands cartesianisme, avec sommaire et table des matières en français.* Amsterdam: Noord-Hollandsche Uitg. Mij.

Trevisani, Francesco (1992): *Descartes in Germania: La ricezione del cartesianesimo nella Facoltà filosofica e medica di Duisburg (1652–1703).* Milan: Franco Angeli.

Trevisani, Francesco (2012): *Descartes in Deutschland: Die Rezeption des Cartesianismus in den Hochschulen Nordwestdeutschlands*. Zürich: LIT.

Uebel, Thomas (2010): "Some Remarks on Current History of Analytical Philosophy of Science". In: Stadler, Friedrich (Ed.): *The Present Situation in the Philosophy of Science*. Dordrecht, Heidelberg, New York, London: Springer, pp. 13–28.

Ueberweg, Friedrich (1866): *Grundriss der Geschichte der Philosophie*. Berlin: Mittler.

Vanpaemel, Geert (1984): "Rohault's *Traité de physique* and the Teaching of the Cartesian Physics". In: *Janus: Revue Internationale de l'Histoire des Sciences, de la Médecine et de la Technique* 71, pp. 31–40.

Vanpaemel, Geert (1985): *De mechanistische natuurwetenschap aan de Leuvense Artesfakulteit (1650–1797)*. Leuven: Katholieke Universiteit.

Vasoli, Cesare (2007): *La dialettica e retorica dell'umanesimo*. Naples: La Città del Sole. (1st ed. 1968).

Verbeek, Theo (1988): *La Querelle d'Utrecht*. Paris: Les Impressions nouvelles.

Verbeek, Theo (1989): "Les passions et la fièvre. L'idée de la maladie chez Descartes et quelques cartésiens néerlandais". In: *Tractrix* 1, pp. 45–61.

Verbeek, Theo (1992a): *Descartes and the Dutch. Early Reactions to Cartesian Philosophy, 1637–1650*. Carbondale, Edwardsville: Southern Illinois University Press.

Verbeek, Theo (1992b): "*Ens per accidens*: le origini della Querelle di Utrecht". In: *Giornale critico della filosofia italiana* 71, pp. 276–288.

Verbeek, Theo (Ed.) (1993a): "Le contexte historique des *Notae in programma quoddam*". In: Verbeek, Theo (Ed.): *Descartes et Regius. Autour de l'Explication de l'esprit humain*. Rodopi: Amsterdam, pp. 1–33.

Verbeek, Theo (1993b): "From 'Learned Ignorance' to Scepticism; Descartes and Calvinist Orthodoxy". In: Popkin, Richard H./Vanderjagt, Arjo (Eds.) (1993), *Scepticism and Irreligion in the Seventeenth and Eighteenth Centuries*. Leiden: Brill, pp. 31–45.

Verbeek, Theo (1993c): "Tradition and Novelty: Descartes and Some Cartesians". In: Sorell, Tom (Ed.): *The Rise of Modern Philosophy: The Tension between the New and Traditional Philosophies from Machiavelli to Leibniz*. Oxford: Clarendon, pp. 167–196.

Verbeek, Theo (1994): "Regius's *Fundamenta Physices*". In: *Journal of the History of Ideas* 55, pp. 533–551.

Verbeek, Theo (1995): "Les cartésiens face à Spinoza: l'exemple de Johannes de Raey". In: Cristofolini, Paolo (Ed.): *The Spinozistic Heresy. The Debate on the Tractatus Theologico-Politicus, 1670–1677 and the Immediate Reception of Spinozism. Seminar Cortona 1991*. Amsterdam, Maarssen: APA-Holland University Press, pp. 77–88.

Verbeek, Theo (1998): "Geulincx, Arnold (1624–69)". In: Craig, Edward (Ed.) (1998): *Routledge Encyclopedia of Philosophy*. London: Routledge, vol. IV, pp. 59–61.

Verbeek, Theo (1999) "Johannes Clauberg: A Bio-Bibliographical Sketch". In: Verbeek, Theo (Ed.) (1999): *Johannes Clauberg (1622–1665) and Cartesian Philosophy in the Seventeenth Century*. Dordrecht: Kluwer, pp. 181–200.

Verbeek, Theo (2000): "The Invention of Nature. Descartes and Regius". In: Gaukroger, Stephen/Schuster, John/Sutton, John (Eds.) (2000): *Descartes' Natural Philosophy*. London: Routledge, pp. 149–167.

Verbeek, Theo (2001): "Notes on Ramism in the Netherlands". In: Feingold, Mordechai/Freedman, Joseph S./Rother, Wolfgang (Eds.): *The Influence of Petrus Ramus*. Basel: Schwabe, pp. 38–53.

Verhoeven, Cornelis (1973): *Het axioma van Geulincx*. Bilthoven: Ambo.

Vermij, Rienk (2002): *The Calvinist Copernicans. The Reception of the New Astronomy in the Dutch Republic, 1575–1750*. Amsterdam: Royal Netherlands Academy of Arts and Sciences.

Viola, Eugenio (1975): "Scolastica e cartesianesimo nel pensiero di J. Clauberg". In: *Rivista di filosofia neo-scolastica* 67, pp. 247–266.

De Vleeschauwer H.J. (1957): *Three Centuries of Geulincx Research: A Bibliographical Survey*. Pretoria: Communications of the University of South Africa.

De Vrijer, Marinus Johannes Antoinie (1917): *Henricus Regius. Een 'cartesiaansch' hoogleraar aan de Utrechtsche hoogeschool*. The Hague: Martinus Nijhoff.

Watson, Richard A. (1966): *The Downfall of Cartesianism, 1673–1712: A Study of Epistemological Issues in Late 17th Century Cartesianism*. The Hague: Martinus Nijhoff.

Watson, Richard A. (1998): *The Breakdown of Cartesian Metaphysics*. Atlantic Islands: Humanities Press International.

Webster, Charles (1969): "Henry More and Descartes: Some New Sources". In: *The British Journal for the History of Science* 4, pp. 359–377.

Wiesenfeldt, Gerhard (2002): *Leerer Raum in Minervas Haus. Experimentelle Naturlehre an der Universität Leiden, 1675–1715*. Amsterdam: Koninklijke Nederlandse Akademie van Wetenschappen.

Williams, Michael (1998): "Descartes and the Metaphysics of Doubt". In: Cottingham, John (Ed.): *Descartes*. Oxford: Oxford University Press, pp. 28–49.

Wilson, Catherine (2008): *Epicureanism at the Origins of Modernity*. New York: Oxford University Press.

Van der Wall, Ernestine (1996): "Cartesianism and Cocceianism: A Natural Alliance?". In: Magdelaine, Michelle (Ed.): *De l'humanisme aux Lumierès, Bayle et le protestantisme*. Paris, Oxford: Voltaire Foundation.

Wohlers, Christian (2002): *Wie unnütz ist Descartes? Zur Frage metaphysischer Wurzeln der Physik*. Würzburg: Königshausen & Neumann.

Wollgast, Siegfried (1988): *Philosophie in Deutschland zwischen Reformation und Aufklärung 1550–1650*. Berlin: De Gruyter.

Wundt, Max (1939): *Die deutsche Schulmetaphysik des 17. Jahrhunderts*. Tübingen: Olms.

Zittel, Claus (2009): Theatrum philosophicum. *Descartes und die Rolle ästhetischer Formen in der Wissenschaft*. Berlin: Akademie.

Index

Epicurus 133
Episcopius, Simon 73
Erasmus, Desiderius 40
essence/*essentia* 62, 65, 71, 87, 89, 90, 92,
    93, 94, 95, 96, 97, 101, 102, 103, 116,
    145, 146, 151, 160, 162, 167, 168, 179,
    180, 183, 187, 198
ethics/*ethica*/moral philosophy/*philosophia
    moralis* 1, 2, 4, 11, 13, 18, 20, 22, 38,
    39, 50, 56, 58, 62, 63, 67, 68, 74–78,
    79, 80, 81, 82, 83, 92, 93, 102, 103,
    104, 105, 113, 118, 124, 125, 126, 173,
    189, 194, 195, 196, 199, 202
evidence/*evidentia* 7, 12, 13, 31, 35, 38, 62,
    68, 71, 83, 84, 85, 95, 99, 100–103,
    104, 112–113, 114, 119, 120, 122, 145,
    148, 152, 155, 156, 172, 174, 175, 177,
    178, 179, 181, 183, 186, 187–193, 194,
    195, 196, 201
experience/*experientia* 3, 4, 5, 7, 12, 13,
    16, 18, 19, 32, 33, 35, 36, 38, 72, 73,
    77, 81, 87, 88, 92, 96, 97, 99, 103,
    104, 107, 110, 111, 118, 123, 129, 130,
    132, 135, 136, 137, 138, 140, 141,
    144, 145, 155, 157, 158, 159, 162,
    164, 169, 170, 175, 177, 178, 179,
    182, 185, 187, 188, 190, 193, 194,
    199, 203
experimental cabinet, s. Leiden *Theatrum
    physicum*
experimental philosophy/experimental
    science/*philosophia experimentalis* 2,
    5, 6, 7, 10, 13, 15, 18, 19, 125, 126, 127,
    129, 130, 132, 135, 138, 141, 158, 159,
    170, 175, 176, 178, 197, 200, 201

fermentation/*fermentatio* 126, 130, 131, 132,
    139, 140
Fernel, Jean 27
Flage, Daniel 12
Fourth Lateran Council 98
France 15, 23, 46, 170
Franeker 6, 23, 71, 106, 108, 127, 159, 170
Frederick the Great 174
freedom/*libertas* 34, 41, 74–75, 93, 94, 95,
    96–97, 103, 152, 155, 179, 184–185
Fullenius, Bernhard 129

Galen 13, 113
– *De locis affectis* 113
Galilei, Galileo 126, 133, 136, 137, 176,
    200, 202
Garber, Daniel 10, 11, 12, 13, 16
Gassendi, Pierre 42, 44, 109, 146
– *Disquisitio metaphysica* 44, 109
Gaukroger, Stephen 16, 35
geometry/*geometria* 44, 71, 83, 96, 107, 108,
    136, 140, 173, 176, 186, 187
Germany 4, 15, 17, 40, 45, 46, 48, 115, 126
Geulincx, Arnold 2, 4, 5, 14, 68, 75, 76–104,
    105, 107, 112, 126, 127, 145, 168, 170,
    199, 200, 202
– *Annotata ad Metaphysicam* 87, 93, 95–101
– *Annotata maiora in Principia philosophiae
    Renati des Cartes* 83, 93
– *De virtute et primis eius proprietatibus* 77, 82
– *Dictata ad Logicam* 79
– *Disputationes metaphysicae* 83, 84
– *Disputationes physicae* 87–95
– *Ethica* 77, 80, 82, 83, 93, 102–103
– *Logica restituta* 79, 80, 86, 93
– *Metaphysica falsa sive ad mentem peripa-
    teticam* 80, 82, 102
– *Methodus inveniendi argumenta* 80,
    82, 84
– *Oratio de abarcendo contemptu* 93
– *Oratio de removendis parergis et nitore
    conciliando disciplinis* 78–80
– *Oratio prima, dicta in auspicio
    Quaestionum quodlibeticarum* 80,
    81, 85
– *Physica vera* 91, 96
– *Quaestiones quodlibeticae* 76, 80, 85, 102
– *Van de hooft-deuchden* 77
Gillespie, Charles 8
Goclenius, Rudolph 58, 59, 116, 118, 119
– *Problemata logica* 116
God/*Deus* 4, 7, 11, 12, 19, 21, 25, 28, 29, 32,
    33, 34, 36, 38, 40, 41, 42, 43, 44, 50,
    53, 54, 55, 57, 58, 59, 60, 61, 62, 63, 66,
    67, 69, 71, 76, 77, 78, 81, 83, 85, 86, 87,
    88, 91, 92, 93, 94, 95, 96, 97, 98, 99,
    100, 101, 102, 103, 104, 110, 116, 117,
    119, 120, 122, 123, 128, 134, 145, 147,